HISTORIC CONTROL

TEXTBOOKS

HISTORIC CONTROL TEXTBOOKS

Edited by

Janos Gertler

INTERNATIONAL FEDERATION OF AUTOMATIC CONTROL

Cover design by Bálint D. Sós

Elsevier Ltd is an imprint of Elsevier with offices at:

Linacre House, Jordan Hill, Oxford OX2 8DP, UK
The Boulevard, Langford Lane, Kidlington, Oxford OX5 1GB, UK
84 Theobald's Road, London WC1X 8RR, UK
Radarweg 29, PO Box 211, 1000 AE Amsterdam, The Netherlands
30 Corporate Drive, Suite 400, Burlington, MA 01803, USA
525 B Street, Suite 1900, San Diego, CA 92101-4495, USA

First edition 2006

British Library Cataloguing in Publication Data
A catalogue record for this book is available from the British Library

Library of Congress Cataloging-in-Publication Data
A catalog record for this book is available from the Library of Congress

For information on all related publications visit our web site at books.elsevier.com

ISBN–10: 0-08-045346-5
ISBN–13: 978-0-08-045346-0

Transferred to digital print on demand, 2007
Printed and bound by CPI Antony Rowe, Eastbourne

Elsevier Internet Homepage – http://www.elsevier.com
Consult the Elsevier homepage for full catalogue information on all books, major reference works, journals, electronic products and services.

All Elsevier journals are available online via ScienceDirect: www.sciencedirect.com

To contact the Publisher
Elsevier welcomes enquiries concerning publishing proposals: books, journal special issues, conference proceedings, etc. All formats and media can be considered. Should you have a publishing proposal you wish to discuss, please contact, without obligation, the publisher responsible for Elsevier's materials and engineering programme:

Jonathan Agbenyega
Publisher
Elsevier Ltd
The Boulevard, Langford Lane
Kidlington, Oxford
OX5 1GB, UK

Phone: +44 1865 843000
Fax: +44 1865 843987
E-mail: j.agbenye@elsevier.com

General enquiries, including placing orders, should be directed to Elsevier's Regional Sales Offices – please access the Elsevier homepage or full contact details (homepage details at the top of this page).

PRESIDENT'S FOREWORD

In 1956, a declaration to create an international organization of automatic control was signed in Heidelberg and, in 1957, the Constitution and By-Laws of IFAC were adopted in Paris. For the past 50 years IFAC has fulfilled its original purpose to promote the science and technology of control in the broadest sense in all systems. IFAC has served all those concerned with the theory and application of automatic control systems engineering and has grown into a truly world-wide organization, now encompassing more than 50 National Member Organizations.

An IFAC 50 Task Force was created to commemorate the 50th anniversary of IFAC. A team of IFAC Advisors has undertaken the planning of several activities for this celebration. As one of several projects headed by Stephen Kahne, creation of a summary of historic control textbooks was undertaken by Janos Gertler, while a celebratory event and an IFAC Foundation are also planned by Rolf Isermann and Pedro Albertos and collaborators.

The IFAC 50 Team, in particular, Janos Gertler, did an excellent job to prepare this wonderful collection of historic control textbooks. The IFAC family deeply appreciates the hard work of many individuals who have made this book possible.

This book also reminds us that we have to be prepared for the next 50 years of IFAC.

February 2006.

Wook Hyun Kwon

President of IFAC
2005-2008

EDITOR'S FOREWORD

The mathematical foundations of control go back to the classical mechanics of the eighteenth and nineteenth centuries, marked by such names as Lagrange, Hamilton and Jacobi, and to mathematical results on stability obtained in the last third of the nineteenth century by Maxwell, Routh, Lyapunov and Hurwitz. Following the celebrated steam engine regulator of James Watt in 1784, controllers were designed for various engineering systems, most notably in power systems, in the late eighteen and early nineteen hundreds. Automatic control as a separate engineering discipline was born during World War II, when fundamental development took pace, in parallel, in the UK and USA, in Germany and Japan, and in Russia.

The objective of this collection is to present the earliest textbooks that grew out of the original development of automatic control, and the many others that followed very soon, in various countries (now all members of IFAC), and in various languages. We set out to collect information on one to four books from each country (nine from the USA), including a brief description of the background, history and contents of the book, a picture of the front page, and copies of one to a few "typical" pages. With the latter, we intended to show pages that contain an equation or figure, easily recognizable to anyone familiar with control, embedded in the text written in one of the many languages and, in some cases, in various scripts.

The present collection contains 62 entries from 21 countries.

Rather than going through the official channels of the National Member Organizations of IFAC, most of the material was collected via a network of personal friends. As many as thirty nine people have contributed to this work, to whom the Editor is greatly indebted. The selection of the books from each country was left to the person invited to contribute. Strict selection rules were not enforced; the number of books does not necessarily reflect the perceived significance of the particular subculture. Also, some of the early entries were lecture notes rather than typeset hardcover books. Finally, the number of pages shown is in no way representative of the perceived importance of the book; while books known only locally may be there with several pages, some of the classics recognized world-wide are present with only a single page.

The Editor wishes to express his special thanks to Stephen Kahne who initiated this project and, as coordinator of the entire IFAC 50 effort, has supported it all the way to its completion.

May examining this volume bring the Reader as much joy as its creation has given the Editor.

February 2006.

Janos Gertler

CONTRIBUTORS

Javier ARACIL
Karl Johan ÅSTRÖM
Jens BALCHEN
Liu BAO
Ruth BARS
Stuart BENNETT
Sergio BITTANTI
Mogens BLANKE
Claudio BONIVENTO
Celso BOTTURA
Plìnio CASTRUCCI
Huitang CHEN
Talha DINIBÜTÜN
Peter DORATO
Wladislaw FINDEISEN
Katsuhisa FURUTA
Josè GEROMEL
Ryszard GESSING
Adolf GLATTFELDER
Alberto ISIDORI
Michel KINNAERT
Heikki KOIVO
Oh-Kyu KWON
Giovanni MARRO
István NAGY
Antti NIEMI
Boris POLYAK
Patrice REMAUD
Walter SCHAUFFELBERGER
Marcel STAROSWIECKI
Vladimir STREJC
Manfred THOMA
Jean-Claude TRIGEASSOU
Paul VAN den HOF
Baiwu WAN
Zhongtuo WANG
Deyun XIAO
Raimo YLINEN
Dazhong ZHENG

CONTENTS

EARLY CONTROL TEXTBOOKS
in
BRASIL

Contrôle Automático – Teoria e Projeto

Plínio Castrucci
EDUSP, S. Paulo, SP, Brazil, 1969

Princípios de Controle e Servomecanismos

Celso Pascoli Bottura
Guanabara Dois, Rio de Janeiro, RJ, Brazil, 1982

Contributed by

Plínio Benedicto de Lauro Castrucci
Universidade de São Paolo

Celso Pascoli Bottura
UNICAMP, Campinas SP

José C. Geromel
UNICAMP, Campinas SP

Sociedade Brasileira de Atomática

Brasil

Contrôle Automático – Teoria e Projeto

Plínio Castrucci
EDUSP, S. Paulo, Brazil, 1969

This textbook derived its existence from lectures given at Escola Politécnica da Universidade de S. Paulo, starting in 1961, to all the electrical engineering students. Chapters 1 to 10 were taught regularly in a two-semester course, during the fifth year of the engineering curriculum. Laboratory sessions for frequency and transient analysis, and compensation design, applied to electrical servos and regulators, were held during the second semester of the course.

A brief list of the chapters, with short comments added, is as follows:

Introduction

1. Feedback and Control

Part I - Linear Systems

2. Analysis of Linear Systems
(transfer function, poles and zeros, frequency response, Routh criterion)

Part II – Linear Feedback Systems

3. Quality Criteria
(closed loop transfer function, stationary and transient errors, sensitivity to parameters)

4. Root Locus Method
(rules, spirule, advantages of a "new root locus" defined in a plane $y = s^{-1}$)

5. Frequency Response Method
(Nyquist criterion, stability margins, Nichols chart)

6. The Central Problem in the Design of Servo-Systems
(lead, lag and general compensators, load compensation, field forcing, etc)

7. Modulated Signal Systems (amplitude modulation and sampling)
(synchros, amplitude modulation with suppressed carrier, sampling and extrapolation, Z-transform, stability)

8. Digital Signal Systems
(binary codes, components, a digital servo)

Part III – Nonlinear Systems

9. Describing Functions
(concept, saturation, dead zone, etc)

10. Phase Plane Analysis
(isoclines, Liénard method, parameter time, singular points, limit-cycles, analysis of on-off servos)

11. Introduction to the General Stability Theory
(state space, relative and asymptotic stability, sign-defined functions, Lyapunov theorems, introduction to domain research)

Part IV – Logical Systems

12. Boolean Algebra
(concept, relay circuits, simplification and Veitch maps)

13. General Method of Design
(design principles: from cause-effect map to sequence diagram to equations to relay circuit)

14. Modern Realizations
(transistor logic, design with NORs and NANDs)

Appendices:

Small Laplace transform list; Multiple loop Nyquist criterion; Measurement techniques; Passive compensation circuits, Tustin's theorem on power gain versus time constant)

Proposed problems

References

Later editions developed into three titles: Linear Control, Nonlinear Systems and Digital Control.

Contributed by

Plínio Benedicto de Lauro Castrucci
Escola Politécnica, Universidade de São Paolo
Brasil

PLINIO
CASTRUCCI

contrôle
automático
teoria e
projeto

EDITORA EDGARD BLÜCHER LTDA.

Brasil

Têrmos recomendados	*Têrmos obsoletos ou errôneos*
Comando v	Entrada, valor desejado, "set point"
Elementos de contrôle G_1	Ampliador, controlador, servo-ampliador
Referência r	Entrada, padrão de referência, valor desejado, "set point"
Sinal ou êrro atuante e	Êrro, correção, desvio
Sistema Controlado G_2	Processo, carga, "plant"
Sistema Indireta/Controlado Z	Processo, carga, "plant"
Variável Controlada c	Saída, variável regulada ou medida
Variável Indireta/Controlada a	Saída, variável regulada
Valor ideal i	Valor desejado

motor é igual à integral da velocidade desde o instante zero até o atual, mais a posição angular inicial. O sistema é portanto um *integrador* eletromecânico:

$$c \cong K v$$

$$\therefore a \cong K \int_0^t v \cdot dt + a_0.$$

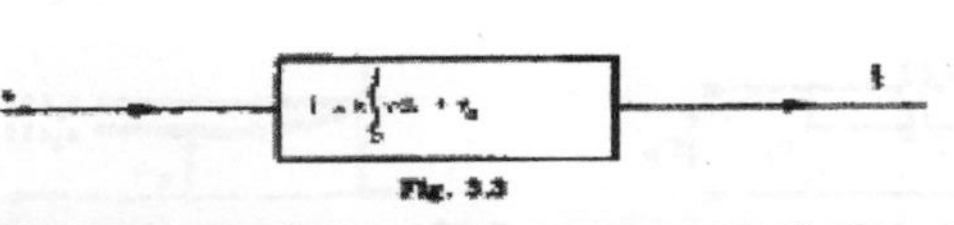

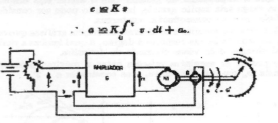

Fig. 3.3

Para melhor compreender a representação geral da Fig. 3.2, façamos a identificação das suas principais variáveis com as do sistema da Fig. 3.3.

A velocidade é a *variável controlada* (c); a posição do eixo é a *variável indiretamente controlada* (a); o *sistema ideal* é o integrador $[K\int_0^t v \cdot dt + a_0]$; o *êrro* do sistema é $y = (i - a)$. O potenciômetro P e a bateria constituem os *elementos da entrada* (A); o eixo do motor à direita do dínamo, e o ponteiro, constituem o sistema Z cuja saída é a variável *indiretamente controlada*. Na Fig. 3.3, não foram indicadas perturbações ao sistema de contrôle.

3.2 RESPOSTA AOS SINAIS DE COMANDO E DE PERTURBAÇÃO

Qualquer resposta de um sistema linear se obtém completamente pela L–transformada inversa do produto $[G(s) \cdot E(s)]$, onde $G(s)$ é a função de transferência correspondente e $E(s)$ é a L–transformada do sinal de entrada. Se houver várias entradas simultâneas, por exemplo referência e perturbações, a resposta total será a soma das correspondentes a cada entrada.

Vejamos a expressão geral da L–transformada da variável controlada, no caso do sistema a realimentação da Fig. 3.1. Temos

$$E = R - B \tag{3.1}$$

$$C = (G_1E + G_2U)G_2 \tag{3.2}$$

$$B = HC \tag{3.3}$$

Com (3.1) e (3.3) em (3.2), eliminam-se B e E, e portanto

$$C = \frac{G_1G_2}{1 + G_1G_2H} R + \frac{G_2G_2}{1 + G_1G_2H} U \tag{3.4}$$

Note-se que a relação entre R e C é semelhante à entre U e C: C é o produto de R (ou U) por uma fração, cujo numerador é o produto das funções de transferência encontradas pelo sinal caminhando de R (ou U) *diretamente* até C, e cujo denominador é a unidade mais o produto das funções que um sinal encontra quando percorre a *malha fechada*. Diz-se, para abreviar, que *o numerador é o "ganho direto"* e *o denominador é a unidade mais o "ganho da malha"*; os "ganhos" aqui são generalizados, isto é, são as funções de transferência.

3.3 CRITÉRIOS DE QUALIDADE EM REGIME ESTACIONÁRIO

Embora o *êrro do sistema* seja pròpriamente o assinalado com y na Fig. 3.2, o *sinal ou êrro atuante* e fornece muita informação sôbre o primeiro: freqüentemente, $Z(s)$, $A(s)$, $G_1(s)$ e $H(s)$ são tais que *o êrro atuante é proporcional ao êrro do sistema*, conforme provaremos no 3º item abaixo.

De grande importância para a análise dos servo-sistemas é saber, no caso de valerem as hipóteses acima, se o êrro tende a zero, a constante ou a ∞. É o que veremos agora, concentrando porém nossa atenção sôbre o *êrro atuante*; ser-nos-á possível estabelecer uma classificação dos sistemas em função dêsse êrro atuante estacionário, por simples inspecção das funções de transferência *da malha aberta*. Repetimos: qualquer conclusão dêsses resultados deve ser interpretada cuidadosamente com respeito ao êrro pròpriamente dito.

É claro também que, no que se segue, é obrigatória a hipótese de estabilidade do sistema.

e o sistema modificado para

$$\frac{200}{(1+400s)\,(1+s)^3} \quad \text{ou} \quad \frac{200\,y^3}{(y+400)(y+1)^3} \cong \frac{y^3}{2\,(y+1)^3}$$

A Fig. 6.1 mostra a compensação obtida, por meio dos gráficos de Bode e do Nôvo Lugar das Raízes.

Fig. 6.1

6.2 COMPENSAÇÃO EM SERIE, COM AVANÇO DE FASE

Trata-se de compensar o servo-sistema por introdução de um bloco em cascata com os outros, já fixados; pelo menos numa faixa de freqüências,

deve ocorrer um avanço da fase dos sinais que atravessam o bloco de compensação. As curvas de Bode da malha original devem-se somar as curvas de Bode do nôvo componente; ao diagrama de polos e zeros originais devem-se acrescentar novos polos e/ou zeros.

Exemplifiquemos com um circuito compensador elétrico, um quadripolo R-C, da função de transferência

$$Gc = \frac{E_2}{E_1} = \frac{1}{a}\;\frac{1+aTs}{1+Ts} \qquad (6.1)$$

$$a = \frac{R_1+R_2}{R_2} \qquad T = \frac{R_1R_2}{R_1+R_2}\,C$$

Supondo que a impedância do gerador que o excita seja desprezível e que a da sua carga seja muito grande, tal circuito pode ser considerado um bloco *isolado*, com as conseqüentes vantagens.

Na Fig. 6.2, temos o circuito e as diversas descrições gráficas que conhecemos; devido ao aspecto da sua resposta a degrau, a qual lembra a derivada do degrau, tal circuito é às vêzes chamado de *derivativo*.

Podem-se provar as seguintes relações entre o máximo defasamento $\varnothing_m$, a freqüência ω_m em que êle ocorre, e as constantes a e T.

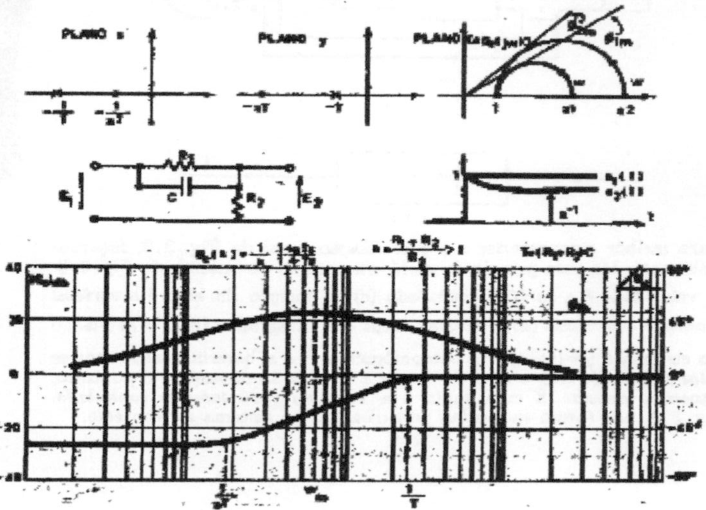

Fig. 6.2

8 — Cattoni

Sejam D_i os menores principais da matriz A:

$$D_1 = |a_{11}| \quad D_2 = \begin{vmatrix} a_{11} & a_{12} \\ a_{21} & a_{22} \end{vmatrix} \quad D_n = \begin{vmatrix} a_{11} & a_{12} & \dots & a_{1n} \\ \dots & \dots & \dots & \dots \\ a_{n1} & \dots & \dots & a_{nn} \end{vmatrix} = |A|$$

Pode-se provar que

1. *Para* v(x) *ser definida positiva é necessário e suficiente que todos os* D_i *sejam* > 0.

2. *Para ser definida negativa, é necessário e suficiente que* $D_1<0$, $D_2>0$, $D_3<0, \dots, D_n(-1)^n > 0$.

3. A substituição dos sinais > em 1. e 2. por ≥ 0 *não leva* às condições para $v(x)$ ser semidefinida positiva; as condições corretas são bem mais complicadas e podem ser encontradas nos tratados de teoria das matrizes.

11.4 OS TEOREMAS DA ESTABILIDADE DE LIAPUNOV [1]

Referem-se à estabilidade local da origem do espaço **y**; enunciemo-los no caso autônomo $\dot{y}=f(y)$ e supostas satisfeitas as condições de existência e unicidade de suas soluções.

1º *Teorema*: a origem é um ponto de equilíbrio estável para o sistema $\dot{y}=f(y)$, *se existe* uma função *definida positiva* $v(y)$ tal que a sua derivada total $\dot{v}(y)$ ao longo das trajetórias do ponto-sistema seja *semidefinida negativa*.

2º *Teorema*: a origem é um ponto de equilíbrio *assintòticamente estável*, se existe uma função *definida positiva* $v(y)$ tal que a sua derivada total $\dot{v}(y)$ ao longo das trajetorias seja *definida negativa*.

Examinemos uma interpretação geométrica de $\dot{v}(y)$ que leva imediatamente a uma justificação dos teoremas acima (Ref. L.3).

Calculemos a derivada total $\dot{v}(y)$ sôbre trajetórias:

$$\dot{v}(y) = \sum_{k=1}^{n} \frac{\partial v}{\partial y_k} \cdot \frac{dy_k}{dt} \tag{11.8}$$

É fácil verificar que esta expressão é o produto escalar de

$$\text{grad } v(y) = \sum_{k=1}^{n} \frac{\partial v}{\partial y_k} \cdot u_k \quad \text{por}$$

$$V(y) = \sum_{k=1}^{n} \frac{dy_k}{dt} \cdot u_k$$

(1) Existe também um Teorema da Instabilidade, de Liapunov (Ref. G.4 e L.2).

onde os u_k são os vetores unitários que definem os eixos do espaço **y** e o $V(y)$ é o vetor-velocidade do ponto sistema nesse espaço.

Sendo o gradiente sempre normal às superfícies de nível da função considerada, isto é, às superfícies definidas por $v(y)$=constante, podemos escrever (11.8) como

$$\dot{v}(y) = V(y) \cdot \text{grad } v(y) = V(y) \cdot n \cdot G \tag{11.9}$$

onde n é a normal à superfície do nível que passa pelo ponto y e G é o módulo do gradiente (número positivo ou nulo).

Concluindo, a derivada total da função $v(y)$ tem sinal algébrico totalmente dependente da projeção do vetor-velocidade do ponto-sistema no ponto y da trajetória sôbre a normal à superfície de nível da função $v(y)$ que passa pelo ponto $\overline{y}$. Ver Fig. 11.4.

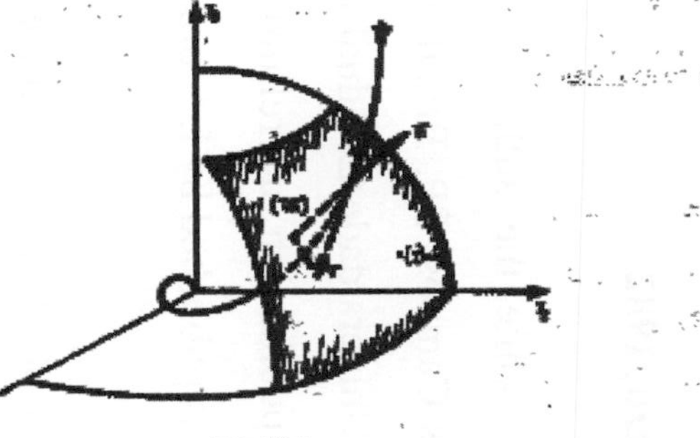

Fig. 11.4

Como a origem é um ponto de mínimo isolado para $v(y)$, forçosamente *existem vizinhanças convenientes* da origem em que $v(y)$ cresce quando $\|y\|$ cresce, isto é, em que *grad* $v(y)$ é dirigida "para fora" das superfícies de nível. Nessas vizinhanças, então, interpreta-se o sinal de $\dot{v}(y)$ assim:

$\dot{v}(y)$	*Projeção do vetor-velocidade sôbre o vetor-normal*	*Significado*
1. > 0	> 0	y "sae" da superfície de nível
2. < 0	< 0	"entra"
3. = 0	= 0	"permanece"

Princípios de Controle e Servomecanismos

Celso Pascoli Bottura
Guanabara Dois, Rio de Janeiro, RJ, Brazil, 1982

Work on this book was started in 1965 while the author was teaching at ITA (Instituto Tecnológico de Aeronáutica) São José dos Campos, SP, Brazil and it was completed in 1972 when he was already was at UNICAMP (Universidade Estadual de Campinas), Campinas, SP, Brazil). After some initial problems with publication, a Preliminary Edition was published at UNICAMP in 1980, and finally the book was published by Editora Guanabara Dois in 1982.

It is a classical control textbook for a sixteen week undergraduate course, with one or two three hour sessions weekly, including both analog and digital control analysis and design with a frequency approach.

Contents:

- Introduction
- Feedback control systems: Linear Analysis
- Feedback control systems: Linear Synthesis
- Multivariable control

Contributed by

Celso Pascoli Bottura
School of Electrical and Computer Engineering, UNICAMP
Campinas SP, Brasil

PRINCÍPIOS
DE CONTROLE E
SERVOMECANISMOS
Bottura

Na Fig. 3.20 apresentamos os diagramas de um compensador lag.

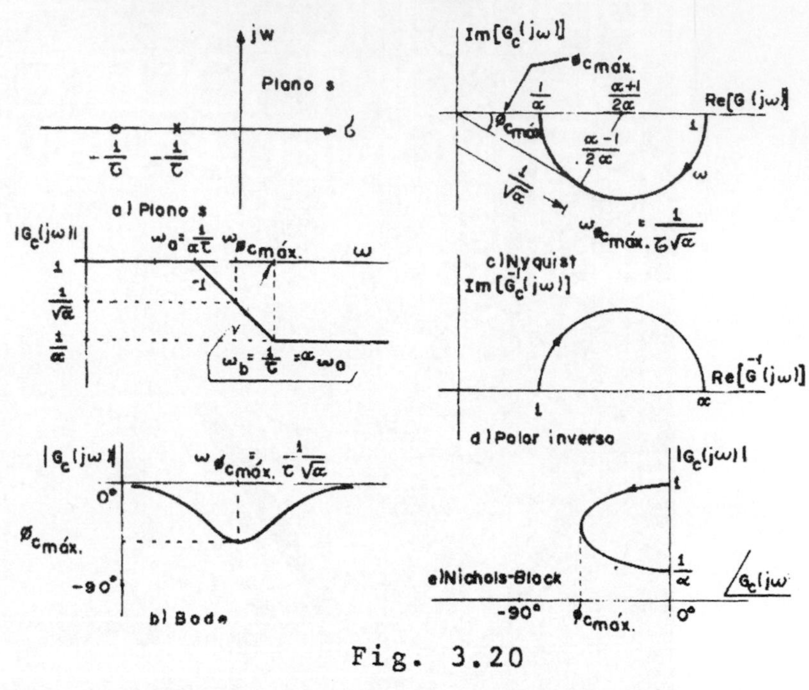

Fig. 3.20

Com o compensador-atraso não alteramos o tipo do sistema. Contudo, é claro que quanto mais próximo da origem do plano s colocamos o pólo real negativo do compensador, mais próximos estaremos da realização do nosso objetivo. Além disto, também é evidente que o zero do compensador não deve ficar muito longe do seu pólo, pelas razões discutidas anteriormente. Portanto, em termos dos diagramas de resposta ou freqüência, deveremos colocar o compensador lag nas baixas freqüências.

Da Fig. 3.20, é imediato que o atr

EARLY CONTROL TEXTBOOKS in the PEOPLE'S REPUBLIC OF CHINA

Fundamental Theory of Automatic Control

Translation from the Russian original by A. A. Voronov
Translated by J. R. Xu, B. W. Wan, Z. T. Wang and S. Q. Zheng
Date of publication: 1957

Fundamentals of Automatic Control

Translation from the Russian original by V. V. Solodovnikov et al.
Translated by Zhongtuo Wang
Date of publication: 1957-59

Servomechanisms

Author: Huitang Chen
Date of publication: 1961

Fundamentals of Automatic Control Theory

Author: Liu Bao
Date of publication: 1963

Material collected and processed by

Xiao Deyun and Zheng Dazhong

Tsinghua University
Beijing, P.R. China

Fundamental Theory of Automatic Control

Russian original written by A. A. Voronov

Chinese translation by Jun-Rong XU, Bai-Wu WAN, Zhong-Tuo WANG and Shou-Qi ZHENG.

Published in 1957.

The author of the book "Fundamental Theory of Automatic Control" is A.A. Voronov from the Soviet Union.. The book was translated into Chinese by four Chinese university lecturers: Jun-Rong XU, Bai-Wu WAN, Zhong-Tuo WANG and Shou-Qi ZHENG.

The second edition of this book in Russian was published in 1954 by the Military Publisher, Ministry of Defense, Soviet Union. The textbook was based on the teaching material the author gave for graduate students and scientific researchers of Moscow Bauman High Technology University and for students of Moscow Molotov Power and Energy Institute in 1948-1952.

The translated manuscript in Chinese of the very book had been used as teaching material for students in the Department of Electrical Engineering, Jiaotong University, Shanghai, since the beginning of 1955 by XU and WAN. This was the earliest control theory course taught in China. The book was published by the Electrical Power Publisher in 1957 with 470 pages and 2 impressions and the total number of copies of the book was about 10,000.

The book in Chinese was widely used as a textbook or reference for undergraduate students in Electrical Engineering as well as others over 15 years. It became one of the most important books for linear control theory in China.

The book was focused on linear control theory. It gave some simple mathematical preparation like Laplace transformation, theory of complex variables for students that were not familiar with. Its outstanding characteristic was that it paid attention to the basic concepts of the control problem. For example, it gave a good explanation to the linearization of a nonlinear plant or element such as prime mover, d.c. generator, two–phase induction motor etc. To explain the real meaning of frequency characteristics the book introduced the concept of harmonic excitation to control systems; to prove the amplitude-phase stability criterion it used the diagram of rotation of each vector denoted by $(j\omega - s_{in})$ in numerator and denominator instead of the strict mathematical proof. Another outstanding characteristic was that it also paid attention to some practical problems, which evidently were very important like the steady state behavior of the control systems, differential equations of plants and elements of different kinds and different nature. The book addressed not only the control problems in electrical engineering but also in mechanical and aviation engineering. Very detailed examples were given to illustrate the theory such as: electronic voltage regulator, servo system of radar station, prime mover speed control by centrifugal governor, carbon-pile voltage regulator etc. Enlightening questions were available for students at the end of each chapter.

Contents

Chapter 3 Plants and Elements Equations in Automatic Control Loop

A. Controlled plants

B. Sensitive elements of controllers

C. Prime movers

D. Amplifiers

E. Stabilizing elements

Chapter 4 Equations, Transfer Functions and Frequency Characteristics of Elements and Systems

Chapter 5 Stability of Linear Dynamic Systems

Chapter 6 Examples to Study Stability of Automatic Control Systems

Chapter 7 Study on Transient Quality of Control Systems

Appendixes

References

Contributed by
Prof. Wan Baiwu
Xi'an Jiaotong University, Xian, China

自动调整理论基础

苏联 阿·阿·伏龙諾夫著

徐俊荣 万百五 王众託 郑守淇譯

电力工業出版社

A Typical Page from the Book "Fundamental Theory of Automatic Control"

Schematic diagram and mathematical description of a selsyn servosystem with amplifier-dc generator-motor set

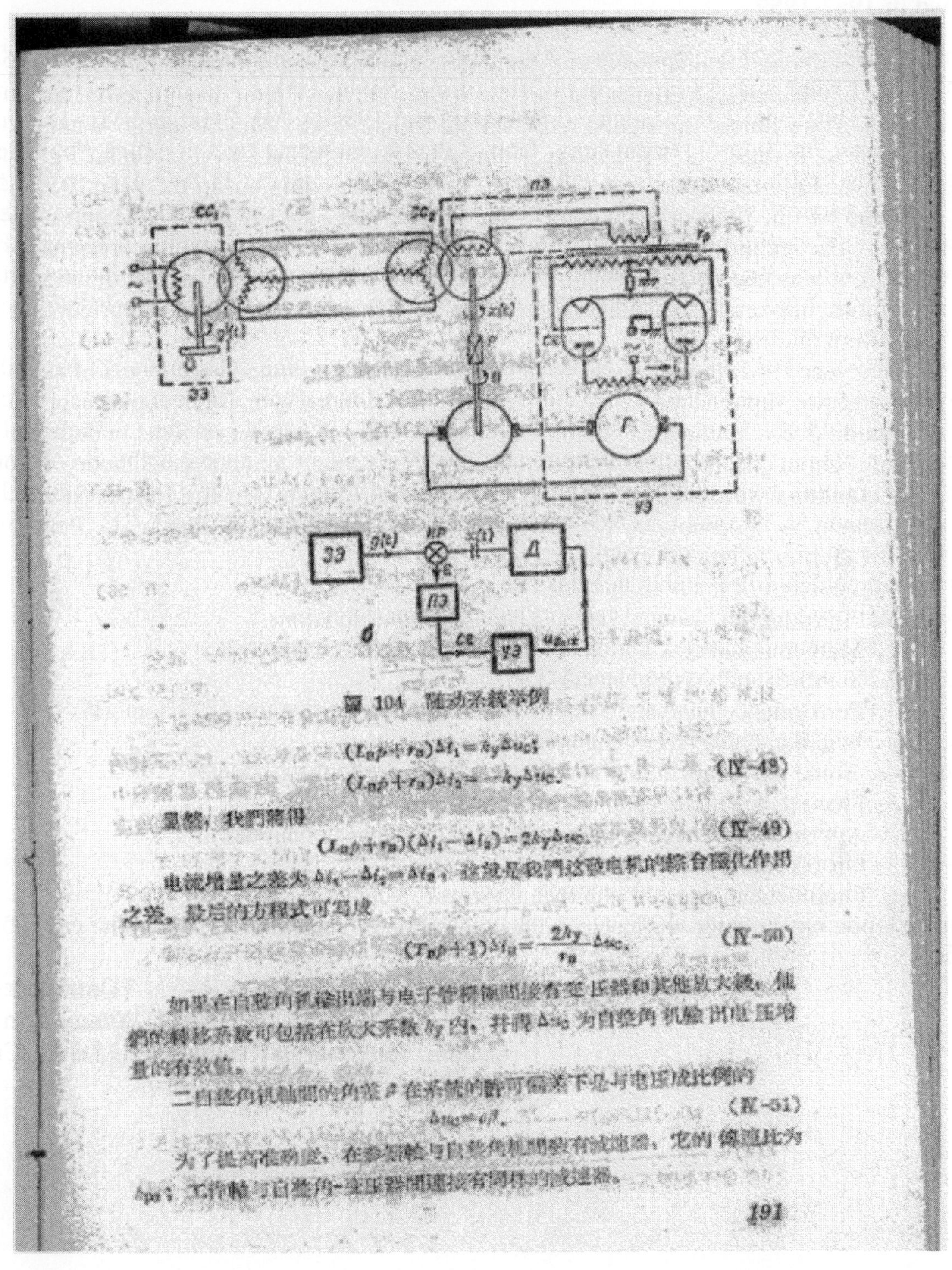

图 104 随动系統举例

$$(L_n p + r_n)\Delta i_1 = k_y \Delta u_c;$$
$$(L_n p + r_n)\Delta i_2 = -k_y \Delta u_c. \qquad (IV-48)$$

显然，我們將得

$$(L_n p + r_n)(\Delta i_1 - \Delta i_2) = 2k_y \Delta u_c. \qquad (IV-49)$$

电流增量之差为 $\Delta i_1 - \Delta i_2 = \Delta i_B$，这就是我們这發电机的綜合磁化作用之差。最后的方程式可写成

$$(T_n p + 1)\Delta i_B = \frac{2k_y}{r_n}\Delta u_c. \qquad (IV-50)$$

如果在自整角机输出端与电子管栅極間接有变压器和其他放大器，他們的轉換系數可包括在放大系數 k_y 內，并視 Δu_c 为自整角机輸出电压增量的有效值。

二自整角机軸間的角差 θ 在系統的許可偏差下是与电压成比例的

$$\Delta u_c = c\theta. \qquad (IV-51)$$

为了提高准确度，在參加軸与自整角机間裝有减速器，它的傳速比为 k_{p1}；工作軸与自整角-变压器間連接有同樣的减速器。

191

Fundamentals of Automatic Control

Russian original written by a collective of authors (V. V. Solodovnikov, Editor)
Chinese translation by Zhongtuo Wang
Published in 1957-1959

The Russian book "Fundamentals of Automatic control" was published in the year 1954 by the Publisher of Mechanical Engineering of the former Soviet Union and imported to China in the year 1955. The Chinese translation work started from 1956 by Mr. Zhongtuo Wang, a young teacher at Dalian Institute of Technology, China. It was published by the China Publisher of Electrical Power Engineering in three volumes. Volume one come out in the year 1957 in 5100 copies, volume two in 1958 in 5100 copies, and volume three in 1959 in 5780 copies. Later, in the year 1960, the second printing of volume one and two come out in 4620 copies each. From the year 1958, it was becoming the main textbook of the course "Theory of Automatic Control" in many famous universities of China. The Chinese translation was also exported to Korean Democratic Republic from 1960.

From the year 1961, the book was assigned as textbook in some departments of several key universities and was republished by the China Publisher of Industry in 7620 copies each volume.

As a standard text, it contains not only the linear theory of control systems in detail, but also some chapters about the problems of nonlinear theory as well as statistical theory of control systems. The authors were distinguished scholars in control theory of the former Soviet Union like M. Aizerman, A. Voronov, A. Krasovsky, A. Lerner, A. Letov, B. Petrov, G. Posperov, V. Solodovnikov (Editor in chief), Y. Zipkin, etc.

The main content of the book includes two parts:

Part one: Introduction. Linear Theory. There are four divisions:

1. Mathematical descriptions of control systems (8 chapters).
2. Stability analysis (6 chapters).
3. Performance analysis. System Synthesis. Discrete systems (7 chapters).
4. Statistical analysis (4 chapters).

Part two: Some problems in nonlinear theory. There are four divisions:

1. Phase space and its applications (6 chapters).
2. Approximate analysis of the periodical process (3 chapters).
3. On-off systems (3 chapters).
4. Graphical analysis of nonlinear systems (2 chapters).

This textbook had a major influence on the education of control theory from the year 1958 to 1966.

Contributed by
Prof. Wang Zhongtuo
Dalian University of Technology, Dalian, China

The Front Covers of the Book "Fundamentals of Automatic Control"

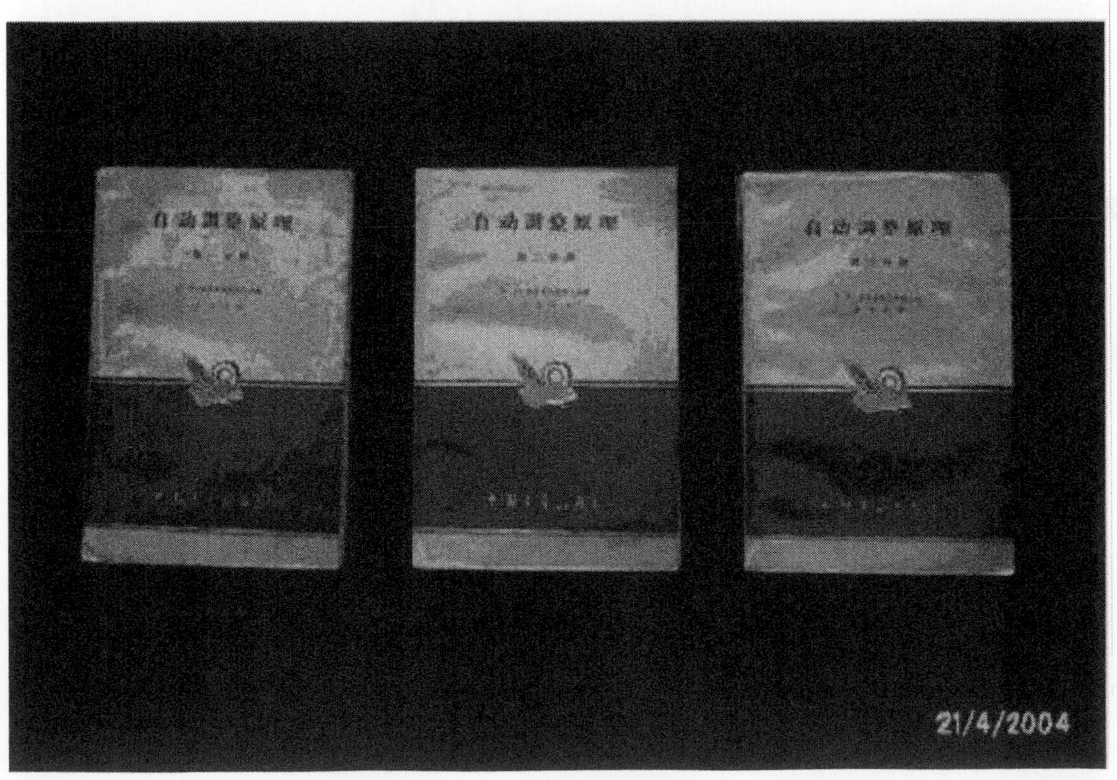

A Typical Page from the Book "Fundamentals of Automatic Control"

The control system of an aircraft compressor and the differential equations describing the dynamics of the system

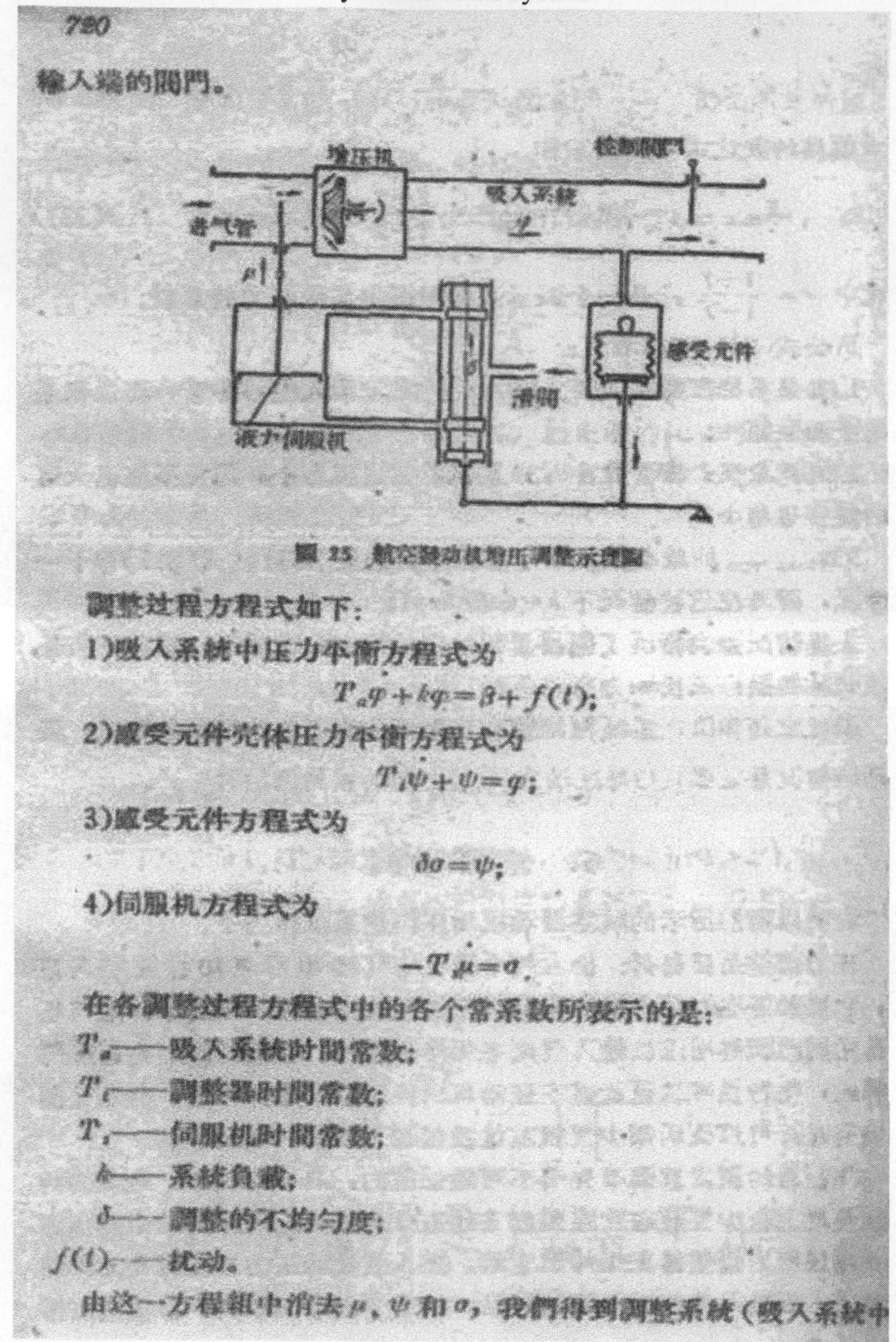

720

輸入端的閥門。

圖 25 航空發动机增压調整示理圖

調整过程方程式如下:

1)吸入系統中压力平衡方程式为

$$T_a\dot{\varphi}+k\varphi=\beta+f(t);$$

2)感受元件壳体压力平衡方程式为

$$T_i\dot{\psi}+\psi=\varphi;$$

3)感受元件方程式为

$$\delta\sigma=\psi;$$

4)伺服机方程式为

$$-T_s\dot{\mu}=\sigma.$$

在各調整过程方程式中的各个常系数所表示的是:

T_a——吸入系統时間常数;

T_i——調整器时間常数;

T_s——伺服机时間常数;

k——系統負載;

δ——調整的不均匀度;

$f(t)$——扰动。

由这一方程組中消去μ、ψ和σ,我們得到調整系統(吸入系統中

Servomechanisms

Author: Huitang Chen
Published in 1961

The author of the book “Servomechanisms” is Mr. Huitang Chen. This book was written based on the author’s lecture notes for the course of “Servomechanisms” in Xi’an JiaoTong university in the fall of 1958. It was published as a national unified textbook by the People’s Educational Press in July, 1961 after being evaluated and recommended by the National Unified Textbooks Committee. Revised twice, this book has 3 editions. By the time of July 1964, when the 3rd edition of the book was printed the 4th time, the book had a total printing of over 11.5 thousand copies.

The history of writing this book is as follows: During the early 1950s, we learned from the Soviet Union and began to set up the specialty of “Electrification for Industrial Enterprises” in China .A course named the “Theory of Automatic Regulation” was arranged in the program. The textbooks we used were those translated from the textbooks of the Soviet Union. For example, the book “Theory of Automatic Regulation” by Voronov was adopted. This book mainly described the basic principles of automatic control. By the time of the mid 1950s, another specialty called “Automation and Telemechanics” was established. In order to strengthen the students’ ability of design and let them grasp the practical systems, the course “Servomechanisms” was included in the program. As the teacher of this course, I deeply felt the deficiencies of that time, namely that we had overemphasized learning from the Soviet Union while ignoring new achievements made by America and the European countries, such as the root-locus method, M-N graph and methods of investigating servomechanism in a statistical context, etc. Therefore, most of the contents in my teaching materials came from the textbooks of America and the European countries. Meanwhile, some advanced methods developed in the Soviet Union were also included, such as time-optimal relay control, time-optimal servomechanisms, synthesis of systems based on the desired frequency response etc. Therefore, this book gave a brand new look to China of that time and was widely used nationwide. It was also very popular with engineers and technicians. It made great contributions by educating many specialists of a generation in automation. Many of those who now are academic leaders in the fields of automatic control remember the influence this book had on them.

Contents

Chapter 6. Time-optimal servomechanism
6-1. Concepts of time-optimal servomechanism
6-2. Determination of the form of optimal process
6-3. Building of time-optimal servomechanism
6-4. Example
Chapter 7. Selection of power element of servomechanism
7-1. Introduction
7-2. Load characteristic
7-3. Torque , speed and power demanded by the load
7-4. Determination of torque, speed and power for the power element
7-5. Motor-load gear ratio selection for total inertia
7-6. Influence of gear ratio to the total inertia
Chapter 8. Experimental methods for research of servomechanism
8-1. Introduction
8-2. Determination of dynamic characteristic using infra-low frequency device
8-3. Determination of system parameters basing on transient diagram by oscilloscope
Chapter 9. Relay servomechanism
9-1. General relay servomechanism
9-2. Linearized relay servomechanism
9-3. Time-optimal relay servomechanism
Chapter 10. Statistical dynamics of Servomechanism
10-1. Necessity of research using statistical method
10-2. Fundamental knowledge of possibility theory
10-3. Stationary random process
10-4. Response of system to the random signal
10-5. Calculation of integral of error spectrum density
10-6. Conclusion
Appendix

Contributed by
Prof. Chen Huitang
Tongji University, Shanghai, China

The Front Cover of the Book "Servomechanisms"

A Typical Page from the Book "Servomechanisms"

An example of a time-optimal servomechanism for an automatic compensator by using nonlinear state feedback of velocity

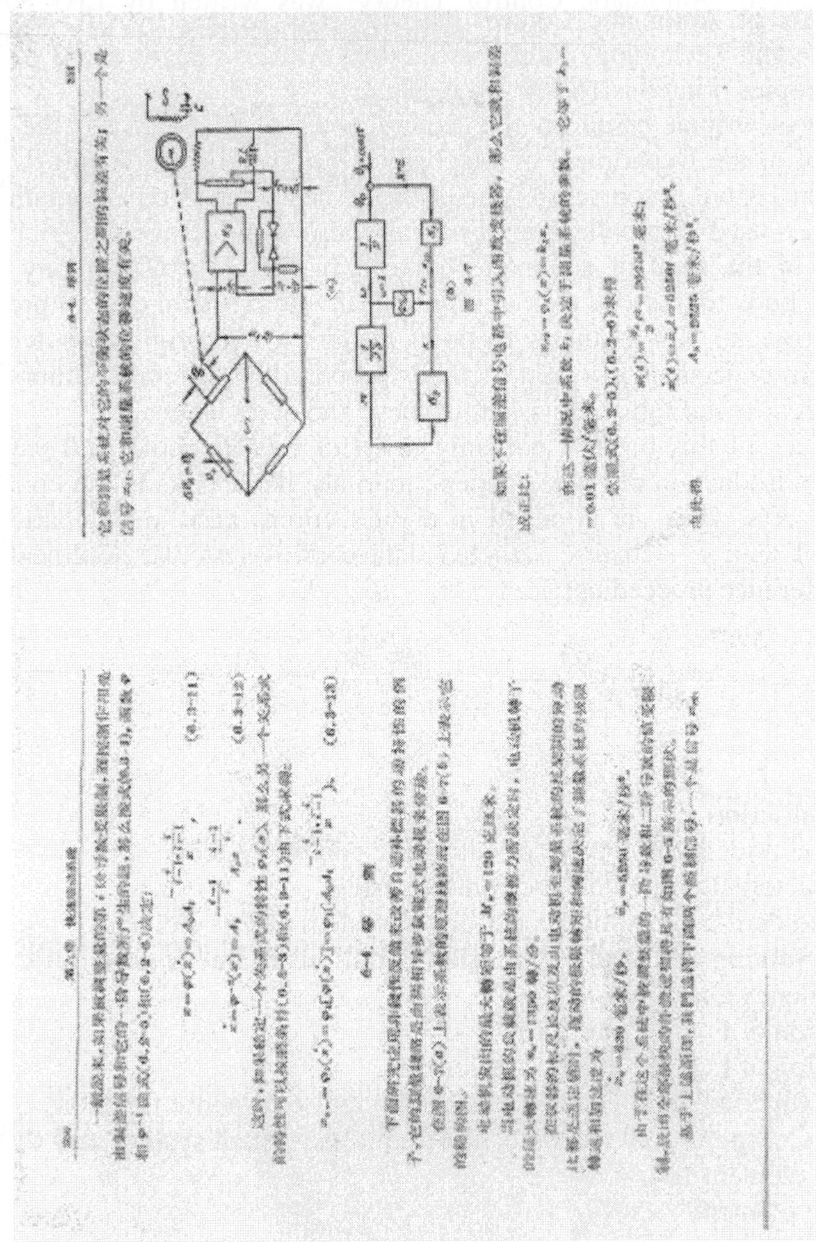

Fundamentals of Automatic Control Theory
Author: Liu Bao
Published in 1963

"Fundamentals of Automatic Control Theory" was written by LIU BAO, published by Shanghai Science and Technology Publisher in 1963 with 568 pages and 3 impressions and the total number of copies printed is 16,500.

This book was written based on the lecture notes of Liu Bao for the senior course on automatic control in the Department of Mechanical Engineering at Tianjin University between 1956 and 1966. In 1956 this course was one of the earliest courses on automatic control taught in China. It had been used as a college textbook and also as reference material by engineers and research people in the field of automatic control. In the early 60's, many Chinese research workers read this book to improve their ability to deal with practical control problems.

This book provided 127 footnotes to point readers to the original sources of theories and rules, to help them understand different methods proposed by different authors and to help them find various references in English, Russian, Chinese and other languages.

The References of this book is not only a list of related books and papers, it is also an explanation and introduction of books, papers, journals, handbooks in the control field from the 40's to the early 60's. They are arranged in eight sections: basic mathematics; linear systems; automatic control theory; statistics; sampled data control systems; nonlinear systems; others; journals and conference proceedings.

Contents

Chapter 4 Stability of automatic control system
4-1 The fundamental idea of stability
4-2 Rouse-Hurwitz stability criterion
4-3 Mihailov stability criterion
4-4 Nyquist stability criterion
4-5 Analyzing stability by log frequency characteristics
4-6 Stability of multi-loop system
4-7 Remainder of stability
4-8 Stability of automatic control system with dead time element
4-9 The structure stability of automatic control system
Chapter 5 Quality analysis of automatic control system
5-1 Quality index for the automatic control systems
5-2 Integral appraisal method
5-3 Relationship between system quality and distribution of zeros and poles of transfer function
5-4 Root locus method
5-5 Frequency method
5-6 Methods to draw different characteristic curves
Chapter 6 Stabilization, correction and synthesis of automatic control system
6-1 Correction of automatic control system
6-2 Fundamentals of servo system synthesis
6-3 The extension of frequency method
6-4 Synthesis of multi-input multi-loop systems
6-5 Synthesis by the distribution of system poles and zeros
6-6 Application of root locus method in system analysis
6-7 Comparison of different synthesis methods
Chapter 7 Analysis and synthesis problem of control system under random process action
7-1 Some fundamental conceptions of probability
7-2 Correlation function and spectral density
7-3 Statistical method to calculate the dynamic characteristics of the controlled object
7-4 Dynamic precision of automatic control system
7-5 Synthesis of system under random process action
Chapter 8 Nonlinear automatic control system
8-1 General statement
8-2 Types of nonlinear characteristics in automatic control system
8-3 Basic idea about phase locus method
8-4 Phase diagram of nonlinear system
8-5 Point turning method
8-6 The method of improving system dynamic quality by use of nonlinear characteristics
8-7 Harmonic balance method
8-8 Analysis of nonlinear system under random inputs
Appendixes
I h-function table
II To find Lm/W+-/ and arg(W+-)from log frequency curves Lm/W/ and arg /W+-/
III Table of P,Q functions

Contributed by
Prof. Liu Bao
Tianjin University, Tianjin, China

The Front Cover of the Book "Fundamentals of Automatic Control Theory"

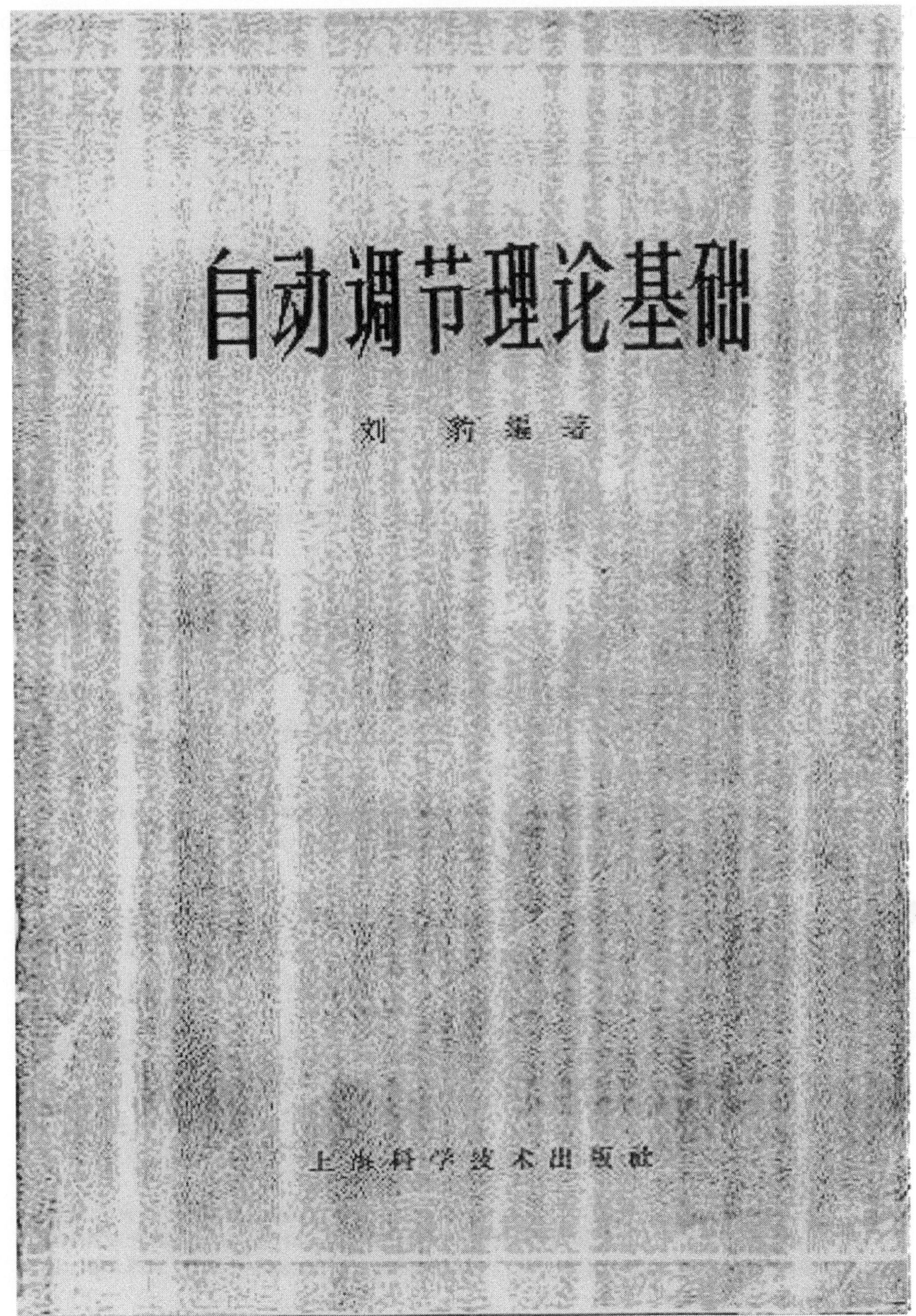

A Typical Page from the Book "Fundamentals of Automatic Control Theory"

A new method to synthesize a control system by drawing frequency characteristics from the system's unit step response

5-6 各种特性曲线的绘制方法 247

也可以用三角形方法代替梯形法❶，其基础是一样的。

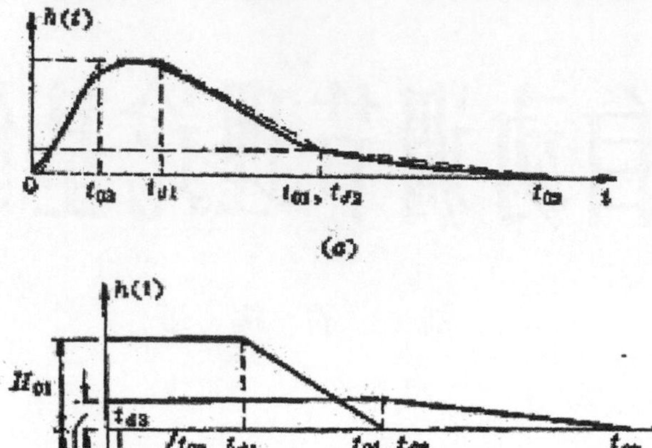

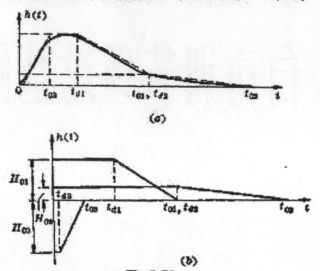

(a)

(b)

图 5-79

(2) 从单位过渡过程繪制频率特性的方法❷ 如已知单位过渡过程如图 5-79 a，以直線段来代替，则这些直線段本身又可以用图 5-79 b 中很多梯形时間特性来表示，亦即

$$h(t) \approx \sum_{i=1}^{n} H_i(t) \qquad (5\text{-}187)$$

式中 $H_i(t)$ 为各梯形时間特性，它的定义如图 5-80 所示，其数学式为

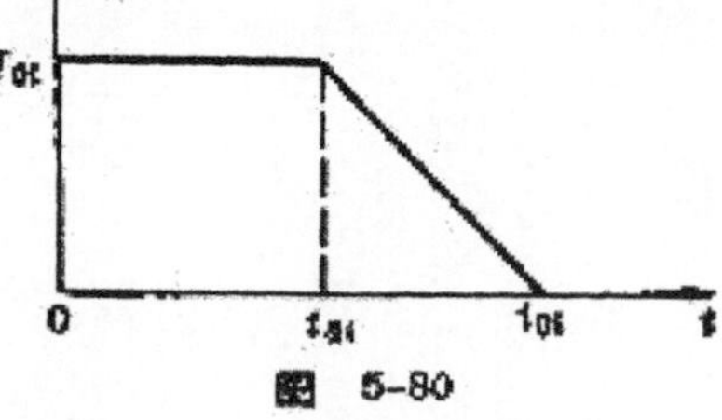

图 5-80

$$\left.\begin{array}{ll} \text{当 } 0<t<t_{di} \text{ 时，} & H_i(t)=H_{0i} \\ t_{di}<t<t_{0i} \text{ 时，} & H_i(t)=H_{0i}\dfrac{t_{0i}-t}{t_{0i}-t_{di}} \\ t_{0i}<t \text{ 时，} & H_i(t)=0 \end{array}\right\} \qquad (5\text{-}188)$$

---

❶ А. А. Воронов, Элементы теории автоматического регулирования, Военное изд., 1954.

❷ 本法基本概念，见刘豹，同时按扰动作用及給定作用来綜合自动調节系統的频率法，天津大学，1961，其具体計算方法及表格由天津大学工业控制仪表专业 1962 年应届毕业生何恩智在著者指导下做成，见何恩智，自动調节系統以频率法綜合的推广，天津大学工业控制仪表专业毕业論文，1962。

EARLY CONTROL TEXTBOOKS

in the

CZECH REPUBLIC

(Czechoslovakia)

Servomechanisms

Trnka Z.
In Czech, SNTL Prague, 1954

Foundations of Automatic Control

Strejc V., M. Šalamon, Z. Kotek and M. Balda
In Czech, SNTL Prague, 1958

Elements of Control Loops

Balda M., V. Strejc and M. Krampera
In Czech, SNTL Prague, 1958

Dynamics of Controlled Systems

Čermák J., V. Peterka and J. Závorka
In Czech, ACADEMIA Prague, 1968

Contributed by

Vladimir Strejc

Czech Technical University
Prague, Czech Republic

Servomechanismy

Trnka Z.
In Czech, SNTL Prague, (1954), pp. 372, Fig.274, Tab. 5.

The book is devoted to the linear theory of positional servomechanisms of electrical and mechanical controlled systems. The author originally wrote the book for students of the department of Electrical Engineering at the Technical University in Prague. These students were familiar with the mentioned systems by that time and could easily accept an extension of lectures to the automatic control of servomechanisms. It was a right premise and, consequently, the book got in a short time a very high popularity. In the USA several books appeared after the Second World War concerning this subject, so that the evolution at the Technical University actually followed a general development.

In the first chapters the ways of formulating the differential equations for electrical and mechanical linear systems are briefly reviewed. The author then designs a most simple linear positional servomechanism, solved first by means of conventional procedures in time domain. Using this example, different solutions are demonstrated, which are deeply elaborated in the next chapters introducing the frequecy characteristic in a complex plain and in logarithmic coordinates. The reader can recognise the main parts of servomechanisms as well as their behaviour when a sudden step change, input speed change or sinusoidal input occurs.

The Laplace transform is briefly presented. The basic positioning servomechanisms with derivative and integral of the control error are solved by means of the Laplace transform. The resulting differential equations are at most of the third order.

The next part of the book is devoted to the elements of servomechanisms. Principles, operation and Laplace transforms of the most common positioning servomechanisms are explained.

The following part of the book presents the synthesis of servomechanisms. The author mentions the stability cost functions which are frequently used and go directly to the desired result. Besides the complete Nyquist cost function, Michailov-Leonhard and Hurwitz criterions, important for a rapid verification of stability, are introduced.

The other part presents the algebra of block diagrams and procedures of designing a control by means of graphs in the complex plain and in logarithmic coordinates. It is mentioned that the procedure of graphs in logarithmic coordinates is for the synthesis of linear systems most rapid. It holds particularly for control loops where it is not necessary to ascertain the stability of arrangements with unstable inner loop.

Another part deals with experimental determination of elements of servomechanisms and of complete systems. The aim is to get responses to arbitrary input time signal with the help of the frequency characteristic. Some numerical examples of linear positional servomechanisms are given, e.g. servomechanism with viscous damping, servomechanism controlled by error and its integral, one with viscous damping corrected by imperfect integrating circuit, modification of the frequency characteristic by differentiating circuit and design of control loop stabilised by feedback.

Non-linear servomechanisms are treated only partly due to the difficulties in the solutions. Only the principle of solutions in the phase plain is mentioned. Parasite changes appearing at the "linear" servomechanisms are discussed.

Prof. Ing. Dr ZDENĚK TRNKA

SERVOMECHANISMY

Schváleno výnosem ministerstva školství ze dne 17. března 1954
č. j. 19.079/54-S/4
jako celostátní vysokoškolská učebnice

PRAHA 1954

STÁTNÍ NAKLADATELSTVÍ TECHNICKÉ LITERATURY

vinutí (obrázek 18,18). Tím se dosáhne stejného účinku, jako u závitu nakrátko.

To platí ovšem jen pro selsyny použité pro přímý přenos úhlu v zapojení podle obrázku 18,16. U selsynů použitých jako polohové transformátory by závity nakrátko rušily a je lépe zde použít rotorů bez vyjádřených pólů, neboť vyjádřená magnetická osa rotoru vyvolává momenty i u selsynů zapojených podle obrázku 18,15 jako polohové transformátory.

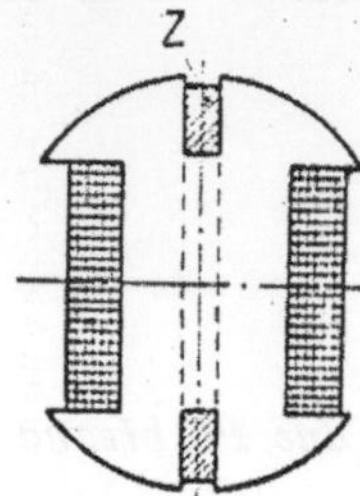

Obr. 18,17. Rotor selsynu se závitem nakrátko (z), který potlačuje příčnou složku pole.

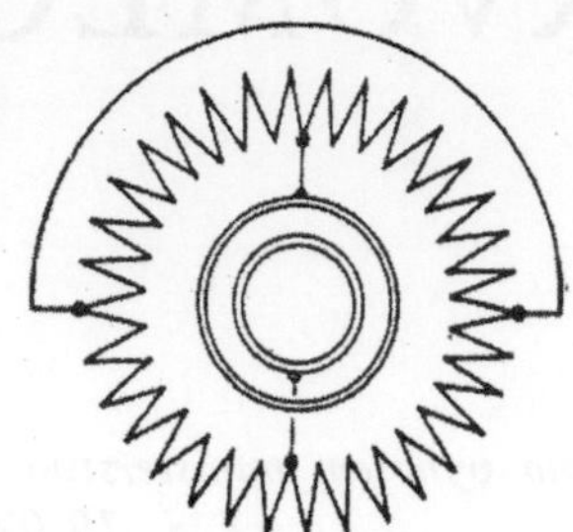

Obr. 18,18. Vinutí rotoru selsynu s nevyjádřenými póly. Diametrálně položené body vinutí jsou spojeny nakrátko.

Označíme-li výchylku selsynu vysilače φ_1 a výchylku selsynu přijimače φ_2, je odchylka poloh obou selsynů $\varphi = \varphi_1 - \varphi_2$ a moment působící na selsyn přijimač je $M = K\varphi$. Pro pohyb selsynu přijimače platí diferenciální rovnice

$$J \frac{d^2\varphi_2}{dt^2} + B \frac{d\varphi_2}{dt} = K\varphi \qquad (18,12)$$

$$J \frac{d^2\varphi_2}{dt^2} + B \frac{d\varphi_2}{dt} + K\varphi_2 = K\varphi_1 \qquad (18,13)$$

kde J je moment setrvačnosti rotoru přijimače, B je konstanta tlumení (indukovanými proudy při pohybu). Máme zde opět diferenciální rovnici harmonického pohybu tlumeného. Rovnici transformujeme a zavedeme opět přirozenou frekvenci ω_n a konstantu poměrného tlumení a (viz vztahy 12,20 a 12,21)

$$(p^2 + 2a\omega_n p + \omega_n^2)\, \Phi_2(p) = \omega_n^2\, \Phi_1(p)$$

$$\frac{\Phi_2(p)}{\Phi_1(p)} = \frac{\omega_n^2}{p^2 + 2a\omega_n p + \omega_n^2} \qquad (18,14)$$

Máme zde stejné vztahy jako u servomechanismu s viskosním tlumením, ale přímé spojení dvou selsynů pro přenos úhlu není servomechanismus, neboť

184

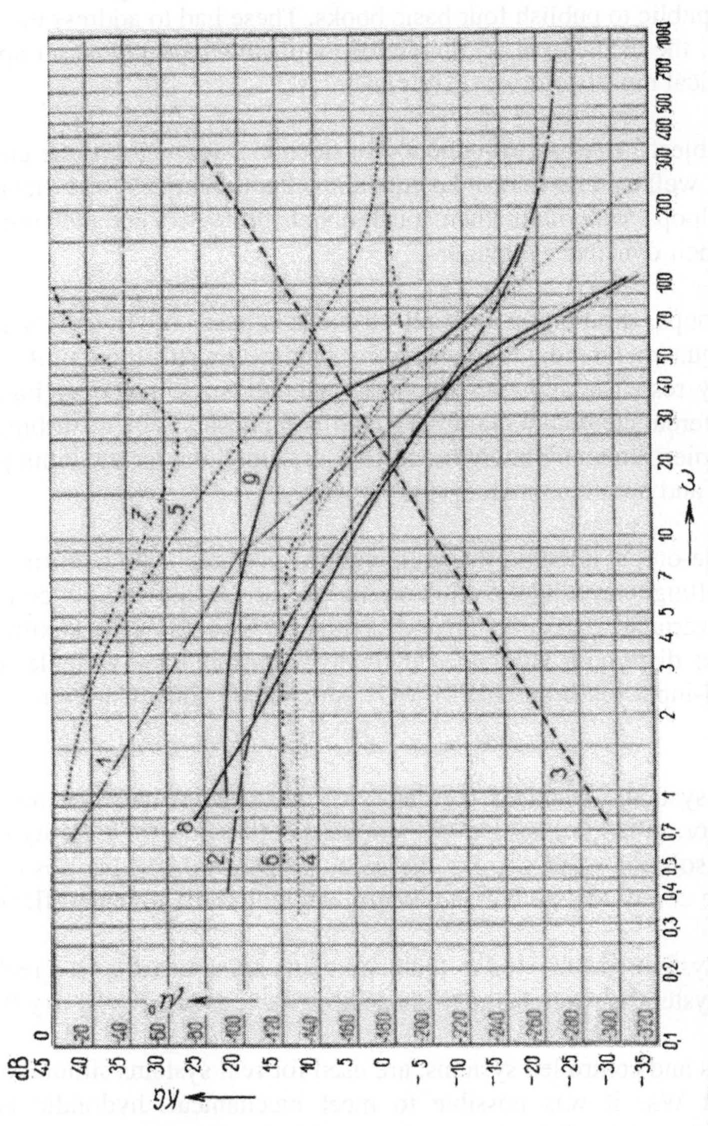

Obr. 28,6. Frekvenční charakteristika servomechanismu stabilisovaného zpětnou vazbou. 1, 2. Původní průběh K_1G_{1dB} a fáze μ_1 nestabilního servomechanismu. 3. Frekvenční charakteristika K_zG_{zdB} členu ve zpětné vazbě. 4, 5. Asymptoty frekvenční charakteristiky a fáze funkce $\mathbf{K_1G_1}(j\omega)\ \mathbf{K_zG_z}(j\omega)$. 6, 7. Průběh absolutní hodnoty a fáze $1 + \mathbf{K_1G_1}(j\omega)\ \mathbf{K_zG_z}(j\omega)$. 8, 9. Výsledná frekvenční charakteristika a fáze.

319

Foundations of Automatic Control

Strejc V., M. Šalamon, Z. Kotek and M. Balda
In Czech, SNTL Prague, (1958), pp. 275, Fig. 214, Tab. 2.

After the first international meeting of experts in automatic control in Heidelberg in 1956, it was decided in the former Czechoslovak Republic to publish four basic books. These had to address the general theory of automatic feedback systems, the elements used in control loops, practical industrial applications and the calculation of the mathematical models of controlled systems.

This book refers to the first subject. After an introduction in Section A describing the control loop and main and auxiliary variables as well as dimension-free equations, the book deals with linear, continuous, discrete and non-linear control loops and with mathematical models of controllers and controlled systems including the measurement of their dynamic behaviours.

In section B, Linear control loops, the reader can find all basic aspects of systems starting with the solution of linear differential equations in time domain and by Laplace transformation. Attention is given to transfer functions, frequency response and step response characteristics including logarithmic Bode plots. Closed control loops differentiate command control and compensation of disturbing variable. All standard stability criterions are mentioned in intelligible form. Cost functions for optimum process control are focused on linear, quadratic and optimum module control actions.

In addition to single-input single-output (SISO) systems applying only four main control variables of the control loop (command, controlling, controlled and disturbing variables), the authors consider branched SISO systems with auxiliary variable measured on the controlled system, with auxiliary controlling variable, with measuring of the disturbing variable and finally with auxiliary variable measured on a controlled system model. Multi-input multi-output (MIMO) systems are for illustration confined to two inputs and two outputs.

Section C, Discrete controlled systems, describes the theory of two-step control and pulse control with step proportional to control error. Attention is paid to their transient responses, stability conditions and optimum responses. Particular solution concerns e.g. the speed of controller valve positioning, process control with transport lag of the controlled system and controllers with derivative and flexible correcting actions.
Section D, Non-linear control systems, is devoted to the main elements appearing in a real control loop, such as saturation, dead-zone, hysteresis and relay non-linearities affected possibly by dry friction.

Section E, Models of controllers and controlled systems, are used for real systems simulation. In the early years after the Second World War it was possible to meet mechanical, hydraulic and pneumatic integrators used for mathematical models of the simulated control loops. The importance and flexibility of electronic elements were soon proven, a development properly recognised in this section.

The last section F, Measurement of dynamical characteristics of control loop elements, by conventional instrumentation facilities, describes briefly the different possibilities. Measurement results serve to calculate the step response or frequency characteristics.

ING. VLADIMÍR STREJC • ING. DR. MIROSLAV ŠALAMON
ING. ZDENĚK KOTEK, kand. tech. věd
DOC. ING. MILAN BALDA, kand. tech. věd

ZÁKLADY TEORIE SAMOČINNÉ REGULACE

PRAHA 1958

STÁTNÍ NAKLADATELSTVÍ TECHNICKÉ LITERATURY

je kvadratická regulační plocha

$$P = \frac{1}{2}\,\frac{\begin{vmatrix} d_0^2 & -(d_1^2 - 2\,d_0\,d_2) & d_2^2 \\ 1 & c_2 & 0 \\ 0 & c_1 & c_3 \end{vmatrix}}{\begin{vmatrix} c_1 & c_3 & 0 \\ 1 & c_2 & 0 \\ 0 & c_1 & c_3 \end{vmatrix}} =$$

$$= \frac{d_0^2\,c_2c_3 + (d_1^2 - 2\,d_0d_2)\,c_3 + c_1d_2^2}{2\,(c_1c_2c_3 - c_3^2)} \tag{21.61}$$

Podobně bychom nalezli kvadratickou regulační plochu pro $n = 4$

$$P = \frac{d_3^2\,(c_1c_2 - c_3)}{2\,c_4\,(c_1c_2c_3 - c_3^2 - c_1^2\,c_4)} +$$

$$+ \frac{c_1(d_2^2 - 2\,d_1d_3) + c_3\,(d_1^2 - 2\,d_0d_2) + d_0^2\,(c_2c_3 - c_1c_4)}{2\,(c_1c_2c_3 - c_3^2 - c_1^2\,c_4)} \tag{21.62}$$

a pro $n = 5$

$$P = \frac{d_4^2\,z_0}{2\,c_5\,z_5} +$$

$$+ \frac{(d_3^2 - 2\,d_2d_4)\,z_1 + (d_2^2 - 2\,d_1d_3 + 2\,d_0d_4)\,z_2 + (d_1^2 - 2\,d_0d_2)\,z_3 + d_0^2\,z_4}{2z_5} \tag{21.63}$$

$$z_0 = c_1c_2c_3 - c_3^2 - c_1^2\,c_4 + c_1c_5$$

$$z_1 = c_1c_2 - c_3$$

$$z_2 = c_1c_4 - c_5$$

$$z_3 = c_3c_4 - c_2c_5$$

$$z_4 = c_2c_3c_4 - c_2^2\,c_5 - c_1^2\,c_4 + c_4c_5$$

$$z_5 = z_1z_3 - z_2^2$$

Je důležité uvědomit si, že fysikální rozměr počítané plochy musí být [sec], neboť hodnota poměrné odchylky regulované veličiny je bezrozměrná. Kontrolou fysikálních rozměrů si při praktických výpočtech můžeme stále ověřovat správnost početního postupu.

Podobně bychom mohli sestavit vztahy pro regulační plochu i pro funkce $J(p)$ vyššího řádu. Naznačený početní postup má tu neobyčejnou výhodu, že není třeba hledat kořeny polynomů $M(p)$ a $N(p)$, jako u jiných početních metod, které jsou právě z těchto důvodů pro praktickou aplikaci neupotřebitelné.

Známe-li vztahy pro kvadratickou regulační plochu, je další určování podmínek pro optimální regulační pochod již snadné.

Zavedeme-li nové označení

$$\left.\begin{aligned} s_0 + r_0 &= u \\ s_1 + r_1 &= v \end{aligned}\right\} \tag{21.64}$$

Elements of Control Loops

Balda M., V. Strejc and M. Krampera
In Czech, SNTL Prague, (1958), pp. 474, Fig. 498, Tab. 40, Encl. 2.

The book describes all the important elements used in industrial control loops. The aim is to show, for each physical or chemical variable, the different sensors applied in practical solutions, together with their mathematical description. After the introduction in section A, dealing with basic knowledge of control, section B concerns measuring devices of controllers.

The first Subsection enumerates sensors for measuring level of liquids, starting with different kinds of floats and ending with the application of radioactive isotopes. Subs. 2 deals with pressure meters based on deformations or on float, piston, diaphragm or on electric resistance. Subs. 3 specifies flow meters for gases, liquids and vapours, by some kind of narrowing in the pipe line, e.g. by orifice gauge, and then measuring the pressure. Other mechanical devices are the Pitot tube and the rotameter. Likewise an electromagnetic induction meter for conductive liquids is mentioned.

Subs. 4 is dedicated to temperature measurements. Characterised are liquid-, gas-, dilatation-, bimetallic- and pressure-type and resistance thermometers. A great attention is paid to thermocouples. Finally heat radioactive sensors are considered.

The next Subs. introduces sensors for the continuous measurement of gas humidity. Mentioned are mechanical hydrometers, psychrometers, electrolytic moisture meters and others. The following Subs. deals with continuous measuring of density. Liquid density sensor may apply a vessel suspended on elastic input and output tubes. Next possibility for gases and liquids is based on measuring of hydrostatic pressure. Another way for liquids is the application of the law by Archimedes. Density of flammable gases can be established by the increase of their kinetic energy. An additional possibility is the measuring of internal friction of gases or measuring of sonic speed of gases and liquids and others. A separate problem is posed by the continuous measuring of viscosity. The respective means are e.g. capillary viscosity meters, rotary meters and those based on free fall of particles in case of liquids.

Next Subs. describes sensors for continuously taking the chemical composition of liquids and gases. A great number of instruments and methods are enumerated starting with measuring the penetration of infrared radiation of characteristic wavelengths, colorimeters, refractometers and others using different physical principles. The last example is based on measuring the changes of electroconductivity of solutions by chemical reactions. Another Subs. provides important information about continuous measuring of pH; the basic approach is followed by introducing the titrimetric curve and chemical equilibrium as the way of measuring pH.

The last Subs. is dedicated to the measuring of revolutions, by mechanical, hydraulic, air-operated and electric speedometers including speedometers based on eddy current.

The short section C informs about the remote transfer of controller variables specifying the simple well-tried electric procedures.

Section D deals with three term controllers, differentiating proportional, integral and derivative actions and all useful combinations, as well as with mechanic, pneumatic, hydraulic, electronic and magnetic amplifiers. The next part is devoted to compensators such as self-acting electromechanical and electronic potentiometers and measuring bridges. At the end, different kinds of feedback are mentioned.

Section E concerns driving mechanisms and control valves. Described are pneumatic, diaphragm and piston drives, solenoid and electromagnetic drives and finally electro-servomotors. The second part of this section describes first the characteristic forms of valve cones and then sizing the valves including the calculation of their pressure drop.

Section F specifies the design of different kinds of continuously acting controllers taking into account the structure of the controlled system itself. The discontinuously acting controllers are represented by on-off and pulse- controllers.

The last part of the book, sections H, I and J concerns controlled systems.

Doc. ING. MILAN BALDA, kand. tech. věd,
ING. VLADIMÍR STREJC a ING. MILOSLAV KRAMPERA

PRVKY REGULAČNÍCH OBVODŮ

PRAHA 1958

STÁTNÍ NAKLADATELSTVÍ TECHNICKÉ LITERATURY

Skleněná trubice je upevněna v kovovém pouzdře pouze spodním koncem, takže tlak může horním spojením působit na trubici i zvenčí. Rotametr má nahoře indukční vysílač, kterého se vůbec u těchto přístrojů hodně používá. Dole je připojena stavoznaková vysokotlaká armatura s ukazatelem polohy plováku, určená pro místní sledování průtoku. Je-li měřená látka neprůhledná, může být místní ukazatel nahoře nad výstupním hrdlem a jeho vnitřní prostor se pak vyplní vhodným plynem.

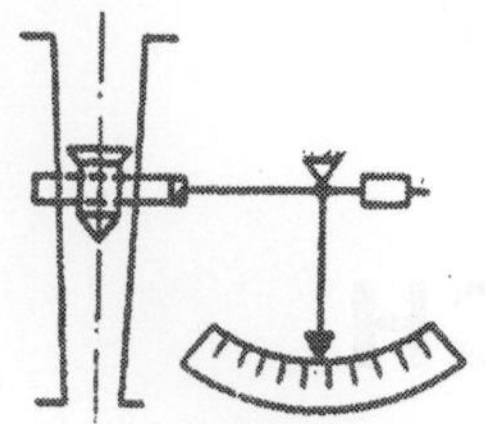

Obr. 11.19. Magnetický přenos polohy plováku rotametru

Rotametrů s vysílačem nelze použít pro tak malé průtoky jako rotametrů bez vysílače a se skleněnou trubicí.

Tlaková ztráta rotametru je přibližně stálá a je v podstatě určována váhou plováku a plochou jeho průmětu do roviny kolmé k ose rotametru. Přírůstek kinetické energie se opět mění na potenciální energii tlakovou jen v nepatrném rozsahu, vzhledem k náhlému velkému rozšíření průřezu po výstupu z mezikruží.

Nepřihlížíme-li k vlivu tření a přítokové rychlosti, můžeme základní rovnici rotametru odvodit velmi jednoduše. Označíme-li váhu plováku, zmenšenou o vztlak, P a plochu průmětu plováku do roviny, kolmé k ose rotametru, F_p, bude stálý diferenční tlak Δp na mezikruží dán vztahem

$$\Delta p = \frac{P}{F_p},$$

z čehož podle Bernoulliho rovnice plyne stálá rychlost průtoku mezerou

$$v = \sqrt{\frac{2g}{\gamma} \frac{P}{F_p}}$$

Označíme-li proměnlivou plochu mezikruží F, bude v rovnovážných stavech plováku platit pro objemový průtok Q

$$Q = F K, \tag{11.27}$$

kde konstanta

$$K = v = \sqrt{\frac{2g}{\gamma} \frac{P}{F_p}}$$

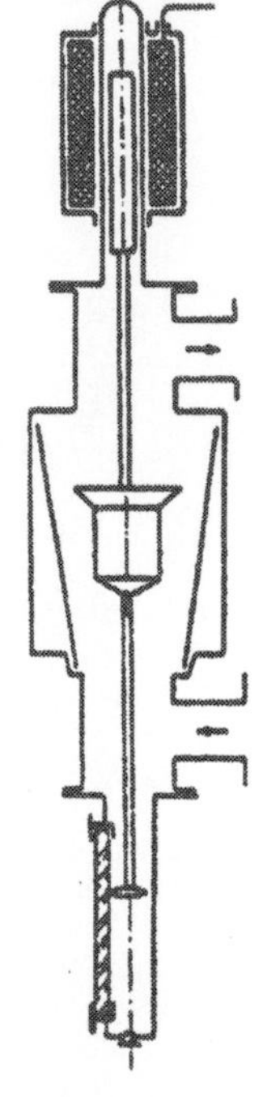

Obr. 11.20. Rotametr s indukčním vysílačem

Ze vztahu (.27) můžeme pro maximální a minimální průtok určit výstupní a vstupní průměr trubice a za použití téhož vztahu kótovat výšky mezi oběma průřezy hodnotami průtoku.

Při praktickém provedení však vliv tření, přítokové rychlosti ani poměrů proudění a u plynů též kompresibility nemůžeme přehlížet. Rozsah a působení těchto vlivů (s výjimkou přítokové rychlosti) závisí na Reynoldsově čísle (viz odst. 11.1), na tvaru plováku, na poměru zúžení a na exponentu adiabatické expanse (jde-li o plyny nebo páry). Přesnou závislost průtoku na poloze plováku tedy můžeme zjistit jedině cejchováním. Již při navrhování rotametrů však s těmito vlivy musíme počítat a řídit se podle dosavadních výsledků praktických měření.

Celkový odpor potenciometru P_0 vypočteme, dělíme-li celkový rozsah v milivoltech, odpovídající rozsahu měřené teploty, procházejícím proudem v miliampérech. Pak

$$P_0 = \frac{\text{EMS}_{\max} - \text{EMS}_{\min}}{i_1}$$

Protože velikost odporu potenciometru je dána použitým vinutím, vypočteme potřebný šunt

$$R_6 = \frac{P\,P_0}{P - P_0}$$

Nyní známe všechny odpory měřicího obvodu až na R_5. Protože však součty odporů obou větví musí být stejně veliké, bude

$$R_5 = (R_2 + R_3) - (R_4 + P_0)$$

Tím je vypočten celý měřicí obvod.

Zbývá určit odpory R_1 a R_7 až R_{10}. Odpor reostatu R_1 volíme tak, aby srazil napětí čerstvého článku (asi 1,7 V) na požadovanou hodnotu asi 1,02 V. Při napájecím prou-

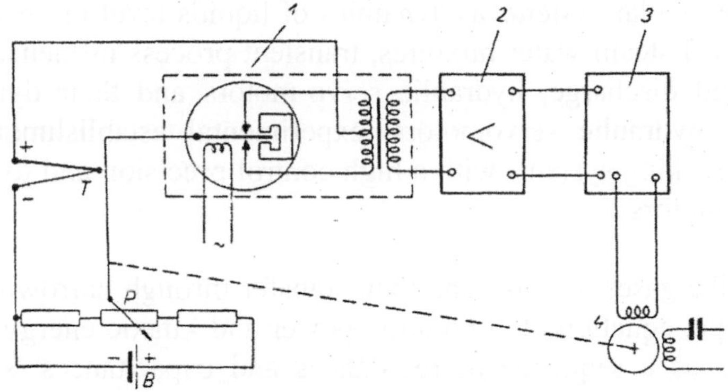

Obr. 27.7. Blokové schema elektronického potenciometru: *1* — vibrátor, *2* — zesilovač napětí, *3* — zesilovač výkonu, *4* — balanční motorek

du 10 mA bude tedy odpor reostatu asi 70 ohmů. Odpor R_7 je tlumicím odporem v obvodu galvanoměru a má vliv na jeho citlivost. Podle velikosti odporu vedení thermočlánku volíme jeho velikost 15 až 30 ohmů; při velmi dlouhém vedení s velkým odporem se cívky R_7 nepoužívá.

Odpor R_{10} se zapojuje při kontrole mechanické nuly přístroje paralelně ke galvanoměru. Má hodnotu 20 až 50 ohmů. Odpory R_8 a R_9 tvoří dělič napětí při standardisaci napětí baterie na ochranu galvanoměru před přetížením. $R_8 = 10$ až 50 ohmů, $R_9 = 0$ až 60 ohmů.

27.22 Elektronické samočinné potenciometry a můstky

Elektronické kompensátory jsou založeny na stejném principu jako přístroje elektromechanické. Místo galvanoměru s mechanickým vyvažovacím ústrojím však používají elektronického zesilovače, který napájí balanční motorek. Balanční motorek se otáčí ve smyslu odpovídajícím polaritě vstupního signálu a otáčí běžcem potenciometru P tak dlouho, až je měřicí obvod opět v rovnováze. S pohybem balančního motorku je svázán pohon ukazatele na stupnici a převody na další členy regulačního

Dynamics of Controlled Systems

Čermák J., V. Peterka and J. Závorka
In Czech, ACADEMIA Prague (1968), pp. 584, Fig. 320, Encl. 1 printed table.

This book is devoted to the procedure of establishing a mathematical model by mathematical-physical analysis. A detailed analysis of electric machines was already carried out in this way, but math. models of thermal systems were elaborated only partly and incompletely. The authors elaborated a large and difficult subject in a form accessible to readers and experts in automatic control opening and facilitating the solution of identification problems. The authors could not present all possible systems. First of all it was needful to concentrate the analysis on typical cases. On the other hand, the reader must be aware of the fact, that no math. model can completely cover reality in all coherence.

Chapter 1 is gives the reasons for math. models calculation.

Chapters 2 to 16 present particular systems as dynamics of liquids level in open tanks, transient states of free level of foamed steam-water mixtures, transient process influenced by controlling organs, nonstationary liquid discharge, hydraulic servo-motors and their dynamic equations, single and double acting hydraulic servomotors, experimental establishment of their time constant, linear hydraulic tracking systems with a high control precision and dynamic behaviour of hydraulic tracking servomotors.

Pressure dynamics of air-like gases in tanks and their transfer through narrowing, nonstationary flow of liquids trough a pipe, equation of continuity, power and kinetic energy, sonic waves in pipeline, pipe presentation as a sequence of resistances and capacitances, simplified single-capacity model.

Dynamics of heat transfer through a thin wall, detailed solution of the heat transfer dynamics and transfer function of the heat flow sensor, heating and cooling of monophasig liquids in case of perfect stirring, mixing heat exchangers, stirring heat ex-changers, recuperating heat exchangers, temperature dynamics of liquid flowing trough a pipe, tubular heat exchanger, transcendental transfer function and its approximation.

Steam-liquid mixtures, dynamics of pressure in drum boilers, level dynamics in closed systems under pressure, dynamics of steam-water zone of a single throughput boiler, the dynamics of engines like steam turbine with a superheater, turbine with warmed up steam, engine speed dynamics.

Nonstationary states of composition determined by physical and physico-chemical actions. Dynamics of plate rectifying columns, equations describing the dynamics of a plate, analytical as well as analogue and digital solution of a dynamical equation set of a column and dynamics of filled rectifying columns and their dynamic and static equations.

Dynamics of chemical reactors working continuously or periodically, tubular reactors and chemical composition in the reactor influenced by temperature changes. Nuclear reactors, kinetic equation of the reactor, simplifications and simulation of nuclear reactors.

JIŘÍ ČERMÁK VÁCLAV PETERKA JIŘÍ ZÁVORKA

DYNAMIKA REGULOVANÝCH SOUSTAV

v tepelné energetice a chemii

ACADEMIA

nakladatelství Československé akademie věd

Praha 1968

16.3 PŘENOSOVÉ FUNKCE REAKTORU

Lineární přenos reaktoru v oblasti nulového výkonu

Všimneme si nyní některých přenosových funkcí reaktoru, především však nejjednoduššího linearizovaného přenosu reaktoru, který je právě v oblasti nulového výkonu. Oblast nulového výkonu znamená, že reaktor je sice kritický, ale v podstatě nedodává teplo chladivu a také, že kinetika kritického reaktoru nepodléhá (v uvažovaném čase) žádným změnám vlivem fyzikálních procesů a změn v moderátoru, palivu, struktuře atd. – předpokládáme tedy téměř „ideální" reaktor. Dále předpokládáme, že podíl zpožděných neutronů μ je malý, (menší než 0,01) a že δk nebude větší nežli maximálně $\mu/2$ a poměrné odchylky $|\Delta N(t)/N_0| \leqq 0{,}05$.

Převedeme-li po úpravě obě rovnice (.49) a (.50) v odchylkovém tvaru z časové oblasti do operátorového tvaru, budou platit tyto vztahy:

$$s\,\Delta N(s) = \frac{\delta k}{T_0} N_0 - \sum_1^m \frac{\mu_i}{T_0} \Delta N(s) + \sum_1^m \lambda_i \,\Delta R_i(s)\,, \tag{.1}$$

$$s\,\Delta R_i(s) = \frac{\mu_i}{T_0} \Delta N(s) - \lambda_i \,\Delta R_i(s)\,. \tag{.2}$$

Dosadíme-li nyní z rovnice (.2) výraz

$$\Delta R_i(s) = \frac{\mu_i}{T_0} \Delta N(s) \left[\frac{1}{s + \lambda_i}\right] \tag{.3}$$

do rovnice (.1), dostáváme vztah

$$s\,\Delta N(s) = \frac{\delta k}{T_0} N_0 + \frac{\Delta N(s)}{T_0} \left[\sum_1^m \frac{\lambda_i \mu_i}{s + \lambda_i} - \sum_1^m \mu_i\right], \tag{.4}$$

respektive

$$\Delta N(s) \left[s + \frac{1}{T_0} \sum_1^m \frac{s\mu_i}{s + \lambda_i}\right] = \frac{\delta k}{T_0} N_0 \tag{.5}$$

a odtud dostaneme po úpravě přenos „ideálního" reaktoru v oblasti nulového výkonu jako vztah mezi přebytkem reaktivity δk a poměrnou změnou množství neutronů $\Delta N(t)/N_0$, v anglosaské literatuře zpravidla značený takto:

$$G(s) = \frac{\Delta N(s)/N_0}{\delta k} = \frac{1}{T_0 s} \left[\frac{1}{1 + \dfrac{1}{T_0} \sum_1^m \dfrac{\mu_i}{s + \lambda_i}}\right]. \tag{.6}$$

EARLY CONTROL TEXTBOOKS
in
DENMARK

Automatisk Kontrol

Jens R. Jensen

Polyteknisk Forenings Forlag, 1956

Notes on the Measurements of Dynamic Systems

Jens R. Jensen

Servoteknisk Forskningslaboratorium, 1958

Contributed by

Mogens Blanke

Technical University of Denmark

Lyngby (Copenhagen), Denmark

Early Automatic Control Education in Denmark

The development of the field in Denmark dates to 1949 when the Civ.Ing. (later Professor) *Jens R. Jensen* held a course in **"Automatisk Kontrol"** for engineers, organized by the Danish Engineering Association (Dansk Ingeniørforening). The lecture notes were the result of developing interest in the area of servomechanisms rapidly followed by industrial demand and research interest in control of machines, industrial mechanisms and processes.

In January 1956, the Danish industry founded the "Servoteknisk Forskningslaboratorium" as an independent research institute funded by industry. With Jens R. Jensen in charge of a few research engineers and a workshop assistant, the small team was busy bringing the latest theoretical developments into practice. Much effort was dedicated to the dynamic modelling of control objects and the measurement of response characteristics. The efforts included equipment to do correlation analysis to measure the dynamic characteristics for example of heat exchangers. Analog computers were built to support simulation, Jens R. Jensen constructed a frequency response calculator stick, cipher computers were brought into work in factory automation. Industrial undertakings included the overall control and servomechanisms used to cut steel profiles at the Burmeister and Wain shipyard.

Teaching of students of the "Den polytekniske Læreanstalt, Danmarks tekniske Højskole" was another of the tasks of the new institute and the activity in research and teaching gradually dominated that of assistance to industry. The institute became a laboratory of the "Danmarks tekniske Højskole" and named "Servolaboratoriet".

The original lecture notes from 1949 were rewritten in 1956 and were used to teach automatic control theory. Subjects included P and PI control for systems with one and two time constants, with integration, systems with time delay, stability analysis and design for stability using the frequency domain and root locations in the complex plane, Routh-Hurwitz criterion, feed-forward compensation. Control was treated for two-variable systems, exemplified by simultaneous temperature and flow control on page 44 and a general controller architecture on page 46. The use of complex zeros to shape phase responses are shown on page 121.

A second book was used to teach measurement of dynamic characteristics, the area that later developed into system identification. The "**Notes on the Measurements of Dynamic Systems**", Parts I and II from 1957 and 1958 respectively, show how correlation analysis was done using thermocouple multipliers. A digital correlation machine was subsequently constructed and the *seven frequency binary test signal* was invented by Jens R. Jensen.

References

Jens R. Jensen: Automatisk Kontrol. 1. udgave 1949.

Jens R. Jensen: Automatisk Kontrol, 2. udgave, 1956. Servoteknisk Forskningslaboratorium, Danmarks Tekniske Højskole.

Jens R. Jensen: Notes on the Measurement of Dynamic Characteristics of Linear Systems. Part I, 1957. Servoteknisk Forskningslaboratorium, Danmarks Tekniske Højskole.

Jens R. Jensen and M. Drost Larsen: Notes on the Measurement of Dynamic Characteristics of Linear Systems. Part II, 1958. Servoteknisk Forskningslaboratorium, Danmarks Tekniske Højskole.

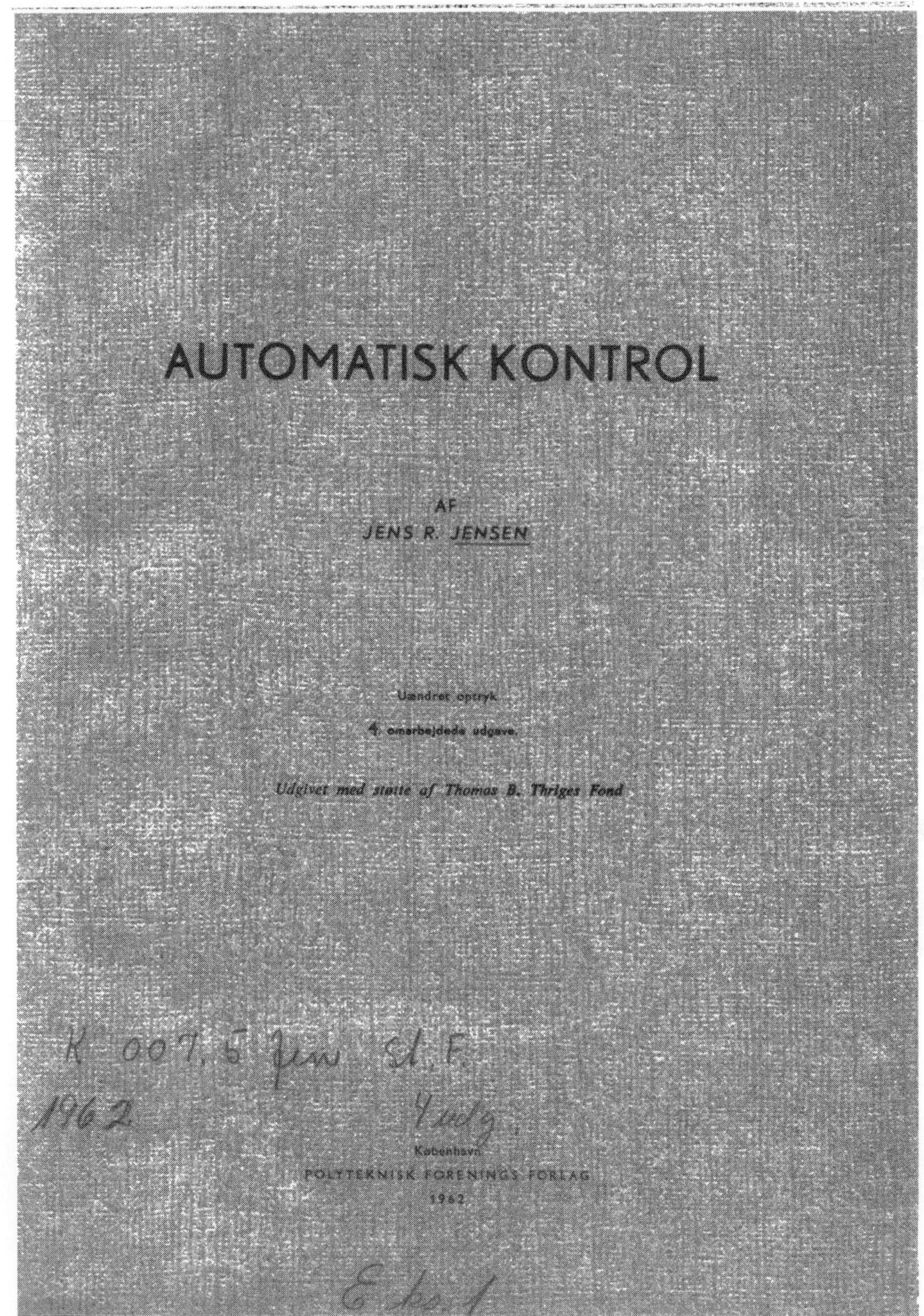

AUTOMATISK KONTROL

AF

JENS R. JENSEN

Uændret optryk

4. omarbejdede udgave.

Udgivet med støtte af Thomas B. Thriges Fond

København

POLYTEKNISK FORENINGS FORLAG

1962

44.

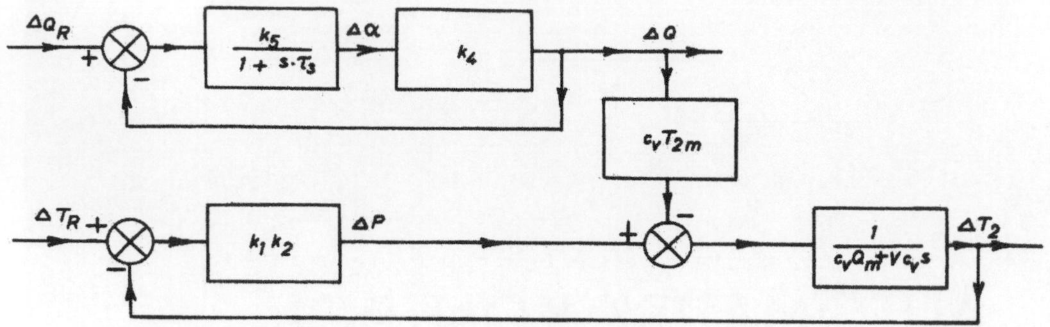

Fig. 39.

Ved hjælp af de i et foregående afsnit angivne regler kan dette diagram omformes ved f.eks. at flytte det andet summationspunkt i nederste sløjfe hen foran det første summationspunkt og regne de to sløjfer om til en enkelt blok

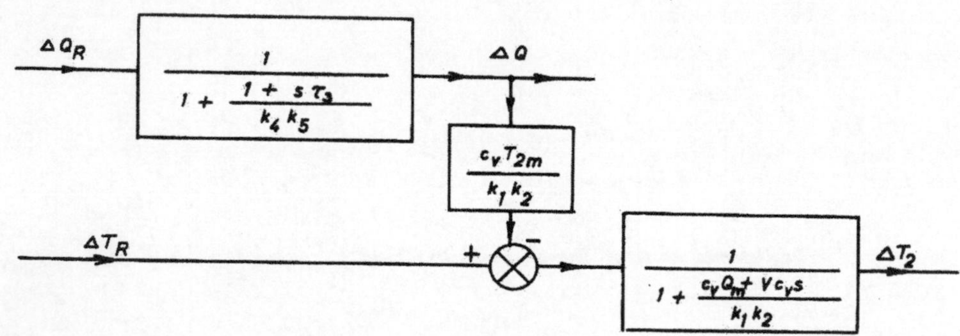

Fig. 40.

som kan pyntes lidt, hvis vi i nederste blok indfører definitionerne i formlerne (20) og (31) og yderligere indfører

$$K_1' = k_4 k_5 \qquad \tau_4 = \frac{\tau_3}{1+K_1}$$

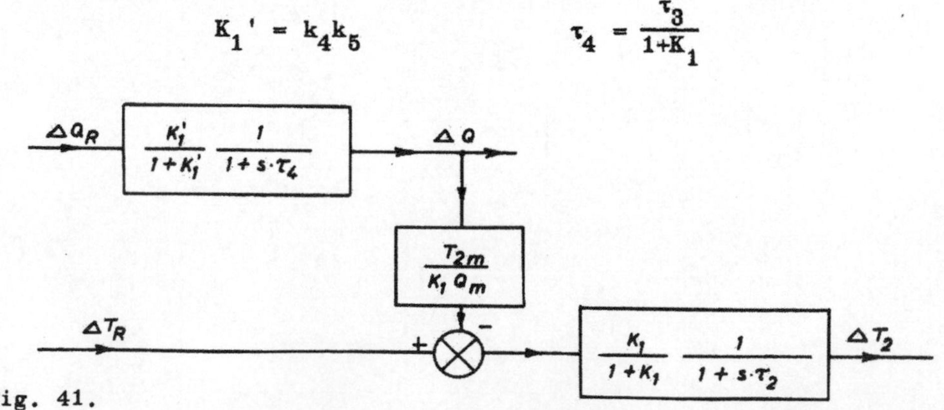

Fig. 41.

anvende en kombination af åben sløjfe teknik (kompensation) og lukket sløjfe teknik (automatisk kontrol).

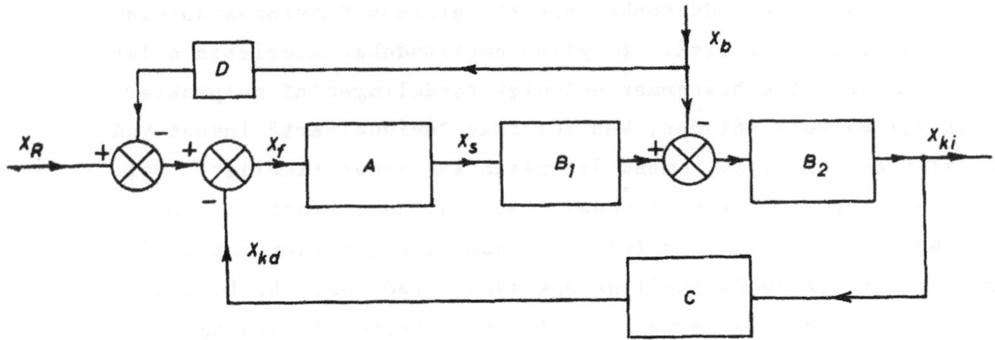

Fig. 42.

Man måler forstyrrelsens størrelse og afleder heraf et signal, hvormed systemet påvirkes et passende sted, f.eks. på det sted, hvor fejlen dannes. Dette afledede signal skal da tilnærmelsesvis have en sådan karakter, at det på det sted i systemet, hvor forstyrrelsen griber ind, udvikler et signal, der netop ophæver forstyrrelsens indflydelse, således at resten af systemet, fortrinsvis den kontrollerede størrelse, ikke mærker noget til forstyrrelsen. Idet forstyrrelsens betydning ifølge fig. 36 er ækvivalent med en vis ændring af referencen, vil dette følgelig kunne opfyldes, når det af forstyrrelsen afledede signal i fig. 42 giver en lige så stor men modsat ændring i referencen, d.v.s. man må tilstræbe

$$D = -\frac{1}{AB_1}$$

I praksis kan denne indstilling næppe holdes helt nøjagtigt over længere tidsrum, men blot den holdes med en beskeden nøjagtighed, f.eks. 20% opnås dog, at den overskydende forstyrrelse, som den automatiske kontrol må modvirke, reduceres til ca. en femtedel, hvilket atter betyder, at de opståede afvigelser reduceres til en femtedel, hvilket meget vel kan være værd at tage med.

Det kan også stilles op på den måde, at det sikkert ofte kan betale sig at anvende både kompensation og automatisk kontrol, idet man da for en given kontrolnøjagtighed kan nøjes med et dårligere kontrolsystem. Specielt kan det måske i tilfælde, hvor kontrolobjektet ikke er dimensioneret med henblik på automatisk kontrol, og hvor det er umuligt at dimensionere et tilstrækkeligt godt automatisk kontrolsystem, være den eneste udvej til opnåelse af en ønsket resulterende nøjagtighed.

Når et netværks nulpunkter og poler er kendte, så er amplitude- og fasekarakteristikkerne kendt (på nær en numerisk faktor på amplitudekarakteristikken). Hvis man indskrænker sig til minimum faseforskydningsnetværk gælder det omvendte også. En given amplitudekarakteristik eller en given fasekarakteristik bestemmer entydigt fordelingen af nulpunkter og poler. At det må være således, kan vel ikke "umiddelbart" indses ved betragtning af f.eks. fig. 24, omend det måske kan synes rimeligt. Det kan imidertid bevises ved hjælp af noget kompleks funktionsteori. Der kan udledes mange formler for den indbyrdes sammenhæng mellem amplitude- og fasekarakteristikker og forskellige egenskaber ved dem. H. W. Bode har beskæftiget sig indgående hermed og bringer en formelsamling og kurver til hjælp ved numeriske beregninger i sin bog, litteraturpunkt 21.

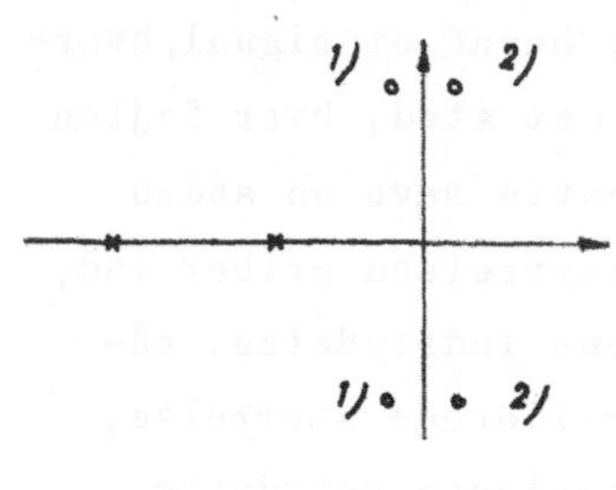

27.

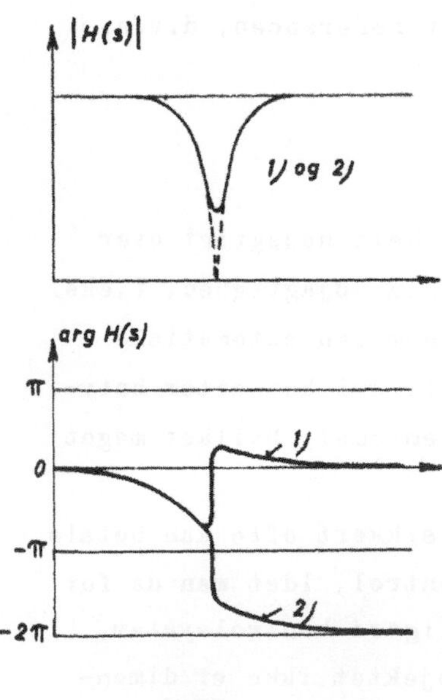

28.

Eksempel: Vi tænker os et netværk med to reelle poler i venstre halvplan og to komplekse, konjugerede nulpunkter. De to nulpunkter tænkes beliggende meget tæt ved den imaginære akse, i tilfælde 1) i venstre halvplan. De to poler og det nederste nulpunkt vil give forholdsvis "langsomme" variationer i både amplitude og fase, når s gennemløber den imaginære akse fra 0 til ∞. Men i det område, hvor ω omtrent er lig den imaginære værdi af nulpunkterne og altså passerer forbi det øverste nulpunkt, vil der komme et stærkt fald og derefter en stærkt stigning i amplitudekarakteristikken, og fasen, der ellers fra begyndelsen er aftagende, vil indenfor et snævert interval af ω vokse omtrent π, se fig. 28, kurverne mærket 1). Vinkeltilvæksten for ω gående fra 0 til ∞ er nul, da de to poler og de to nulpunkter i venstre halvplan ophæver hinanden i denne henseende. I andet tilfælde 2) tænker vi os de to nulpunkter flyttet lige så langt til højre for den imaginære akse. Det giver samme amplitude-

Danmarks tekniske Højskole

Servoteknisk forskningslaboratorium

Notes on Measurement of Dynamic Characteristics of Linear Systems

Part II

Laboratorieingeniør Jens R. Jensen
Civilingeniør M. Drost Larsen

COPENHAGEN DENMARK
1958

Measurement of a System of a Break Frequency of 0.0008 c/s

The arrangement is shown in Fig. 11.

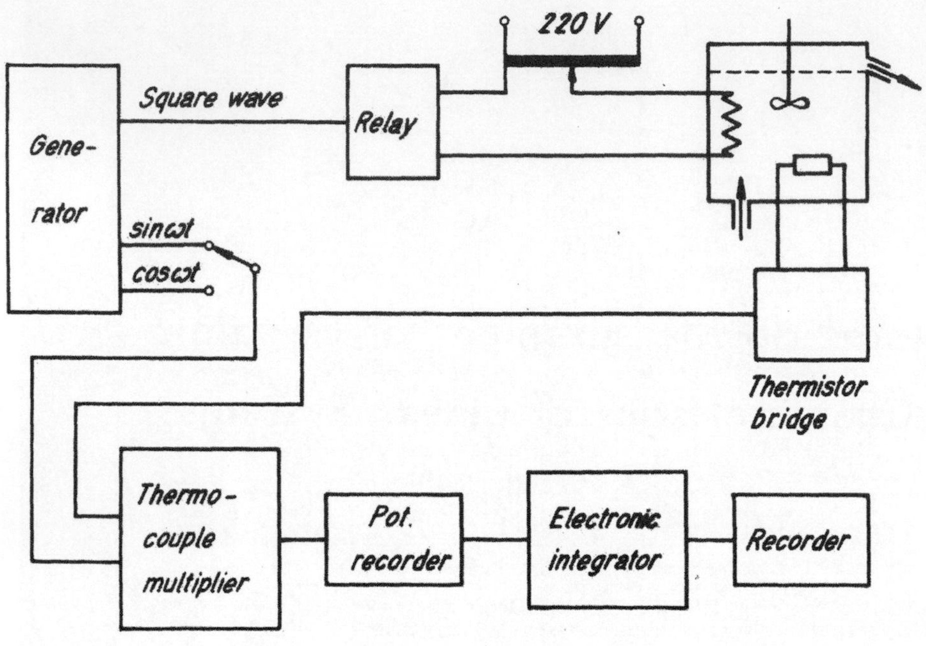

Fig.11.

The system consists of a container through which water flows at a constant rate. The water is electrically heated, and the temperature is measured by means of a thermistor bridge.

Since it is not a simple matter to vary the heating power sinusoidally, a square wave input is used. The reference voltage is a sine wave in phase with or 90° out of phase with the first harmonic of the square wave. The higher harmonics of the square wave are regarded as noise.

The output of the multiplier as well as the integral of the output are recorded. The registrations obtained correspond to Figs. 6 and 7 on pages 9 and 10 in Part I, and samples of the registrations are shown in Figs. 12 and 13.

For demonstration purposes only three points of the

EARLY CONTROL TEXTBOOKS
in
FINLAND

Basics of Servomechanisms

Erkki Laurila
Published in 1954

Theoretical Background of Control Engineering, Vols. 1-2

Hans Blomberg
Helsinki University of Technology, 1963 .

Contributed by

Antti J. Niemi, Heikki Koivo, Raimo Ylinen

Helsinki University of Technology
Helsinki, Finland

Background

Instruction and research of control engineering was initiated in Finland by Dr. Erkki Laurila (1913-98), Professor of Technical Physics at the Helsinki University of Technology, who from the academic year 1946-47 onwards devoted a part of his lectures to theoretical fundamentals of control. The other parts consisted mainly of methods of industrial measurements and control applications, and of fine mechanics which were related to his earlier work in the Finnish instrument industry. A Laboratory of Technical Physics serving the listed subject areas was also planned and soon set up, for experimental illustration of related phenomena and industrial controllers, and for individual constructions related e.g. to thesis works.

The 3rd year students of control engineering were already familiar with electron tube amplifiers, and Prof. Laurila could therefore use these for introduction into the concept of feedback and, further on, into that of stability and transients of feedback loops. Since the relay and proportional controllers dominated the practical applications of automatic control, they were also presented, and the models of the latter type were then augmented with reset (integral) and rate (derivative) elements. Simplified models of some basic industrial processes were taken for representations of the control objects. Transfer function models of control components and loops were introduced by him at the end of the 40's, after the integral transformations had been included in the mathematics courses.

During his first years of instruction, Prof. Laurila expanded the part of control engineering further and introduced electronics as another subject of lectures, while fine mechanics and instrument constructions, except industrial controllers and servo mechanisms, were taken over by another teacher. His lectures on control were occasionally accompanied by written sheets, and existing American and German textbooks were recommended for wider information. He obviously preferred to bring first his earlier and parallel studies on electronics of solid matter and rarefied gases to the form of two textbooks in Finnish, which were published in 1948 and, with Mr. L. Saari, in 1950.

Another pioneer in control engineering education in Finland is Dr. Hans Blomberg (1919-). He was appointed Professor of Theoretical Electrical Engineering at Helsinki University of Technology in 1956 with duties in Swedish, which is the other official language in Finland. Control engineering was at that time a new field of interest to electrical engineers and Professor Blomberg decided to specialize in it. Over the next decade his laboratory, called the Automatic Control Laboratory, was the centre of control engineering education and research in Finland.

Basics of Servomechanisms

Erkki Laurila
Published in 1954

Erkki Laurila's textbook on control, entitled Basics of Servomechanisms, was published in 1954. It contained four chapters: 1. Introduction, 2. Mathematical and experimental methods of servomechanisms, 3. Linear servomechanisms, and 4. Nonlinear servomechanisms. Chapter 1 introduces the concept of automatic control as a feedback system. Two motivating examples are used: temperature control in a water tank and position servo as a typical linear servomechanism. In the latter case basic concepts of steady-state error and effect of damping are discussed. In the second chapter Laplace transform is introduced as a tool to treat linear, time-invariant differential equations. This leads to the introduction of the transfer function concept and further, to frequency domain presentations in the form of Bode and Nyquist diagrams.

Chapter three forms the core of the textbook. Linear servomechanisms are first characterized, then their stability issues are analyzed using e.g. Nyquist criterion. For design purposes, phase margin and gain margin are introduced as measures of degree of stability. To stabilize and improve the system performance, phase-lag and phase-lead networks are treated and examples are given to illustrate the design steps. Electric networks are used as basic compensators. In addition to servos, also pneumatic and hydraulic actuators are discussed and their transfer functions derived. Here and elsewhere, Laurila points out analogies between different systems. The chapter concludes with a discussion on elimination of disturbances. Linear servomechanism is used as a case study and block diagrams strongly support the analysis.

Chapter four studies nonlinear servomechanisms. The motivation comes from common nonlinear actuators like relays, relays with hysteresis and saturating actuators. Phase plane and describing function are explained as basic analysis tools. The use of describing function to specify self-oscillation and also as a design tool to avoid oscillation is briefly explained. The last topic in the book is sampled-data servosystems. These arise naturally from the idea of using contactor type of servosystems – digital computers are not mentioned. The treatment of the topic is rather cursory.

Although Erkki Laurila's textbook is only 100 pages long, he is able to summarize the key ideas and techniques in a concise manner. The reader is introduced to a new field of control engineering and is easily convinced of the usefulness of the approach.

VALTION TEKNILLINEN TUTKIMUSLAITOS

Tiedoitus 128

SERVOTEKNIKAN PERUSTEET

ERKKI LAURILA

HELSINKI 1954

verrannollinen. Takometripiirin aikavakio oletetaan häviävän pieneksi.

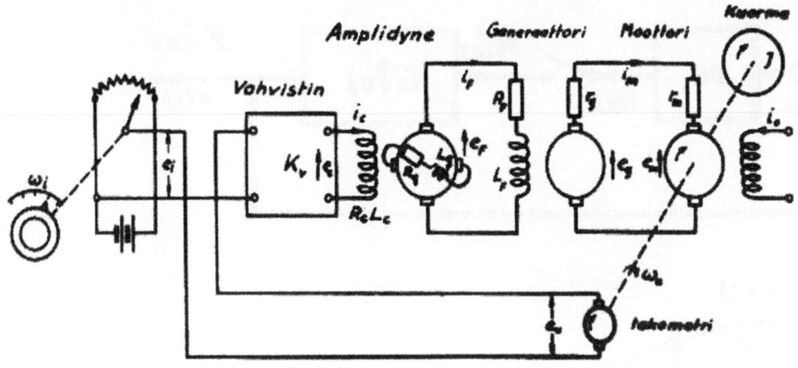

Kuva 45

Kuvaan merkityillä suureilla on seuraavat arvot:

$K_v = 12$	$e_F = 2700(V/A)\cdot i_c$	$r_g = 0,002\,\Omega$
$R_c = 80\,\Omega$	$R_F = 7,75\,\Omega$	$r_m = 0,010\,\Omega$
$L_c = 8$ Hy	$L_F = 24,4$ Hy	$i_{m\,max} = 1250$ A
$R_q = 1,06\,\Omega$	$e_g = 18,5(V/A)\cdot i_F$	$e_t = 1\ (V/rad\ s^{-1})\cdot\omega_u$
$L_q = 0,318$ Hy	$e_m = 40(V/rad\ s^{-1})\cdot\omega_u$	$J = 1800\ kgm^2/rad^2$

vääntömomentti $T = 36(Nm/rad\ A)\cdot i_A$

Voimme kirjoittaa eri suureitten välillä vallitsevat yhtälöt suoraan L-muunnettuina:

$$E_c = K_v(E_i - E_t)$$

$$I_c = \frac{E_c}{R_c + sL_c} = \frac{E_c}{R_c(1+T_1 s)} \qquad T_1 = \frac{L_c}{R_c} = 0,1\ s$$

$$E_F = \frac{k'_F I_c}{R_q + sLq} = \frac{k_F}{1+T_2 s} I_c \qquad T_2 = \frac{L_q}{R_q} = 0,3\ s$$

$$I_F = \frac{E_F}{R_F + sL_F} = \frac{E_F}{R_F(1+T_3 s)} \qquad T_3 = \frac{L_F}{R_F} = 3,2\ s$$

$$E_g = kgI_F$$

77

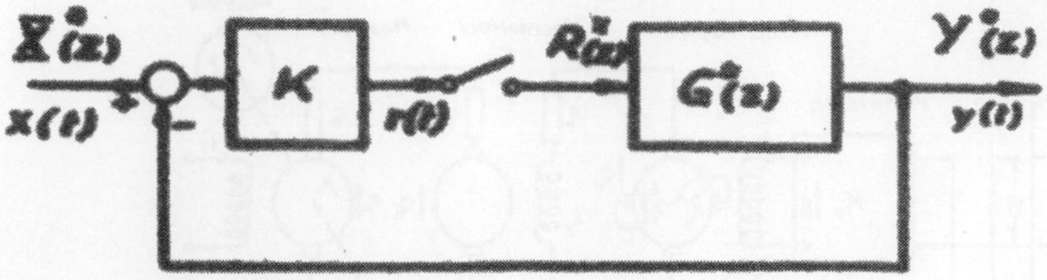

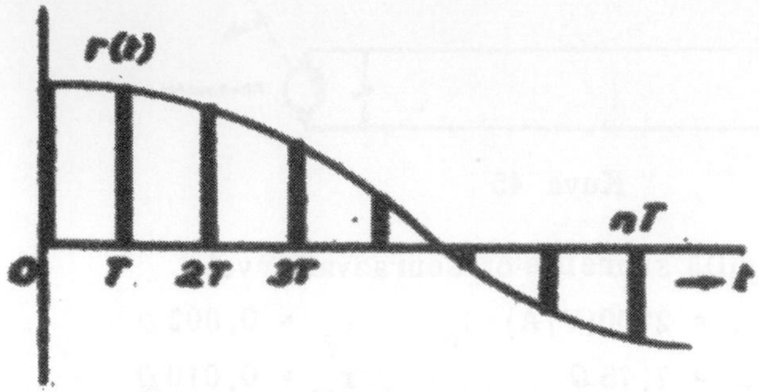

Kuva 60

telymetodiikka kehitetty jo huomattavan pitkälle, ja se muistuttaa suuresti yleispiirteissään lineaaristen servosysteemien käsittelytapaa. Tarkoituksenmukaisuussyistä käytetään L-muunnoksen ohella näissä probleemoissa ns. z-muunnosta. Aikamuuttujan r(t) z-muunnos saadaan lausutuksi L-muunnoksen avulla

$$R^*(z) = L\left[r(t)\,\delta_T(t)\right]$$

jolloin L-muunnoksen lausekkeessa esiintyvä e^{-sT} korvataan z:lla. Kaavassa esiintyvä $\delta_T(t)$ on

$$\delta_T(t) = \sum_{n=-\infty}^{+\infty} \delta(t - nT)$$

z-muunnoksen käyttöön perustuen voidaan näytejonosysteemien käsittelyssä turvautua lineaaristen systeemien yhteydessä käytettyjä graafisia menetelmiä muistuttaviin menetelmiin.

Sikäli kuin näytejonon aikaväli T on hyvin pieni verrattuna ohjattavan systeemin määrääviin aikavakioihin, voidaan servosysteemiä käsitellä lineaarisena. $R^*(Z)$:n vastaava ai-

99

Theoretical Background of Control Engineering, Vols. 1-2

Hans Blomberg
Helsinki University of Technology, 1963 .

Because Professor Blomberg gave his lectures in Swedish but most of the students were Finnish speaking, he started already in the very early phase to use duplicates of his lecture notes in teaching. Those lecture notes were collected to a book Hans Blomberg: Theoretical Background of Control Engineering, Vols. 1-2, Helsinki University of Technology, 1963. This book was theoretically oriented; later on Professor Blomberg and his staff moved to more general systems theory.

The whole set of these lecture notes was never used as a textbook of the basic course in control engineering, because this was taught by other teachers and Professor Blomberg himself gave lectures in advanced methods of control engineering.

Volume 1 of the book contains four chapters. Chapter one is an introduction to control systems and their properties starting from some physical control systems. Then in chapter two the concepts of transfer function and block diagram algebra are introduced. Chapter three gives more examples of real control systems, in particular of those of electrical circuits and motors. Finally, in chapter four the discussion of properties of control loops is started from static error coefficients. The volume has also an appendix with introduction to Laplace transform.

Volume 2 is a direct continuation of chapter four in volume 1 and contains only one chapter and two appendices. The consideration of static errors is first completed. Then stability and instability definitions as well as their connection to location of poles are introduced. The traditional stability tests, like Routh-Hurwitz test and Nyquist test with several examples are presented. Also the construction and use of root locus for control design are considered. Finally, the properties of control systems in time domain using step responses and frequency responses as well as responses to stochastic inputs are discussed.

The first appendix of volume 2 contains a comprehensive introduction to Fourier transform and spectral theory of deterministic and stochastic signals and the second appendix presents the basic stability definitions of Ljapunov.

TEKNISKA HÖGSKOLAN

Kompendium n:o 169

REGLERINGSTEKNIKENS TEORETISKA GRUNDER

Del 1

Prof. Hans Blomberg

Svensk-finsk ordförteckning

av

Dipl.ing. Seppo Rickman

Helsingfors 1963

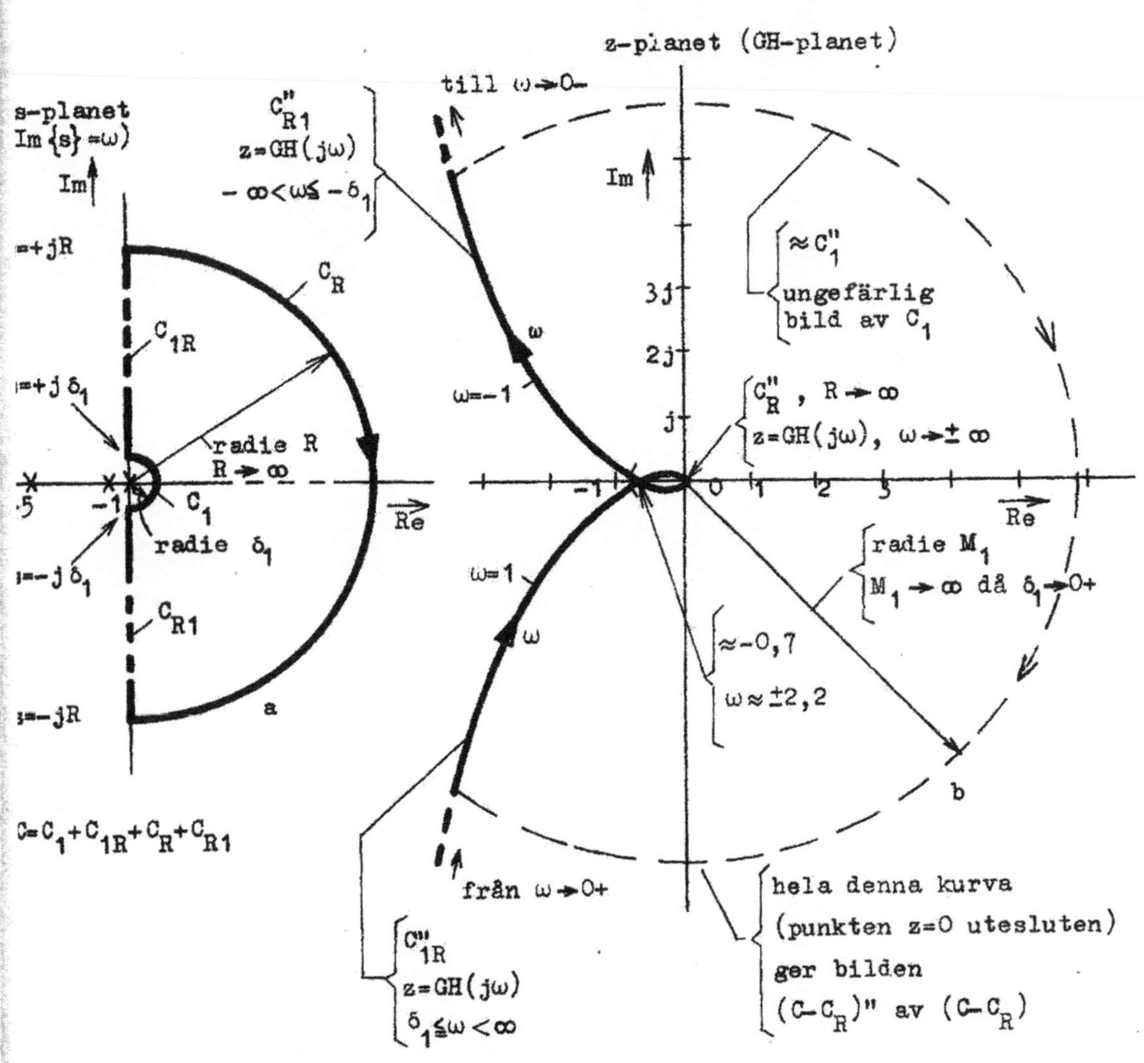

Fig. 4.8.3-2

Tillämpning av Nyquists stabilitetskriterium, exempel 2. Konturen C (bild a) i s-planet är avbildad över funktionen

$GH(s) = \frac{20}{s(s+1)(s+5)}$ i z-planet. Bilden b visar resultatet.

EARLY CONTROL TEXTBOOKS
in
FRANCE

Technologie et calcul pratique des systèmes asservis

P. Naslin
Dunod, 2nd edition, 1958

Méthodes Modernes d'Etude des Systèmes Asservis

J.-C. Gille, P. Decaulne, M. Pelegrin
Dunod, 1960

Commande optimale des processus

R. Boudarel, J. Delmas, P. Guichet
Dunod, 1967

Contributed by

Marcel Staroswiecki
Ecole Polytechnique Universitaire de Lille
France

Jean-Claude Trigeassou and Patrice Remaud

Ecole Supérieure d'Ingénieurs de Poitiers
France

Michel Kinnaert

Université Libre de Bruxelles
Belgium

Background

In 1960, the classical theory of linear control systems was well established, following developments in Germany and in the United States during World War II. In France, it was disseminated through textbooks originally written in French, like

- M. Bonamy, *Servomécanismes, Théorie et Technologie* (Masson, 1957),
- J.-C. Gille, P. Decaulne, M. Pelegrin, *Théorie et Technique des Asservissements* (Dunod, 1956), *Théorie et Calcul des Asservissements* (Dunod, 1958),
- P. Naslin, *Les Systèmes Asservis* (Ed. de la Revue d'Optique, 1951), *Technologie et Calcul Pratique des Systèmes Asservis* (Dunod, 1958),

or translated from American English

- H. Chestnut, R. Mayer, *Servomécanismes et Régulation* (Dunod, 1956)

and from German

- W. Oppelt, *Manuel des Régulations Techniques* (Eyrolles, 1957).

On a related subject, the book

- M. Pelegrin, *Machines à calculer électroniques arithmétiques et analogiques* (Arithmetic and analog electronic computers) (Dunod, 1959)

deals mainly with the application of the first electronic computers to closed loop systems used in aircrafts, or used to simulate the dynamics of these complex systems. *M. Pelegrin,* French aeronautics engineer, was also a graduate of MIT (1949) in the domain of electronic computing. From 1954, he was responsible for teaching computers at Ecole Nationale Supérieure de l'Aéronautique. This book reproduces his teachings for young aeronautics engineers, also specialists of automatic control for aircrafts. The two types of computers used at that time are presented, analog computers necessary for simulation purpose, and arithmetic computers, in order to perform complex computations or to incorporate them in closed loop systems. Though the technologies presented are all obsolete nowadays, this book gives evidence on the natural links between automatic computing and control in the fifties. Moreover, it gives numerous examples of computers manufactured by French companies of this time. Since 1950, France was able to produce the analog and digital computers needed by the army and industry, thanks to several companies, like "Société d'Electronique et d'Automatismes" (SEA), and "Compagnie des Machines BULL". It is also important to note that these companies had been created by (or developed with the help of) pioneering control engineers (like *P. Naslin* and *H.F. Raymond*).

Technologie et calcul pratique des systèmes asservis (Technology and practical design of feedback systems)

P. Naslin
Dunod, 2nd edition, 1958

About the book

This text book originates from a lecture given in 1953 by P. NASLIN to the Belgian Society of Mechanical Engineers. The text of this lecture was first published in France by DUNOD. Then, this first edition was enlarged, mainly with the numerous papers of the author published in the French journal "Automatismes", created in 1956. This second edition, published in 1958, constitutes a reference text book on automatic control, for the interest of the different technical examples presented, as well as for the pedagogy of the author in order to convince French engineers of the relevance of this new science.

This book is divided into three main parts:

About the author

P. NASLIN, graduate of "Ecole Polytechnique" and "Ecole Supérieure d'Electricité", was a pioneer in the introduction of automatic control in France after 2nd World War. He greatly contributed, with the help of his teachings in engineering schools, his papers in technical journals, his lectures and his numerous books, to the diffusion of a modern approach to deal with closed loop systems, towards an audience composed of engineers from various origins. As he stated in the foreword of this book, his main objective was to demonstrate the fundamental functional identity between the controllers of steam engines and the negative feedback of electronic amplifiers. The originality of this book, one of the first to be published in France, is to be based on numerous examples, taken from different technological domains, in order to demonstrate the universality of the feedback concept.

Moreover, P. NASLIN deliberately decided to simplify theoretical developments, in order to facilitate the understanding of this new system methodology. Thus, he uses essentially the *p* Heaviside operator for the writing of differential equations and his demonstrations are mainly based on the frequency response. In all his other books, he made a great effort to be comprehensible to a large audience, by favouring approximate (but rigorous) techniques and methods in order to avoid sophisticated mathematics. Thus, he developed original techniques, particularly for the synthesis of feedback systems. He is the creator of the method "Polynômes normaux à amortissement réglable" (normalized polynomials with tunable damping), also called in France "Naslin polynomials" (comparable to the normalized polynomials of Whiteley or those of Graham and Lathrop).

P. NASLIN is the autor of several textbooks in automatic control like: "Les régimes variables dans les systèmes linéaires et non linéaires" (Dynamics of linear and nonlinear systems) published by DUNOD in 1962 (500 pages) and "Introduction à la commande optimale" (DUNOD). He also wrote two textbooks dealing with discrete event systems: "Circuits logiques et automatismes à séquences" which was published three times by DUNOD, and "Construction des machines séquentielles industrielles" with co-author P. GIRARD in 1973, where an original method for the analysis of sequential systems is presented, this approach being a precursor of French normalized GRAFCET (based on graph theory and Petri nets). He is also the author of a book on electronic computers "Principes des calculatrices numériques automatiques", which is the fruit of his experience in the development of the French computer industry.

Moreover, he is not only the brillant author of numerous books in all fields of control, he has also been responsible for a reference collection on automatic control published by DUNOD: "La bibliothèque de l'automaticien", composed of 40 titles. Thus, he is at the origin of the publication in French of the major books related to control, originating from the USA and the USSR.

Contributed by
Jean-Claude Trigeassou and Patrice Remaud
Ecole Supérieure d'Ingénieurs de Poitiers
86022 Poitiers Cedex France

TECHNOLOGIE ET CALCUL PRATIQUE DES SYSTÈMES ASSERVIS

(RÉGULATEURS ET SERVOMÉCANISMES)

PAR

P. NASLIN

Ancien élève de l'École Polytechnique et de l'E. S. E.
Ingénieur militaire en chef de l'Armement
Professeur à l'École nationale supérieure de l'Armement
à l'École supérieure d'Électricité
et à l'Institut supérieur des Matériaux et de la Construction mécanique

DEUXIÈME ÉDITION
ENTIÈREMENT REFONDUE
ET AUGMENTÉE

PARIS

92, RUE BONAPARTE (6e)
1958

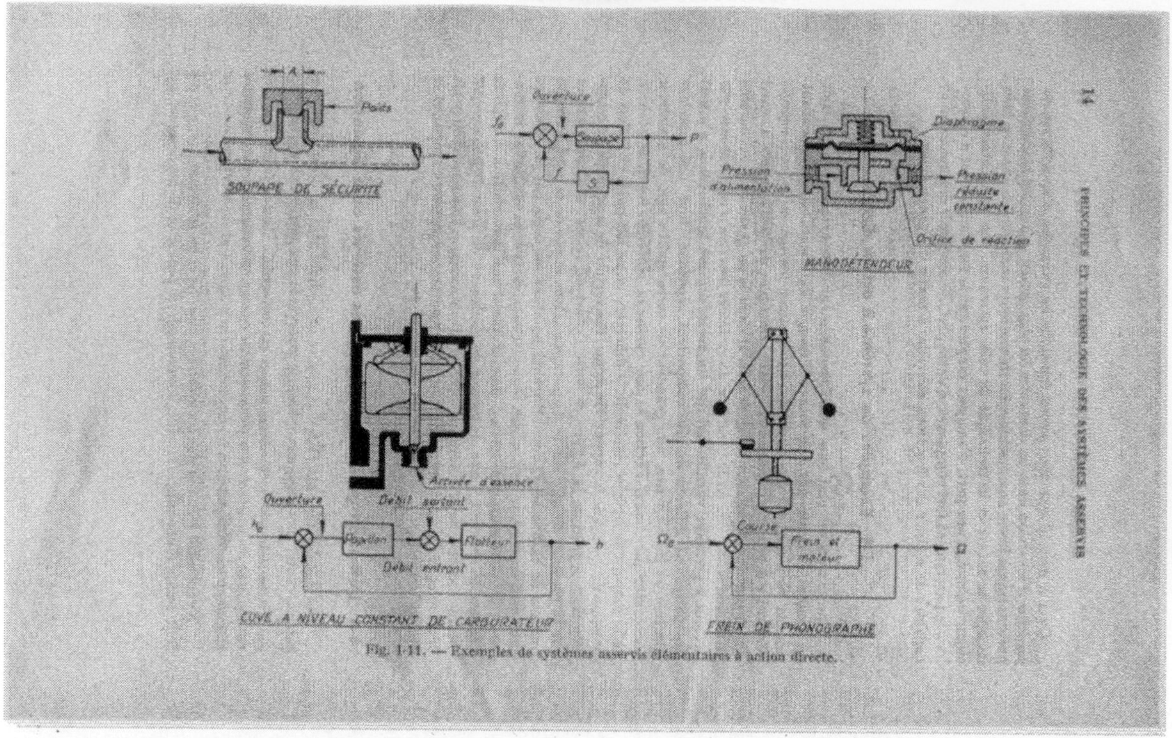

The functional identity of different closed loop systems based on the feedback concept

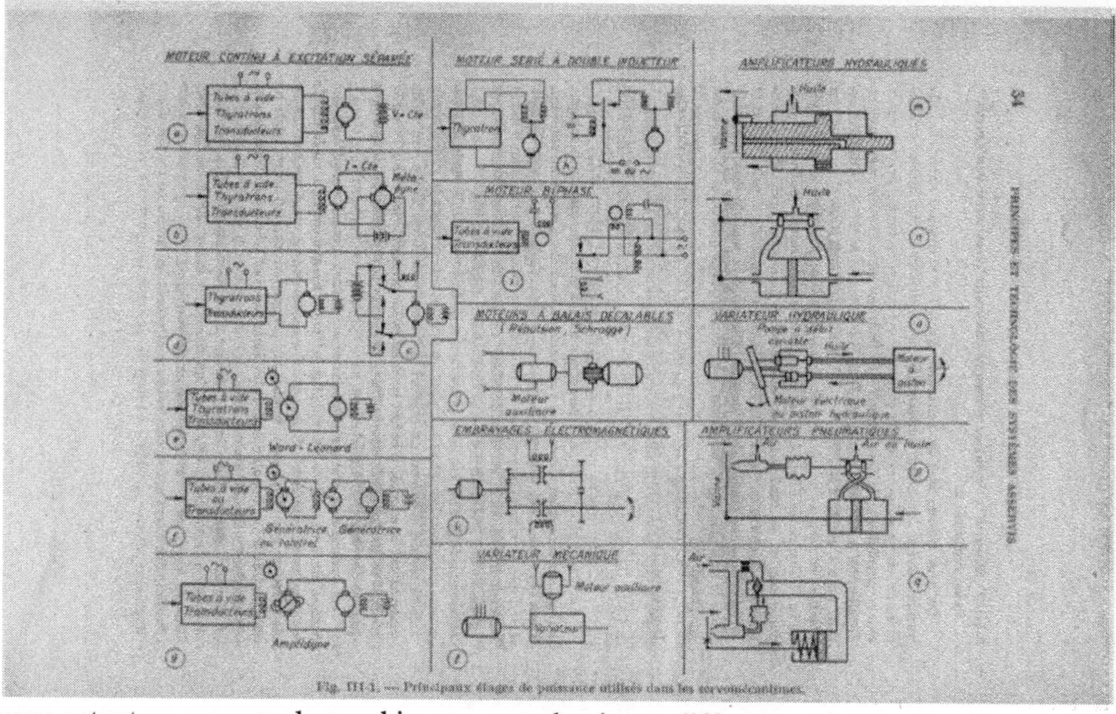

Power actuators commonly used in servomechanisms: different technologies (electricity, mechanics, hydraulics, pneumatics).

En désignant, avec les indices convenables, par $1/R$ les débits dans les étranglements par unité de différence de pression et par $1/C$ la variation de pression

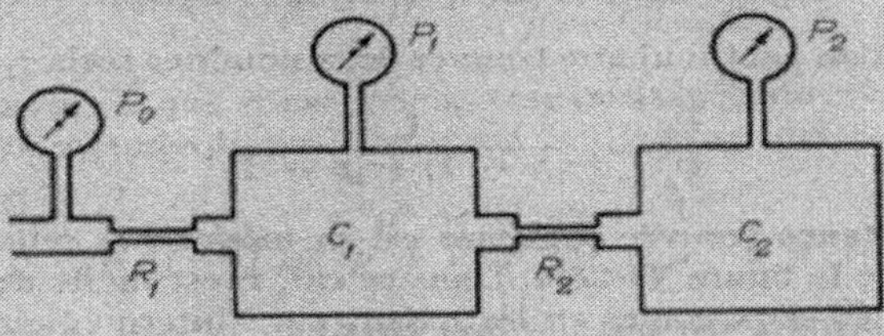

Fig. VI-27.

dans les récipients par unité de débit d'air, le processus est décrit par les 2 équations suivantes :

$$\frac{dP_1}{dt} = \frac{P_0 - P_1}{R_1C_1} - \frac{P_1 - P_2}{R_1C_2},$$

$$\frac{dP_2}{dt} = \frac{P_1 - P_2}{R_2C_2}.$$

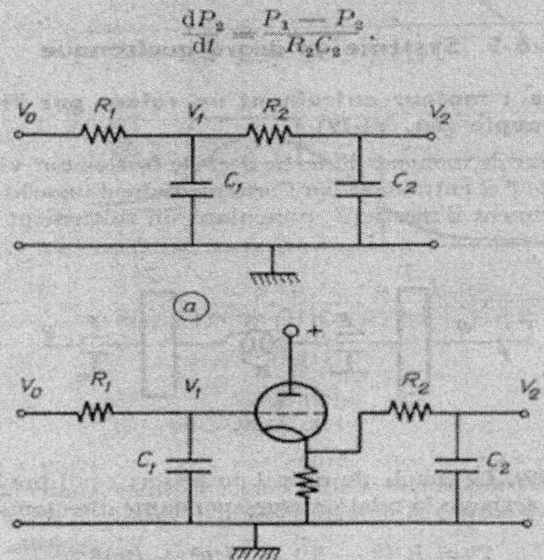

Fig. VI-28. — Cascade de deux étages du premier degré (a) avec couplage et (b) sans couplage.

Cascading of first order systems is presented (originating from electricity or pneumatics, with eventual coupling)

Considérons par exemple le système de la figure VI-36, doué d'une chaîne de réaction principale de fonction de transfert unité et d'une chaîne de réaction secondaire β_2. On commencera par analyser le système $\mu_2\beta_2$. Si cette boucle ne comporte pas d'éléments instables, sa fonction de transfert ne présentera pas de pôle à partie réelle positive et le nombre de tours dans le

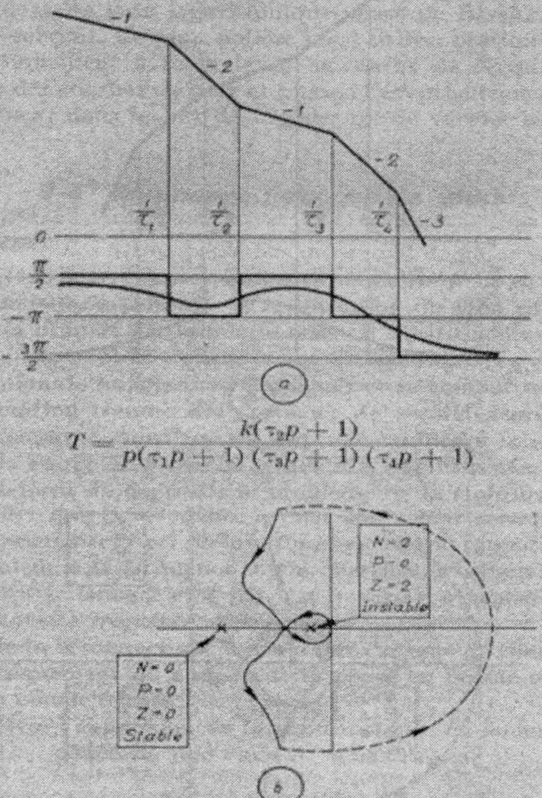

Fig. VII-16. — Application du critère de Nyquist-Cauchy à un système composé uniquement d'organes stables.

sens direct de sa courbe de Nyquist-Cauchy autour du point critique nous fournira le nombre de pôles de sa transmittance en boucle fermée, à parties réelles positives. Ce nombre de pôles est égal au nombre de tours que doit faire dans le sens inverse la courbe de Nyquist-Cauchy de la boucle principale autour du point critique pour que le système complet soit stable.

Use of Nyquist stability criterion

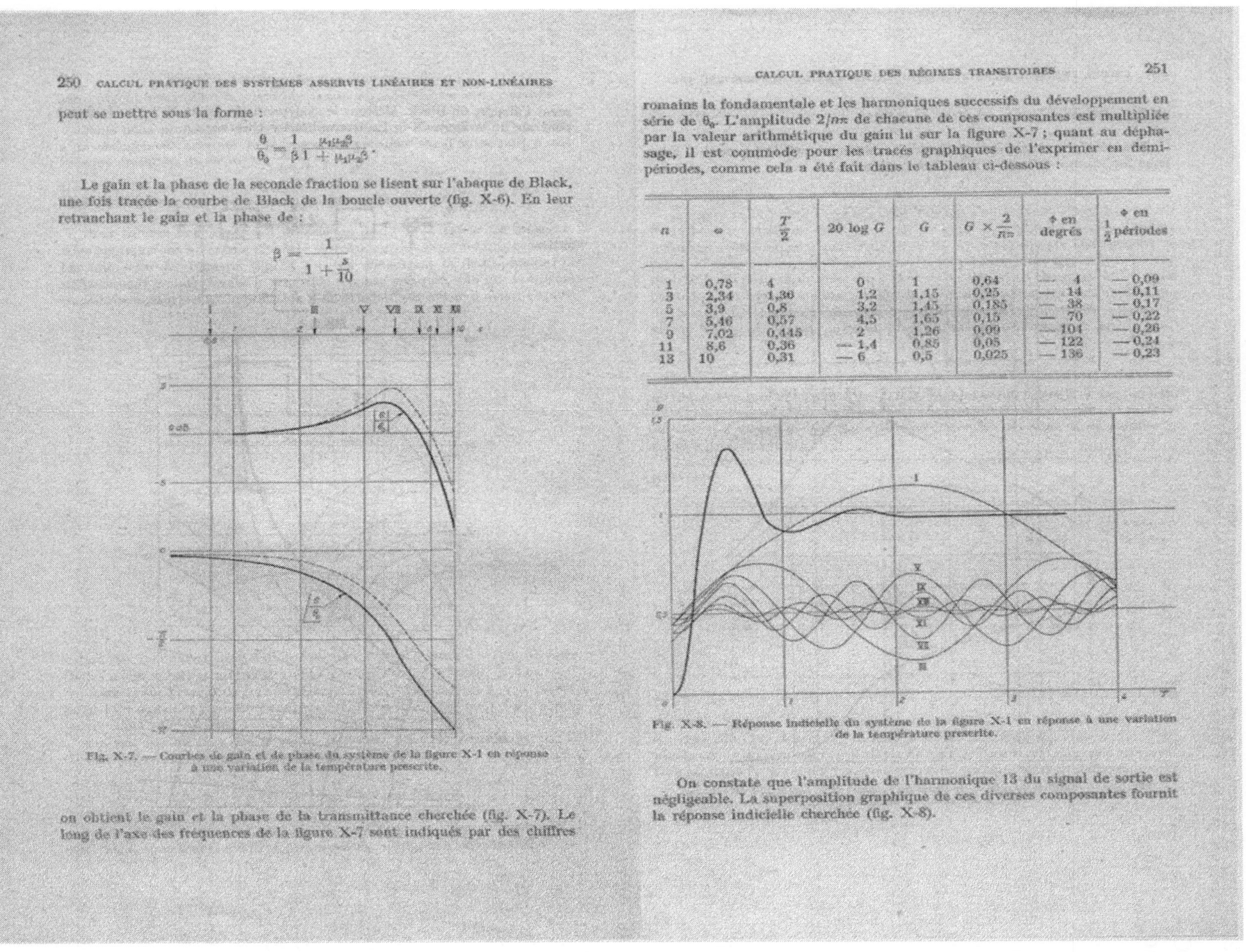

peut se mettre sous la forme :

$$\frac{\theta}{\theta_0} = \frac{1}{\beta} \frac{\mu_1\mu_2\beta}{1 + \mu_1\mu_2\beta}.$$

Le gain et la phase de la seconde fraction se lisent sur l'abaque de Black, une fois tracée la courbe de Black de la boucle ouverte (fig. X-6). En leur retranchant le gain et la phase de :

$$\beta = \frac{1}{1 + \frac{s}{10}}$$

Fig. X-7. — Courbes de gain et de phase du système de la figure X-1 en réponse à une variation de la température prescrite.

on obtient le gain et la phase de la transmittance cherchée (fig. X-7). Le long de l'axe des fréquences de la figure X-7 sont indiqués par des chiffres

romains la fondamentale et les harmoniques successifs du développement en série de θ_0. L'amplitude $2/n\pi$ de chacune de ces composantes est multipliée par la valeur arithmétique du gain lu sur la figure X-7 ; quant au déphasage, il est commode pour les tracés graphiques de l'exprimer en demi-périodes, comme cela a été fait dans le tableau ci-dessous :

n	ω	$\frac{T}{2}$	20 log G	G	$G \times \frac{2}{n\pi}$	ϕ en degrés	ϕ en $\frac{1}{2}$ périodes
1	0,78	4	0	1	0,64	— 4	— 0,09
3	2,34	1,30	1,2	1,15	0,25	— 14	— 0,11
5	3,9	0,8	3,2	1,45	0,185	— 38	— 0,17
7	5,46	0,57	4,5	1,65	0,15	— 70	— 0,22
9	7,02	0,445	2	1,26	0,09	— 101	— 0,26
11	8,6	0,36	— 1,4	0,85	0,05	— 122	— 0,24
13	10	0,31	— 6	0,5	0,025	— 136	— 0,23

Fig. X-8. — Réponse indicielle du système de la figure X-1 en réponse à une variation de la température prescrite.

On constate que l'amplitude de l'harmonique 13 du signal de sortie est négligeable. La superposition graphique de ces diverses composantes fournit la réponse indicielle cherchée (fig. X-8).

These two pages illustrate the approximate determination of the step response of a closed loop using frequency response.

Méthodes Modernes d'Etude des Systèmes Asservis

J.-C. Gille, P. Decaulne, M. Pelegrin
Dunod, 1960

The specific aim of the book by Gille, Decaulne and Pelegrin was to collect later research results, which extended the classical ones and opened new application areas, and which were only available through separate and fragmented papers in journals and conferences, generally written in English, in order to make them available to a general technical audience under a synthetic and didactic form. Its content was used by the authors for lectures at Ecole Nationale Supérieure de l'Aéronautique de Paris, and at Graduate level at the Faculté des Sciences de Québec.

The foreword of the book carefully states the mathematical background which is required from the reader: complex numbers, complex exponential, Laplace transform, decomposition of rational fractions. Matrix notation is used exceptionally, in very limited occurrences. The book is divided into four independent sections.

Section 1 presents the new analysis techniques associated with poles and zeros of the transfer function. These techniques, due to W. Evans, E. Guillemin and J. Truxal, are not believed by the authors to replace the classical harmonic approach, but they are credited with offering a new view of the problem, and stated as a nice complement to classical approaches. Closed loop analysis of linear mono-variable systems and the synthesis of a closed loop regulator are the main topics addressed in Section 1.

Section 2 addresses the statistical analysis of closed loop systems. Stochastic inputs are presented, and the design and optimization of linear systems is addressed. The methods presented in Section 2 follow from the works of C. Shannon and N. Wiener, Y. Lee and G. Newton in the United States, V. Solodovnikov in USSR, and M. Pélegrin in France.

The third section gathers several extensions of the harmonic approach to multivariable systems and to sampled data systems. Arguments given by the authors to justify this extension are related with the practical implementation of closed loop control, by using analogue or digital computers. The section relies on the works by F.-H. Raymond (France), M. Colomb, E. Udsin, J. Ragazzini, L. Zadeh, W. Linvill, L. Sittler and J. Truxal (USA), Ja. Cypkin (USSR).

The last section is devoted to non-linear problems, the methods selected according to their interest for practical applications of closed loop systems, and graphical analysis being preferred to algebraic approaches. Transient behavior, first harmonic, second order state space, on-off regulators, and stability theory (Poincaré, Ljapunov) are the main topics. Research on non-linear systems was already very active, and Section 4 builds on results from USSR (Ja. Cypkin, B. Naumov, N. Bautin, A. Lëtov, M. Ajzerman), Germany (W. Hahn, E. Pestel and K. Magnus), England (P. Nikiforuk and J. West) and France (B. Hamel).

The book contains a glossary in five languages (French, English, Russian, German and Spanish).

Contributed by

Marcel Staroswiecki
Ecole Polytechnique Universitaire de Lille

TECHNIQUES DE L'AUTOMATISME

MÉTHODES MODERNES D'ÉTUDE DES SYSTÈMES ASSERVIS

PAR

J.-C. GILLE
Ingénieur de l'Air
Master of Arts

P. DECAULNE
Ingénieur de l'Air
Master of Science

M. PELEGRIN
Ingénieur de l'Air
Docteur ès Sciences

DUNOD
PARIS
1960

TABLE DES MATIÈRES

QUATRIÈME PARTIE

ASSERVISSEMENTS NON LINÉAIRES

2.3.2 Asservissements à stabilité conditionnelle.

D'autres systèmes asservis, au contraire, sont stables pour certaines plages de variation du gain en boucle ouverte, c'est-à-dire lorsque k satisfait à des conditions telles que

$$k_o' < k < k_o'' \,, \quad \text{avec } k_o' \neq 0.$$

Ils sont dits *conditionnellement stables*. Le cas le plus important de tels systèmes est celui des asservissements *instables en boucle ouverte*, c'est-à-dire dont la fonction de transfert en boucle ouverte a des pôles à partie réelle positive. On peut donner comme exemples les avions instables pilotés, les systèmes à boucles superposées quand l'asservissement interne est instable, etc...

Pour ces systèmes, au moins une branche du lieu des pôles part pour $k = 0$ d'un point du demi-plan de droite. Elle coupe alors l'axe imaginaire ($k = k_o'$) et présente une portion stable. Si k augmente encore, le lieu repasse dans le demi-plan de droite (pompage).

L'exemple 2.2.5 ci-dessus était conditionnellement stable dans le cas d'un avion instable (voir fig. 2-22).

On peut imaginer des asservissements à stabilité conditionnelle possédant plusieurs plages de stabilité (cf. fig. 2-23). Ils se rencontrent assez rarement en pratique.

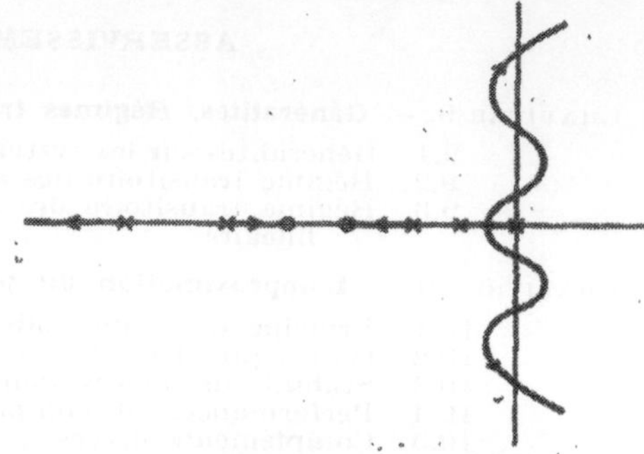

FIG. 2-23. — Asservissement à stabilité conditionnelle complexe.

2.3.3 Autres cas possibles.

D'autres cas peuvent se rencontrer [1].

Ainsi, il peut arriver qu'une branche entière du lieu d'Evans, de $k = 0$ à k infini, soit située dans le demi-plan de droite. On a alors affaire à un système instable quel que soit son gain en boucle ouverte : c'est un cas particulier de ce que les auteurs soviétiques

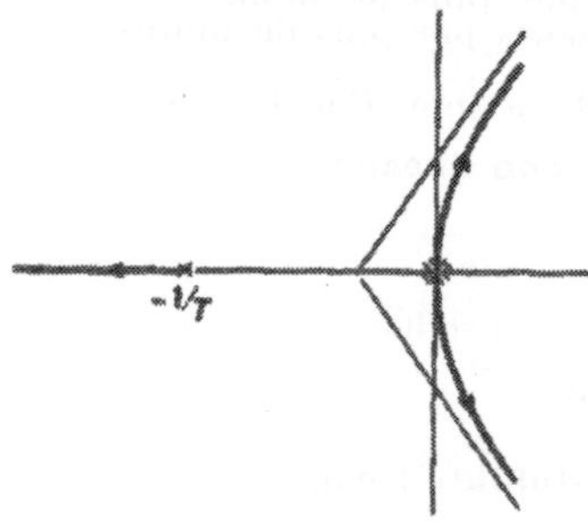

FIG. 2-24. — Lieu d'Evans d'un asservissement intrinsèquement instable.

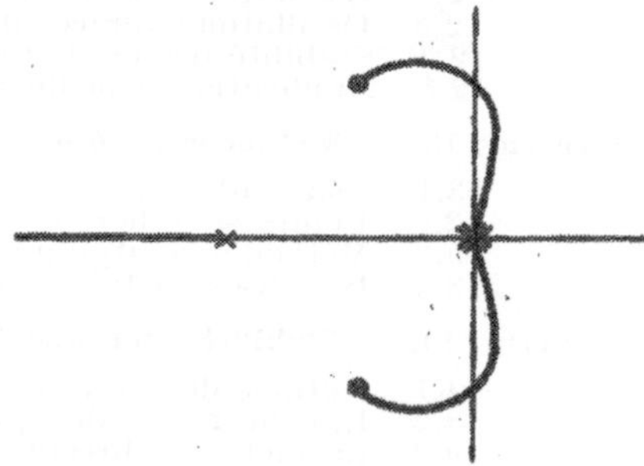

FIG. 2-25. — Asservissement à stabilité paradoxale.

[1] Pour plus de détails sur ces cas complexes, voir *Théorie et Calcul des Asservissements*, § 14.3.2 (pp. 255-259).

7.1.3 Transformée de Laplace d'une fonction du temps échantillonnée.

Si on ne s'intéresse qu'aux valeurs de la sortie $s(t)$ aux instants d'échantillonnage 0, T, 2 T,... (Fig. 7-7) et non à la connaissance complète de $s(t)$, il est possible d'obtenir une relation simple entre l'entrée échantillonnée et la sortie échantillonnée.

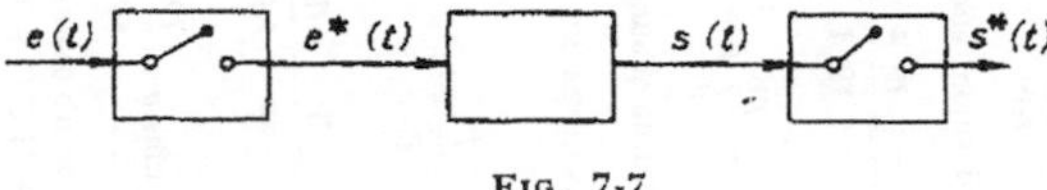

Fig. 7-7

Par application du principe de superposition au système linéaire $H(p)$ on sait qu'on peut obtenir la réponse $s(t)$ par sommation des réponses du système pour chaque impulsion. D'où, en appelant $h(t)$ la réponse du système à une impulsion unitaire [1] :

$$s(t) = e(0)\,h(t) \qquad 0 \leqslant t < T$$
$$s(t) = e(0)\,h(t) + e(T)\,h(t - T) \qquad T \leqslant t < 2T$$
$$s(t) = e(0)\,h(t) + e(T)\,h(t - T) + e(2T)\,h(t - 2T) \qquad 2T \leqslant t < 3T$$

etc., et aux instants d'échantillonnage :

$$s(0) = e(0)h(0)$$
$$s(T) = e(0)h(T) + e(T)h(0)$$
$$s(2T) = e(0)h(2T) + e(T)h(T) + e(2T)h(0)$$
$$\cdots\cdots\cdots$$
$$s(nT) = \sum_{m=0}^{n} e(mT)h[(n - m)T] \,.$$

Or, d'après l'équation (7-2) :

$$S^*(p) = s(0) + s(T)\,e^{-Tp} + s(2T)\,e^{-2Tp} + \cdots = \sum_{n=0}^{\infty} s(nT)\,e^{-nTp}$$

D'où :

$$S^*(p) = \sum_{n=0}^{\infty}\sum_{m=0}^{\infty} e(mT)h[(n - m)T]\,e^{-nTp}.$$

On peut vérifier à partir de cette expression que :

$$S^*(p) = [e(0) + e(T)\,e^{-Tp} + \cdots]\,[h(0) + h(T)\,e^{-Tp} + \cdots] = E^*(p)\,H^*(p)$$

où $H^*(p)$ est la transformée de Laplace par échantillonnement associée à la réponse impulsionnelle $h(t)$ ou, ce qui revient au même, à la fonction de transfert $H(p)$. En d'autres termes, *en ne considérant que des transformées de Laplace et fonctions de transfert pulsées : la transformée de la sortie d'un système linéaire pour une entrée échantillonnée est simplement la transformée de l'entrée multipliée par la fonction de transfert du système*, exactement comme dans le cas des systèmes continus :

$$S^*(p) = E^*(p)\,H^*(p). \qquad (7\text{-}6)$$

[1] La fonction $h(t)$ est la *réponse impulsionnelle* du système. Rappelons que sa transformée de Laplace est la fonction de transfert du système :

$$H(p) = \mathscr{L}\,h(t)$$

et que $h(t)$ est la dérivée de la réponse unitaire $q(t)$.

11.2.2 Asservissement par plus ou moins avec frottement visqueux, sans retard ni seuil.

A) Équations.

Soit le système décrit au § 11.2.1, mais supposons à la sortie l'existence d'un frottement visqueux de telle manière qu'on ait

$$J \frac{d^2s}{dt^2} + f \frac{ds}{dt} = h \operatorname{sign}\left(\varepsilon + \lambda \frac{d\varepsilon}{dt}\right) \qquad (h > 0),$$

soit

$$\tau \frac{d^2s}{dt^2} + \frac{ds}{dt} = K \operatorname{sign}\left(\varepsilon + \lambda \frac{d\varepsilon}{dt}\right),$$

en posant

$$\tau = \frac{J}{f} \quad , \quad K = \frac{h}{f}.$$

B) Régime transitoire en régulateur.

Dans le cas du fonctionnement en régulateur ($e \equiv 0$), on aura

$$J \frac{d^2\varepsilon}{dt^2} + f \frac{d\varepsilon}{dt} = -h \operatorname{sign}\left(\varepsilon + \lambda \frac{d\varepsilon}{dt}\right),$$

ou

$$\frac{d^2s}{dt^2} + \frac{1}{\tau}\frac{ds}{dt} = -\frac{K}{\tau} \operatorname{sign}\left(s + \lambda \frac{ds}{dt}\right).$$

En prenant d'abord le signe moins pour la fonction sign :

$$\frac{d^2s}{dt^2} + \frac{1}{\tau}\frac{ds}{dt} = \frac{K}{\tau}$$

d'où, après intégration, en posant $y = ds/dt$:

$$y = K - (K - y_0)\, e^{-t/\tau}.$$

On a donc affaire à une mise en vitesse, avec une constante de temps τ, vers une vitesse de saturation K.

Posant $s = x$ on a, par une nouvelle intégration :

$$x = Kt - \tau(K - y_0)(1 - e^{-t/\tau}) + x_0.$$

Cette équation montre que toutes les trajectoires se déduisent les unes des autres par translation en x, suivant la valeur de x_0.

La trajectoire qui passe par l'origine a notamment pour équations paramétriques :

$$(x_0 = 0 \quad , \quad y_0 = 0 \quad \text{pour} \quad t = 0)$$

$$\begin{cases} x = Kt - \tau K(1 - e^{-t/\tau}) \\ y = K(1 - e^{-t/\tau}) \end{cases}$$

On voit que la trajectoire *admet comme asymptote l'horizontale* $y = K$. Son équation cartésienne est

$$x + \tau y + \tau K \operatorname{Log}\left(1 - \frac{y}{K}\right) = 0.$$

On peut voir comme en 11.2.1 que les trajectoires se répartissent en deux familles de concavités opposées (suivant la valeur de la fonction sign), les courbes d'une même famille se déduisant l'une de l'autre par translation parallèle à l'axe des x. On a tracé (Fig. 11-14) les deux courbes de concavité opposée qui passent par l'origine. Le passage d'une caractéristique à l'autre a lieu sur la **droite de commutation** Δ :

$$\varepsilon + \lambda \frac{d\varepsilon}{dt} = 0.$$

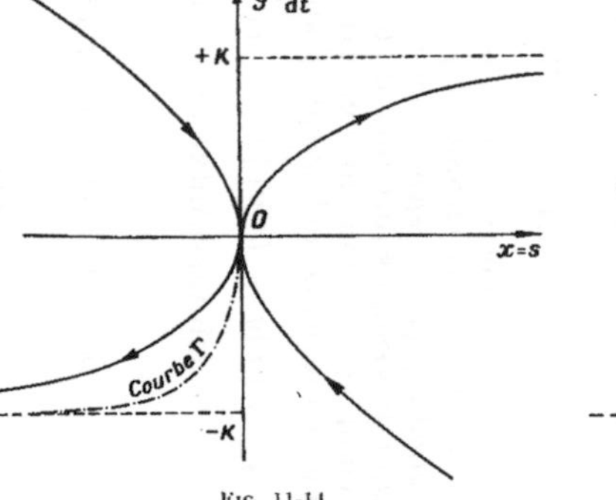
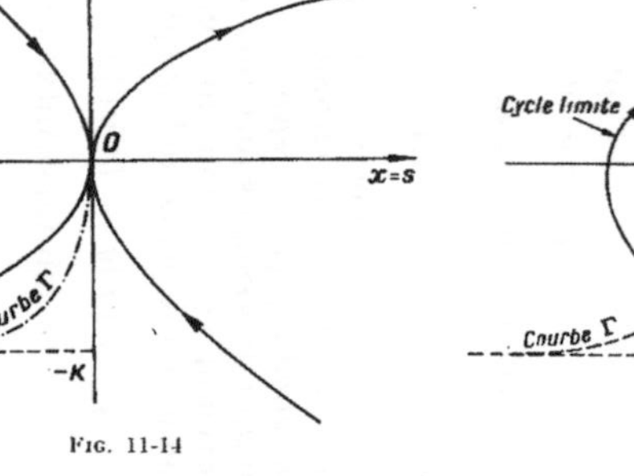

Fig. 11-14

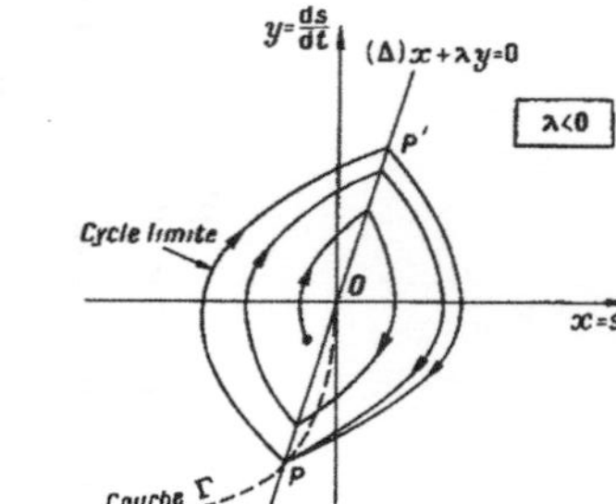
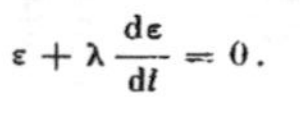

Fig. 11-15

Graphiquement (Fig. 11-15) la présence de l'asymptote a pour conséquence que, si λ est **négatif**, les trajectoires de phase commencent par diverger mais sont limitées pour *t* infini par un *cycle asymptotique* ou *cycle limite* constitué par deux arcs de ces caractéristiques se relayant sur la droite de commutation. Ce cycle correspond à une **solution périodique** unique qui ne dépend pas des conditions initiales.

Pour un système donné possédant une valeur déterminée du paramètre λ il existe sur la droite de commutation Δ un point unique P situé sur le cycle limite. Si on part de conditions initiales situées sur Δ placées plus loin de l'origine que P, la commutation suivante sera plus proche de l'origine ; si, au contraire, on part de conditions (sur Δ) plus près de l'origine que P, les commutations successives seront plus éloignées : dans les deux cas on tend vers le cycle limite. Par contre, si les conditions initiales correspondent exactement au point P, on se trouve d'emblée sur le cycle limite : les commutations suivantes se font en P' (symétrique de P par rapport à l'origine) puis en P, etc.

Lorsque le paramètre λ varie le lieu des points P (conditions initiales correspondant à une solution périodique) est une certaine courbe Γ, d'équations paramétriques :

$$\begin{cases} x = K\left[\tau \operatorname{th} \dfrac{T}{2\tau} - \dfrac{T}{2}\right] \\ y = -K \operatorname{th} \dfrac{T}{2\tau} \end{cases}$$

où T est la moitié de la période de la solution correspondante. Cette courbe Γ est à compléter par symétrie par rapport à l'origine.

La courbe Γ est le **lieu de Hamel** du système (nous la retrouverons au chapitre suivant, § 12.2.1 à 12.2.3, où ces résultats seront généralisés). Elle est tracée sur les

Commande optimale des processus

R. Boudarel, J. Delmas, P. Guichet
Dunod, 1967

This is a series of four books devoted to a new approach in the design of control problems, namely by using optimal control. Optimization approaches and Variation Calculus are known for long, and early applications to Control were developed by Hall and Sartorius, Wiener and followers. However, the practical interest of optimization based design for control applications rose drastically after World War II, when handling of constrained and non-linear problems became really possible. The enormous amount of papers dealing with optimal control, their heterogeneity, and the very small number of synthesis books for an engineering public justify the undertaking of writing this series, where the accent is put on a synthetic approach which highlights the main ideas, and develops in parallel continuous and digital approaches, while treating in detail the problems associated with numerical computation. The authors rely on the expertise gained by the Research Group on Optimization of the Centre d'Etudes et de Recherches en Automatismes at Ecole Nationale Supérieure de l'Aéronautique, and on the huge experience developed through many industrial and military contracts in aerospace and naval applications. They had also been teaching the material of their work in fairly prestigious institutions: Ecole Nationale Supérieure de l'Aéronautique (Paris), Ecole Nationale Supérieure d'Electrotechnique, Electronique, et Hydraulique (Toulouse), Institut National des Sciences Appliquées (Toulouse), Département de Génie Electrique de l'Université Laval (Québec).

The four books of the series are :

1 : *Concepts fondamentaux de l'Automatique* (Fundamental Concepts in Control), which sets the basic models, properties and tools (state space representation and basic system properties, stochastic variables, Markov processes),

2 : *Programmation non linéaire*, (Nonlinear Programming), which addresses finite dimensional optimization, with applications in parameter identification and digital control,

3 : *Programmation dynamique* (Dynamic Programming), centered on control problems, with special emphasis on stochastic processes and computational aspects,

4 : *Méthodes variationnelles* (Variational Methods), which parallels classical Variation Calculus with Pontryagin's maximum principle.

Each of the three last volumes is illustrated with a number of completely solved application problems.

Table of Contents

Volume 1 – Fundamental Concepts in Control

Part I Deterministic concepts

8.3 Sensitivity with respect to parameters
8.4 Linear systems

Chapter 9 Stability
9.1 Introduction
9.2 Notion of stability
9.3 Mathematical definition of stability
9.4 Ljapunov functions
9.5 Linear time invariant systems
9.6 Example

Part II Stochastic concepts

Chapter 10 Introduction
10.1 From deterministic to stochastic systems
10.2 Concepts associated with stochastic systems
10.3 The optimization problem formulation

Chapter 11 Representation of a stochastic continuous time signal
11.1 The notion of stochastic function
11.2 First order statistics
11.3 Second order statistics
11.4 Correlation functions
11.5 Correlation spectrum
11.6 White noise
11.7 Transmission of a stochastic signal through a linear element
11.8 Multivariable signal

Chapter 12 Representation of a stochastic discrete time signal
12.1 Definition of a stochastic discrete signal
12.2 First order statistics
12.3 Second order statistics
12.4 Correlation spectrum
12.5 Particular signals
12.6 Transmission of a stochastic signal through linear elements

Chapter 13 Discrete Markov processes
13.1 Markov process with discrete states
13.2 Markov process with continuous states
13.3 Study of the stochastic linear recurrent system

Chapter 14 Continuous Markov processes
14.1 Markov process with discrete states
14.2 Markov process with continuous states
14.3 Study of the stochastic linear differential system
14.4 Discretization of a continuous process

Contributed by

Marcel Staroswiecki
Ecole Polytechnique Universitaire de Lille

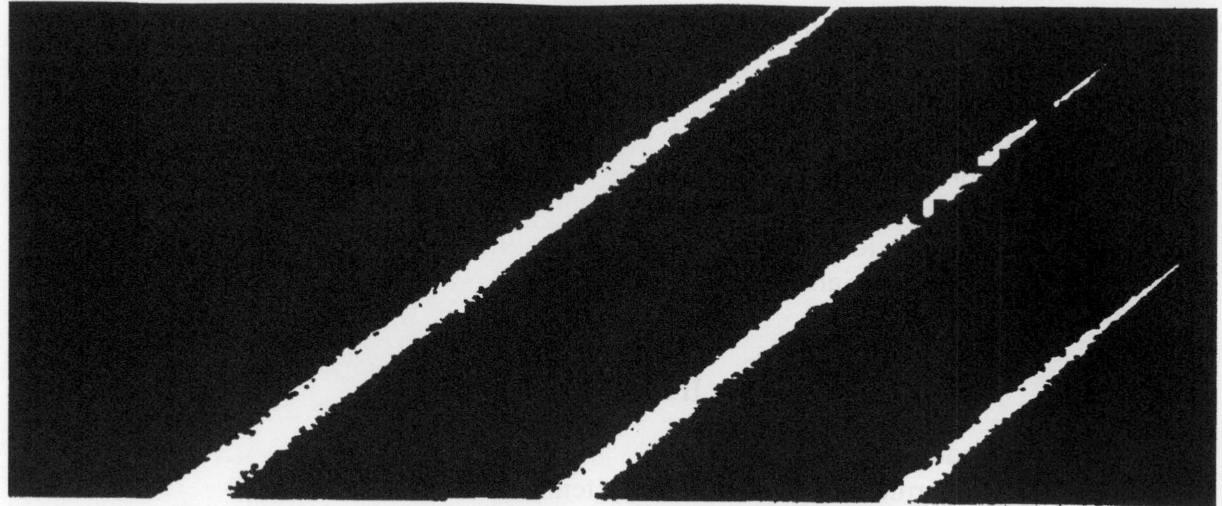

**COMMANDE
OPTIMALE
DES PROCESSUS**

TECHNIQUES DE L'AUTOMATISME

Collection publiée sous la direction de J.-Ch. GILLE et M. PÉLEGRIN

COMMANDE OPTIMALE DES PROCESSUS

TOME I

Concepts fondamentaux de l'automatique

PAR

R. BOUDAREL
Ingénieur A. M. et E. S. E.

J. DELMAS
Ingénieur E. N. S. E. E. H. T., Master of Sciences

Centre d'Études et de Recherches en Automatisme

P. GUICHET
Ingénieur A. M. et E. S. E.
Master of Sciences
Cie internationale pour l'Informatique

PRÉFACE DE

J.-Ch. GILLE
Département de génie électrique, Université Laval, Québec
Professeur honoraire, École nationale supérieure de l'Aéronautique

DUNOD
PARIS
1967

II.2.1 Représentation par une équation d'état

a) STRUCTURE

Considérons comme forme standard :

$$\begin{cases} \dot{x}(t) = F[x(t), e(t), t], \\ s(t) = G[x(t), t], \end{cases} \tag{3}$$

où la première équation vectorielle est un système différentiel du premier ordre, avec les hypothèses suivantes :

G est une fonction univoque,

F est une fonction continue et satisfaisant à la condition de Lipschitz.

N.B. — Rappelons que F satisfait à la condition de Lipschitz s'il existe une constante k positive telle que, pour t et $e(t)$ fixés, on ait :

$$\|F[x_1(t), e(t), t] - F[x_2(t), e(t), t]\| < k\|x_1(t) - x_2(t)\|,$$

quels que soient x_1 et x_2 appartenant à l'ensemble des états admissibles.

La variable $x(t)$ des équations (3) a la propriété du vecteur d'état. En effet, nous savons que pour un $x(t_0)$ et un $e(t_0, t)$ donnés, la solution de l'équation différentielle $x(t)$ est unique, moyennant les hypothèses précédentes (théorème de Cauchy); et la sortie $s(t)$ est alors définie d'une façon unique si G est univoque. Aussi nous appellerons (3) équation d'état du système en précisant « sous forme différentielle » si on veut la distinguer de la forme (2).

b) INTÉRÊT

Une classe très importante de processus physiques peut être représentée par un modèle mathématique du type (3). Le principal avantage d'une telle représentation est d'avoir apporté une unité dans l'étude de ces systèmes; le lecteur déjà un peu familier avec l'automatisme sait que les problèmes de stabilité et d'optimisation par exemple sont actuellement toujours abordés, d'une manière théorique, à partir de cette représentation.

Un autre intérêt est de correspondre explicitement avec le schéma de simulation comme indiqué figure II.2 où les doubles traits indiquent qu'il s'agit de plusieurs variables.

c) NON-UNICITÉ DE LA REPRÉSENTATION

Il est à peu près évident qu'un système peut être représenté par différentes équations ayant la même structure que (3) mais où les vecteurs d'état sont différents. Il suffit pour cela de considérer un nouveau vecteur d'état $y(t)$ lié à $x(t)$ par une fonction biunivoque dérivable :

$$x(t) = T[y(t)].$$

Les équations (II.3) s'écrivent :

$$\begin{cases} \dot{x} = F[T[y(t)], e(t), t] = \left[\dfrac{\partial T}{\partial y}\right]^{T} \dot{y}(t) = T_y^T \dot{y}(t), \\ s(t)_i = G[T[y(t)], t]^i. \end{cases}$$

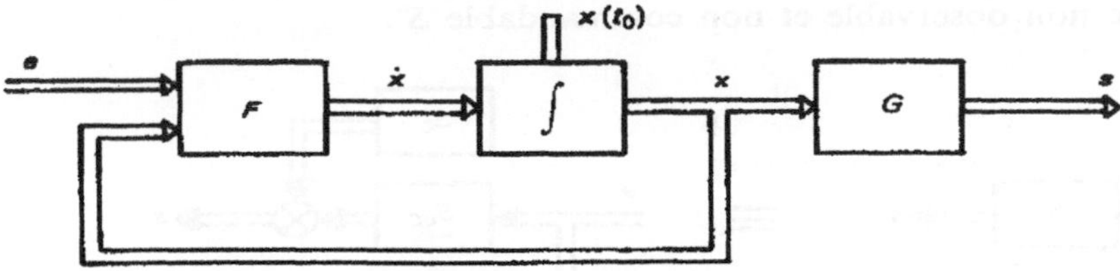

Fig. II.2. — Simulation d'une équation d'état

D'après les hypothèses sur T, la matrice des dérivées partielles $[\partial T/\partial y]$ existe (dérivabilité) et admet une inverse (biunivocité) ; les équations précédentes peuvent se mettre sous la forme :

$$\begin{cases} y(t) = \mathscr{F}[y(t), e(t), t], \\ s(t) = \mathscr{G}[y(t), t], \end{cases}$$

qui est analogue à (3).

Pour la représentation d'un processus par un modèle mathématique du type (3), le choix du vecteur d'état sera guidé par la simplicité des équations et par la signification physique des composantes du vecteur d'état.

II.2.2 États particuliers

Certains états particuliers jouent un rôle important.

Nous les appellerons :

— *état maintenable*

C'est un état tel qu'on puisse trouver une entrée $e(t)$ qui le laisse inchangé sur un intervalle de temps. Ces états satisfont la relation implicite :

$$\dot{x} = F[x(t), e(t), t] = 0. \qquad (4)$$

A un domaine d'entrées admissibles, correspond alors un domaine d'états maintenables.

VI.1.3 Intérêt de ces notions

Si on considère une équation vectorielle d'état S quelconque de la forme (1) on peut toujours la décomposer en quatre parties comme représenté figure VI.1 :

— une partie observable et commandable S_{0c},
— une partie observable et non commandable S_0,
— une partie commandable et non observable S_c,
— une partie non observable et non commandable S'.

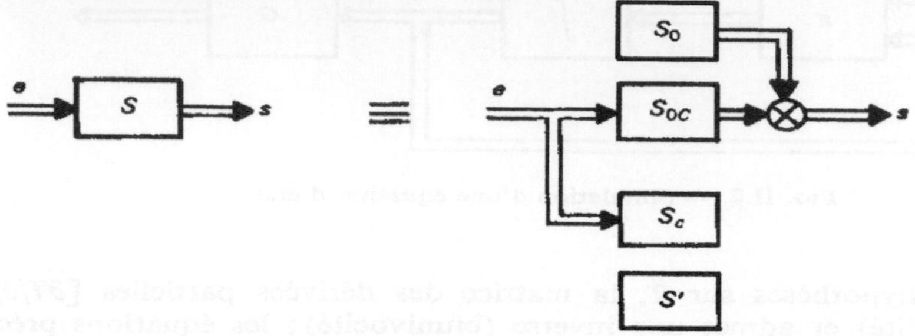

Fig. VI.1. — Décomposition canonique d'un système

Exemple

Dans l'équation d'état suivante :

$$\begin{cases} \dot{x}_1 = x_1 + 2x_2 + x_3 + e_1 + e_2 \\ \dot{x}_2 = -x_2 + x_4 + e_1 + 2e_2 \\ \dot{x}_3 = -2x_3 \\ \dot{x}_4 = x_4 \end{cases} \tag{2}$$

$$s = x_2 + x_4$$

il est évident que :

— x_1 est commandable et non observable,
— x_2 est commandable et observable,
— x_3 est non commandable et non observable,
— x_4 est non commandable et observable.

La notion d'observabilité est importante dans le problème du choix d'un modèle du type (1) pour représenter l'évolution d'un processus réel. En effet l'état d'un système est, comme nous l'avons vu, un concept abstrait; sur un

EARLY CONTROL TEXTBOOKS
in
GERMANY

Die selbsttätige Regelung (Feedback Control)

A. Leonhard
Springer Verlag, 1940, 1949, 1957

Dynamik selbsttätiger Regelungen
(Dynamics of Automatic Feedback Control)

R.C. Oldenbourg and H. Sartorius
R. Oldenbourg Verlag, 1944, 1949, 1951

Kleines Handbuch technischer Regelungsvorgänge
(Compact Handbook of Engineering Control Processes)

W. Oppelt
Verlag Chemie, 1954, 1956, 1960, 1964, 1972.

Contributed by

Manfred Thoma

University of Hannover
Germany

Die selbsttätige Regelung (Feedback Control)

A. Leonhard
Springer Verlag, 1940, 1949, 1957

The first edition of the book **A. Leonhard: Die selbsttätige Regelung in der Elektrotechnik (Automatic Feedback Control in Electrical Engineering)** was published in 1940 by Springer-Verlag. Already two years later it was out of print. However, in 1944 a reprinted version of the book was published in the USA. Instead of printing a new edition, A. Leonhard suggested at that time to the publisher to write a new more general control engineering book which is not restricted to the field of electrical engineering. At the beginning of 1944, the new version of the manuscript was ready to be forwarded to the publisher. However, due to war and postwar circumstances the edition with the title: **Die selbsttätige Regelung – Theoretische Grundlagen mit praktischen Beispielen (Automatic Feedback Control – Theoretical Fundamentals with Practical Applications)** was delayed until 1949. The delay, however, gave the editor the chance to include newly published results, like the Nyquist's Stability Criterion and the Logarithmic Plots (Bode Diagrams). The second revised and last edition of altogether 376 pages and 319 illustrations was published 1957.

At that time, most of the published books in Control Engineering paid increased attention to the mathematical part. In contrast, A. Leonhard's book rested much more on practical applications; in other words, the application of different mathematical methods was used to solve engineering problems (physical modeling).

The content of the last edition of the book is split into four major parts (A-D):

<u>Part A</u> considers fundamentals like different feedback control structures and objectives. Besides, the classical behavior of linear, time-invariant control components is presented in order to enable the reader to understand the basic principles and to apply them. In addition, the principal behavior of control components with dead-time is discussed.

<u>Part B</u> considers the design of mainly single input control system structures. Time responses and sinusoidal responses are used to characterize the transient behavior of feedback control systems. In addition, the reaction of control systems with discontinuous nonlinear control elements is studied.

The stability of feedback control systems is the subject of <u>Part C</u>. Different stability criteria (Routh-Hurwitz, Nyquist and Logarithmic Plots) of linear closed loop systems are presented. In case there are simple nonlinear elements in the loop, the stability analysis is to represent a nonlinear device by a describing function which is easier to handle.

The last <u>Part D</u> treats the problem how to choose the parameters of a closed loop system in order to improve the control behavior. Particular attention is given to the minimization of the transient response of linear feedback control systems with a step input. The optimal system is one that has the minimum control area. In that case the value of the control area is given by an integral error.

Of great practical value are the engineering examples – accompanied by considerable discussion – in each chapter. Some of them serve to extend the scope of the material of the text; others show design applications.

The characteristics in the time and frequency domain of linear, continuous-time controller and control components are tabulated in the appendix. The compilation of the tables includes, in the time domain, ordinary differential equations, their solutions and step responses and, in the frequency domain, transfer functions and polar plots.

The book addressed to practicing engineers and advanced students quickly became a standard in the field.

Die selbsttätige Regelung

Theoretische Grundlagen mit praktischen Beispielen

Von

Professor Dr.-Ing. A. Leonhard

Stuttgart

Zweite neubearbeitete Auflage

Mit 319 Abbildungen

Springer-Verlag

Berlin / Göttingen / Heidelberg

1957

Ermittlung des scheinbaren Gesamthubes in Abb. 8 angedeutet. Die Gl. (22) gilt nun genau so wie für z/H_z auch für w/H_w, wir müssen uns nur darüber klar sein, daß dabei H_w verschieden ist, je nachdem in welchem Bereich der Kurvenbahn wir arbeiten. Bei der Behandlung eines wirklichen Regelvorganges ist selbstverständlich dieser Gesamthub, auf den der Kolbenweg damit bezogen wird, von großer Bedeutung.

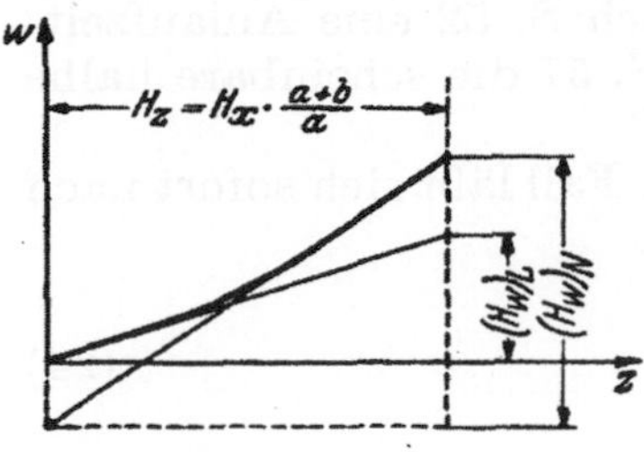

Abb. 8. Zusammenhang zwischen der Stellung des Rückführpunktes (*I*, Abb. 5) und der des Arbeitskolbens

Regelglieder, bei denen der Zusammenhang zwischen Ein- und Ausgangsgröße durch die allgemeine Schwingungsgleichung (13b) dargestellt werden muß, wie bei den beiden behandelten Beispielen, treten verhältnismäßig selten auf. Meistens liegen die Verhältnisse so, daß sich die allgemeine Gl. (13b) durch Wegfall eines oder auch zweier Glieder auf der linken Seite vereinfacht. Die wichtigsten Fälle sollen nun besprochen werden.

b) Weg-Geschwindigkeitssteuerung, Regelglieder mit Ausgleich. Wir betrachten zunächst nochmals die beiden Anordnungen (Abb. 1 und 5), vereinfachen sie aber etwas. Bei Abb. 1 vernachlässigen wir die Selbstinduktion im Ankerkreis, so daß die elektromagnetische Zeitkonstante $T_b = 0$ wird. Den mittelbaren Regler (Abb. 5) vereinfachen wir dadurch,

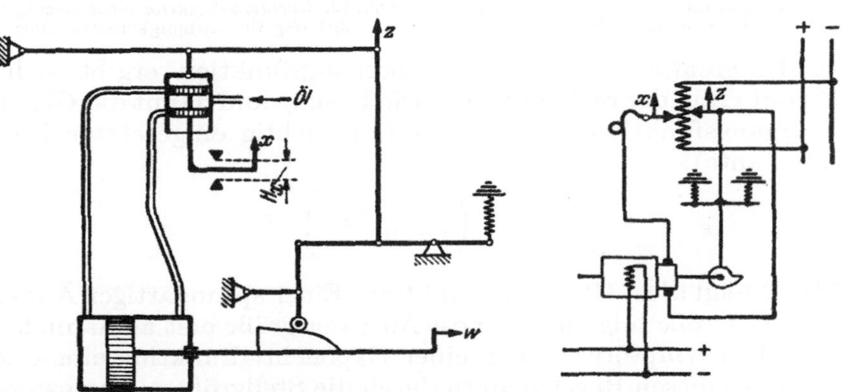

Abb. 9. Mittelbarer Öldruck-Regler als Beispiel eines Regelgliedes mit Weg-Geschwindigkeitssteuerung (Regelglied mit Ausgleich)

Abb. 10. Regler mit Elektro-Steuermotor

daß wir nach Abb. 9 vom Steuerkolben, der vom Meßwerk verstellt wird, unmittelbar den Arbeitskolben steuern. Das gleiche Ersatzschema gilt auch für die Anordnung (Abb. 5), wenn die Verstellgeschwindigkeit des Hauptsteuerkolbens sehr groß ist, so daß $T_y \ll T_z$ wird. Abb. 10 zeigt ein entsprechendes Verstellwerk aber mit Elektro-Steuermotor. In beiden Fällen — Nebenschlußmotor mit vernachlässigbar kleiner Ankerinduktivität, Verstellwerk des indirekten Reglers mit Rückführung nach

Abb. 9 und 10 — können wir in der Differentialgleichung (13b) $T_b = 0$ setzen, so daß sich die Restgleichung ergibt:

$$\alpha' \, T_a + \alpha = \varepsilon. \tag{23}$$

Bei der Anordnung nach Abb. 1 ist T_a nach S. 52 eine Anlaufzeitkonstante, bei der nach Abb. 9 entsprechend S. 57 die scheinbare halbe Schlußzeit des Kolbens.

Der Frequenzgang für den jetzt einfacheren Fall läßt sich sofort nach Gl. (14) angeben, in der $T_b = 0$ gesetzt wird:

$$\mathfrak{F} = \frac{\vec{\alpha}}{\vec{\varepsilon}} = \frac{1}{j\,\omega\,T_a + 1} \tag{24}$$

mit der Kurve (Abb. 11), einem Halbkreis.

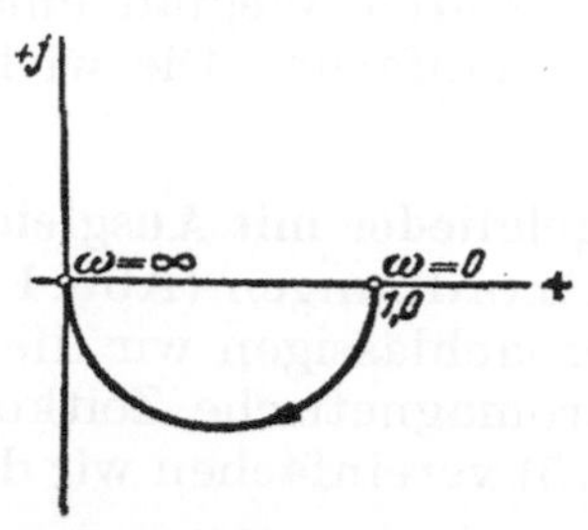

Abb. 11. Frequenzgang eines Verstellgliedes mit Weg-Geschwindigkeitssteuerung

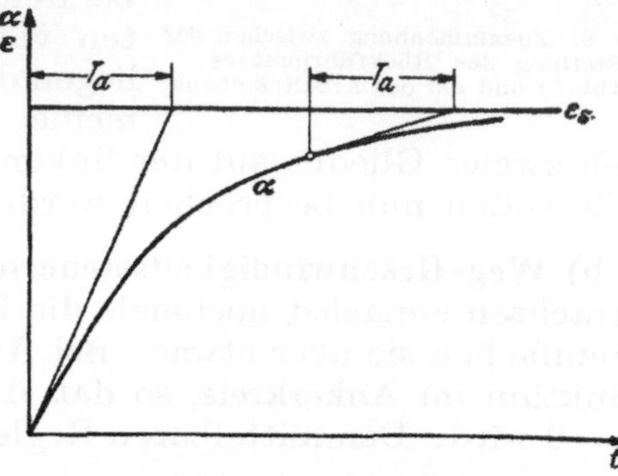

Abb. 12. Übergangsfunktion eines Regelgliedes mit Weg-Geschwindigkeitssteuerung

Die Differentialgleichung der Übergangsfunktion ergibt sich aus Gl. (23) mit ε_s auf der rechten Seite. Die Lösung und damit die Gleichung der Übergangsfunktion selbst lautet (mit richtig eingesetzter Integrationskonstante):

$$\alpha = \varepsilon_s \left(1 - \mathrm{e}^{-\frac{t}{Ta}}\right) \tag{25}$$

Abb. 12 zeigt diese Übergangsfunktion. Einer sprungartigen Änderung der Eingangsgröße folgt also hier die Ausgangsgröße erst allmählich nach und erreicht asymptotisch nach einer Exponentialfunktion einen neuen Wert. Da bei diesen Regelgliedern durch die Stellgröße ε, wenigstens bei mechanischen Anordnungen, sowohl die Stellung oder der Weg α als auch die Verstellgeschwindigkeit der Ausgangsgröße α' beeinflußt wird, können wir von einer *Weg-Geschwindigkeitssteuerung* sprechen. Man heißt solche Verstellglieder auch Glieder „*mit Ausgleich*".

Als *Ausgleichswert* (q) eines Regelgliedes bezeichnet man das Verhältnis von Eingangsgrößenänderung (y) zu der durch sie im Beharrungszustand bewirkten Ausgangsgrößenänderung (x), wobei mit den tatsächlichen Größen, nicht mit den bezogenen zu rechnen ist. Es wird also

$$q = \frac{y}{x} \tag{25a}$$

und damit

$$\varphi_{(n+1)0} = \mu_n + (\varphi_{n0} - \mu_n)\, e^{-\frac{T_c}{T_1}} = \mu_n (1 - D_1) + \varphi_{n0} D_1 . \qquad (25)$$

Dabei ist φ_{n0} die Regelgröße zu Beginn der Schrittzeit n $(t = 0)$ und $\varphi_{(n+1)0}$ am Ende, also am Anfang der Schrittzeit $n + 1$. Für $e^{-\frac{T_c}{T_1}}$ ist D_1 gesetzt.

Da nach Gl. (19) auch der Zusammenhang zwischen $\mu_{(n+1)}$ und μ_n bekannt ist, nämlich

$$\mu_{(n+1)} = \mu_n - \varkappa\, \varphi_{(n+1)0} \qquad (26)$$

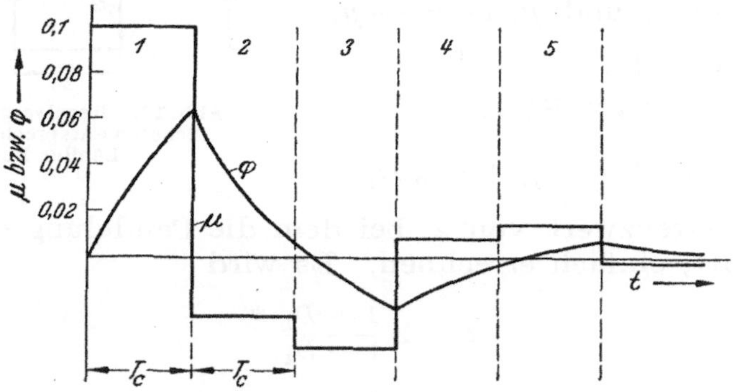

Abb. 16. Regelvorgang bei einer Regelung. Schema entsprechend Abb. 15

kann die Integration schrittweise durchgeführt werden entsprechend Abb. 16.

$$\left.\begin{aligned}
\varphi_{20} &= \mu_1 (1 - D_1) + \varphi_{10} D_1\\
\mu_2 &= \mu_1 - \varkappa\, \varphi_{20}\\
\varphi_{30} &= \mu_2 (1 - D_1) + \varphi_{20} D_1\\
\mu_3 &= \mu_2 - \varkappa\, \varphi_{30}\\
&\cdot \quad \cdot \quad \cdot \quad \cdot\\
&\cdot \quad \cdot \quad \cdot \quad \cdot\\
&\cdot \quad \cdot \quad \cdot \quad \cdot\\
\varphi_{n0} &= \mu_{(n-1)} (1 - D_1) + \varphi_{(n-1)0} D_1\\
\mu_n &= \mu_{n-1} - \varkappa\, \varphi_{n0} .
\end{aligned}\right\} \qquad (27)$$

Bei der Abb. 16 ist für $\frac{T_c}{T_1} = 1{,}0$ und für $\varkappa = 2$ angenommen, so daß $D_1 = e^{-\frac{T_c}{T_1}} = e^{-1} = 0{,}37$ wird. Bei Beginn des Vorganges soll eine Störung aufgetreten sein, die einer Abweichung der bezogenen Steuergröße $\mu_1 = 0{,}1$ entspricht. Der Bezugswert für μ ist so gewählt, wie oben angegeben ($\mu = 1$ gibt letztlich auch $\varphi = 1$).

Der Abb. 16 ist zu entnehmen, daß bei dem angenommenen Wert für $\varkappa$ bzw. für die bezogene wirksame Verstellzeit (Gl. 21) $\frac{T_s}{T_1} = \frac{T_c}{\varkappa T_1}$

$= \frac{1}{2} = 0{,}5$ der Regelvorgang gut gedämpft verläuft. Während aber bei einer Regelung nach Schema Abb. 15 und Verwendung eines *stetigen* Reglers die Regelschwingung unabhängig von der gewählten Verstellzeit immer gedämpft verläuft, besteht jetzt mit Schrittregler bei zu kleiner Verstellzeit, also zu großem Wert für $\varkappa$ die Gefahr von Pendelungen, also von Labilität. Die Grenze der Stabilität ist dann erreicht, wenn entsprechend Abb. 17 Regelgröße φ und Steuergröße μ bei jedem Schritt ihr Vorzeichen wechseln, ihre Größe aber beibehalten. In diesem Fall wird

$$\varphi_{(n+1)0} = -\varphi_{n0} \quad \text{und} \quad \mu_{n+1} = -\mu_n\,.$$

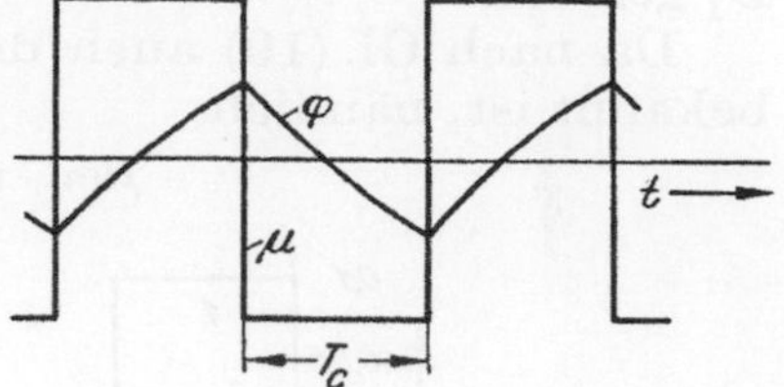

Abb. 17. Regelvorgang bei einem Schema entsprechend Abb. 15. Labiler Betrieb.

Da aber nach Gl. (27)

$$\mu_{n+1} = \mu_n - \varkappa\, \varphi_{n+1}$$

und

$$\varphi_{(n+1)0} = \mu_n (1 - D_1) + \varphi_{n0} D_1$$

läßt sich der Grenzwert von $\varkappa$, bei dem die Pendelung entsprechend Abb. 17 auftritt, einfach errechnen. Es wird

$$\varkappa = 2\,\frac{1 + D_1}{1 - D_1}\;{}^{*} \tag{28}$$

bzw.

$$\left(\frac{T_s}{T_1}\right) = \frac{T_c}{\varkappa_g\, T_1} = \frac{T_c}{T_1}\,\frac{1 - D_1}{2\,(1 + D_1)} = \frac{T_c}{T_1}\,\frac{1 - e^{-\frac{T_c}{T_1}}}{2\left(1 + e^{-\frac{T_c}{T_1}}\right)}\,. \tag{29}$$

Abb. 18 zeigt die Grenzwerte von $\left(\frac{T_s}{T_1}\right)_g$, die nicht unterschritten werden dürfen, abhängig von $\frac{T_c}{T_1}$.

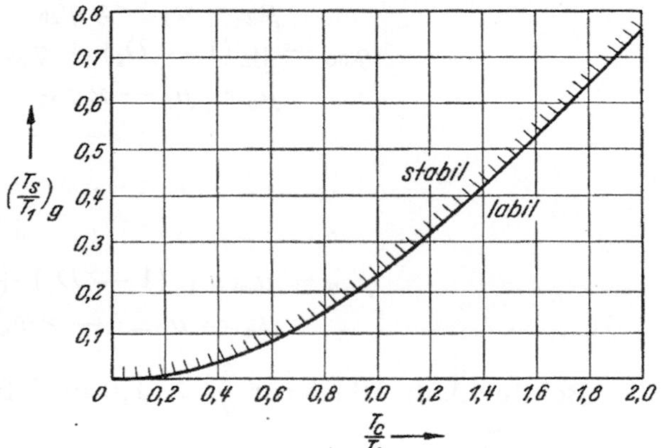

Abb. 18. Grenzwert für bezogene Stellzeit bei einer Regelung entsprechend Abb. 15

Schrittregler werden häufig so ausgeführt, daß die Schritthöhe nicht stetig mit der Regelabweichung wächst, sondern in Stufen. Die Schräge der Wippen 6 in Abb. 13 ist in diesem Fall in mehrere Stufen unterteilt, so daß die Schritthöhe für verschiedene Abweichungsbereiche konstant bleibt. Damit bleiben aber $\varkappa$ und T_s nicht konstant, sondern verlaufen abhängig von der Regelabweichung

* Dieses Ergebnis kann auch mit Hilfe der Differenzrechnung [*39*], [*89*] auf die aber nicht eingegangen werden soll, gefunden werden.

Dynamik selbsttätiger Regelungen (Dynamics of Automatic Feedback Control)

R.C. Oldenbourg and H. Sartorius
R. Oldenbourg Verlag, 1944, 1949, 1951

The first edition of the classic **R.C. Oldenbourg and H. Sartorius: Dynamik selbsttätiger Regelungen, 1. Band (Dynamics of Automatic Feedback Control, Volume 1)** was already published in January 1944 by the R. Oldenbourg Verlag. Actually, the manuscript was more or less completed already in 1942. Since the book was rather quickly sold out a reprinted version was made available in 1949. A pirate edition was even published in 1948 in the USA by the American Society of Mechanical Engineering. In addition, in 1949 a Russian version was on the market in the Soviet Union. Finally, the second and last edition, published in 1951, was translated into Japanese in 1953. This demonstrates clearly that the contributions of R.C. Oldenbourg and H. Sartorius in the field of automatic control were international recognized rather early.

The first volume is characterized by the subtitle: General and Mathematical Basic Foundation, Continuous and Discontinuous Control, Nonlinearities. The authors actually intended to write a second volume on optimal control problems. However, for some reason, the second volume never materialized.

The 258 pages and 118 illustrations of the last edition are split into six chapters (I-VI).

Chapter I provides a short introduction to control systems. The purpose of this chapter is to describe the general approach, to give definitions and to formulate the mathematical background.

A lengthy presentation in Chapter II deals with the structural and mathematical description of feedback control systems. For linear, time-invariant systems, three methods are derived based on differential equations, on frequency responses, and on transfer functions. Mathematically there is no fundamental difference between the three methods. However, depending on the standpoint and/or the practicability point of view, they have to be considered differently. The idea was to present a general way to characterize the transient behavior of all kind of automatic control systems, independent of the equipment and the technological peculiarities.

Chapter III is dedicated to continuous-time processes. Besides a classification into proportional (static) and integral (astatic) compensators, the authors also address the delay-time of simple systems with distributed parameters. Substantial attention is paid to the improvement of the response of feedback compensated systems.

In Chapter IV systems with simple nonlinear characteristics are investigated. Mainly based on practical mechanical equipments, the time response and the stability behavior of two cases are considered. The first one is a servomechanism with threshold and the second one with hysteresis, caused by components with friction. Since a closed form solution does not exist, the integration of the differential equation has to be done stepwise. Note that the describing function was not really known at that time and therefore was not mentioned.

Two types of non-continuous control systems are studied in Chapter V. The authors considered first the transient behavior of feedback systems with bang-bang controllers. The main part is dedicated to step (sampling) control actions. In this case, linear difference equations are used in order to characterize the dynamics of these sampled data systems. The stability domain is already characterized by the inside of the unit circle. The authors also pointed out already that, with the help of a bilinear transformation, the Routh-Hurwitz criterion could be applied for stability determination.

The last Chapter VI considers the appropriate application of diverse control strategies by comparing them in particular continuous and sampled data control systems.

The aim of the book was to derive and present various mathematical methods for analysis and synthesis purposes of feedback control systems. But very little attention was paid to engineering applications. Therefore the book met with the approval of control scientists. However, the text contained a number of ideas that were new at the time and were pointing into future directions.

DYNAMIK SELBSTTÄTIGER REGELUNGEN

1. BAND

ALLGEMEINE UND MATHEMATISCHE GRUNDLAGEN

STETIGE UND UNSTETIGE REGELUNGEN

NICHTLINEARITÄTEN

VON

DR.-ING. RUDOLF C. OLDENBOURG

UND

DR.-ING. HANS SARTORIUS

MIT 112 BILDERN UND EINER TAFEL

2. AUFLAGE

VERLAG VON R. OLDENBOURG

MÜNCHEN 1951

anlage versuchsmäßig vorliegt. Die Ergebnisse werden gewöhnlich in Form von *Integralgleichungen* erhalten. Hier erweist sich nun die Laplacetransformation als ein äußerst wertvolles Hilfsmittel, da mit ihr die Auswertung dieser Integralgleichungen ohne Schwierigkeit nahezu mechanisch erfolgen kann.

§ 38 Die Stabilität des Regelvorganges

Die Übergangsfunktion des aufgetrennten Regelkreises sei bekannt und werde mit $\varphi(t)$ bezeichnet.

Mit Hilfe des Duhamelschen Integrals (§ 17) ist es möglich, den zeitlichen Verlauf der Ausgangsgröße $z_1(t)$ anzugeben, der von einer beliebigen Zeitfunktion $z(t)$ verursacht wird [Gleichung (17. 8)]. Es ist:

$$z_1(t) = \frac{d}{dt}\int_0^t \varphi(\xi) z(t-\xi)\, d\xi. \tag{38. 1}$$

Soll sich nun im Stabilitätsgrenzfall eine einmal eingeleitete, etwa als sinusförmig angenommene Zustandsänderung gerade von selbst aufrechterhalten, so muß bei wieder geschlossenem Regelkreis:

$$z = -z_1 \tag{38. 2}$$

sein, also:

$$z(t) = -\frac{d}{dt}\int_0^t \varphi(\xi) z(t-\xi)\, d\xi. \tag{38. 3}$$

Diese nun in Gestalt einer Integralgleichung erscheinende Stabilitätsbedingung sagt aber nichts anderes aus als die bereits gefundenen Bedingungen. Man kann sich davon überzeugen, wenn man auf beide Seiten der Gleichung (. 3) die Laplacetransformation anwendet:

$$z(p) = -\mathfrak{L}\left(\frac{d}{dt}\int_0^t \varphi(\xi) z(t-\xi)\, d\xi\right). \tag{38. 4}$$

Nach dem *Faltungssatz* der Laplacetransformation (Anhang Ia) wird Gleichung (.4):

$$z(p) = -p \cdot \mathfrak{L}[\varphi(t)] \cdot z(p). \tag{38. 5}$$

Nach Gleichung (16. 3) ist die Laplacetransformierte der Übergangsfunktion nichts anderes als der durch p dividierte komplexe Frequenzgang:

$$\mathfrak{L}[\varphi(t)] = \mathfrak{F}(p)/p. \tag{38. 6}$$

Damit nimmt Gleichung (.4) die Form der bekannten Stabilitätsbedingung des Regelvorganges an:

$$\mathfrak{F}(p) + 1 = 0. \tag{38. 7}$$

§ 39 Der Regelverlauf

Zur Ermittlung des vollständigen Regelverlaufes ist es erforderlich, daß die Übergangsfunktion des Teiles der Regelstrecke bekannt ist, der zwischen *Stör-* und *Meßstelle* liegt. Wir wollen sie mit $\varphi_z(t)$ bezeichnen. Dann ist die Aus-

Damit geht die Stammgleichung (.30) in folgende dreiwertige Funktion über:

$$y^3 - 3\,h\,y + 2 = 0, \qquad (65.\,34)$$

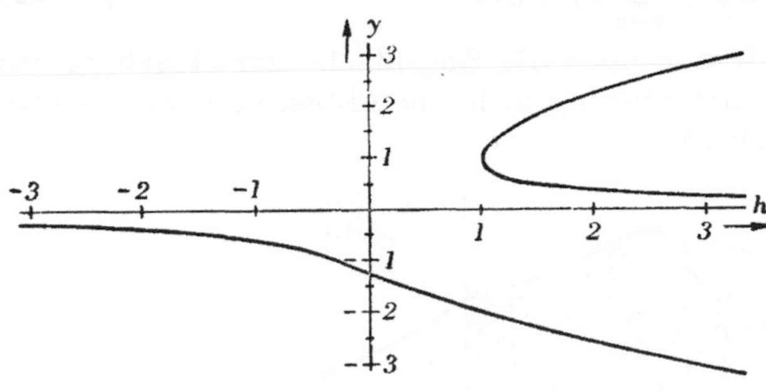

Bild 61: Die dreiwertige Funktion: $y^3 - 3\,hy + 2 = 0$

die in Bild 61 dargestellt ist. Es ist hieraus zu ersehen, daß y und damit Δ nur dann durchwegs reelle Werte annehmen, wenn $h \geqq 1$ (65. 35) wird, da k stets reell sein soll. Aus den Gleichungen (.32) und (.35) folgt als Bedingung für drei reelle Wurzeln:

$$3\,k^2 \gtreqless (A^2/3 - B)\,. \qquad (65.\,36)$$

Diese Ungleichung ist offenbar nur dann erfüllbar, wenn

$$A^2/3 - B \geqq 0\,. \qquad (65.\,37)$$

Beachtet man nun, daß für $A^2/3 = B$ nach Gleichung (.36) $k = 0$ sein muß, so folgt aus (.37) und (.31)

$$C \leqq (A/3)^3 \qquad (65.\,38)$$

als weitere notwendige (aber nicht hinreichende!) Bedingung dafür, daß sämtliche Wurzeln der Stammgleichung reell werden[1]).

Der zweite Teil unserer Aufgabe, nämlich die Entscheidung, wann die Fläche [Gleichung (.28)] ein Minimum wird, gestaltet sich sehr einfach, wenn man in einer räumlichen Darstellung die Fläche als Funktion der beiden Parameter A und C aufzeichnet (Bild 62). In der AC-Ebene ist die Kurve $C = (A/3)^3$ eingezeichnet. Da für A und C die Bedingung (.38) erfüllt sein muß, und außerdem beide Konstanten positiv sind, kommt nur der Bereich der AC-Ebene für unsere Betrachtung in Frage, der zwischen der Grenzkurve und der A-Achse liegt. Längs der Grenzkurve ist die Fläche:

$$F/(M\,T_z) = (A - 1)/(A/3)^3,$$

längs der A-Achse:

$$F/(M\,T_z) = \infty\,.$$

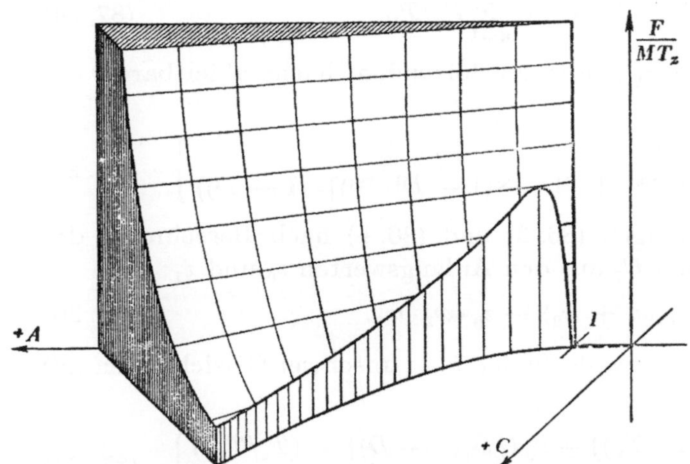

Bild 62: Zur Ableitung der Bedingung für kleinste Regelfläche für den Regelkreis nach Bild 60 *(Flächenrelief)*

---

[1]) Die notwendige und hinreichende Bedingung wird bekanntlich von der *Diskriminante* geleistet, welche für den Fall der Gleichung (.30) lautet:

$$A^2\,B^2 + 18\,ABC - 4\,A^3\,C - 4\,B^3 - 27\,C^2 \geqq 0.$$

mation in den Grenzen von 1 bis ∞, also aus Gleichung (.21):

$$F/T_C = \sum_{n=1}^{\infty} F_n/T_C. \tag{87.23}$$

Als einfaches Beispiel wollen wir nun noch die Regelfläche der Regelung des § 86 für den Fall ermitteln, daß die Störung nicht im Abtastmoment ($\tau = 0$), sondern im Zeitpunkt $\tau = \tau_{St}$ erfolgt.

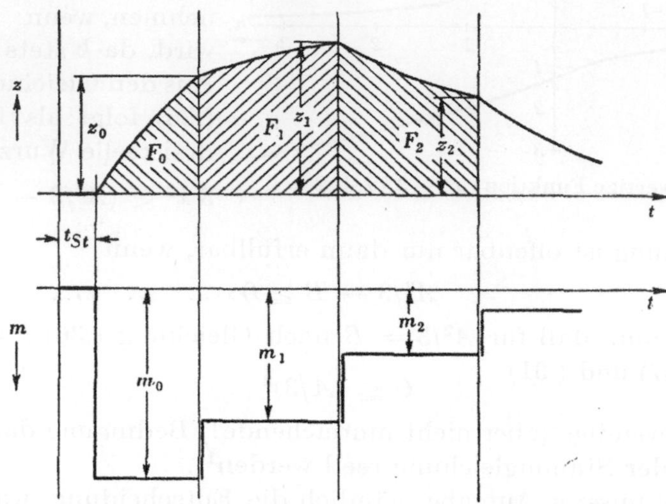

Bild 94: Zur Bestimmung der Fläche bei beliebigem Störzeitpunkt

Wir wählen die in Bild 94 gezeigte Intervalleinteilung. Dann ist die Regelfläche:

$$F/T_C = F_0/T_C + \sum_{n=1}^{\infty} F_n/T_C. \tag{87.24}$$

Für die Anfangswerte z_0 und z_1 gelten die folgenden, leicht ableitbaren Bestimmungsgleichungen:

$$\left.\begin{aligned} z_0 &= M(1 - D^{1-\tau_{St}}) \\ z_1 &= M(D(1 - D^{1-\tau_{St}}) + [1 - S(1 - D^{1-\tau_{St}})] \cdot (1 - D)) \end{aligned}\right\} \cdot \tag{87.25}$$

Weiter folgt mit den Gleichungen (86. 3) und (86. 4) nach Berechnung der Summationskonstanten C_1 und C_2 aus den Anfangswerten z_0 und z_1:

$$z_2 = z_1(w_1 + w_2) - z_0 \cdot w_1 \cdot w_2. \tag{87.26}$$

Für die Flächen der Anfangsintervalle findet man unschwer [Gleichungen (.6) und (.7)]:

$$\left.\begin{aligned} F_1/T_C &= z_1([1/(1-D)] - (T_z/T_C)) - z_0([D/(1-D)] - (T_z/T_C)) \\ F_2/T_C &= z_2([1/(1-D)] - (T_z/T_C)) - z_1([D/(1-D)] - (T_z/T_C)) \end{aligned}\right\} \tag{87.27}$$

und

$$F_0/T_C = \int_0^{1-\tau_{St}} M(1 - D^{\tau})\,d\tau = M[1 - \tau_{St} - (T_z/T_C)(1 - D^{1-\tau_{St}})]. \tag{87.28}$$

Kleines Handbuch technischer Regelungsvorgänge
(Compact Handbook of Engineering Control Processes)

W. Oppelt
Verlag Chemie, 1954, 1956, 1960, 1964, 1972.

Already in the Spring of 1947, W. Oppelt published his first booklet of 118 pages on feedback control systems. It had the title: **Grundgesetze der Regelung (Principles of Control)**. It was printed by the Wolfsburger Verlagsanstalt under war and postwar circumstances as "Notdruck". In other words the paper and the printing were of very low quality. The same publishing house did put out in 1949 a second booklet of 144 pages with the title: **Stetige Regelungsvorgänge (Continuous Control Processes)**. Though the manuscript was forwarded to the publisher together with that of the first published book, publication took for the above mentioned reasons two years longer. While the first booklet did deal with basic mathematical methods, the second booklet did concentrate on applied matters. Hence these two booklets formed the basis for the well known book: **Kleines Handbuch technischer Regelungsvorgänge (Compact Handbook of Engineering Control Processes)** which was for the first time published in 1954 by Verlag Chemie. Altogether there were five improved and extended editions (1954, 1956, 1960, 1964, and 1972) and one improved reprint in 1967. The fifth and last edition had 770 pages and almost 950 illustrations. Beyond that, the handbook was translated into Czech, French, Hungarian, Polish, Rumanian, and Russian. Even the two booklets were translated into Japanese.

The concern of all of Oppelt's books was to make the new and interdisciplinary thinking wider known to applied engineers. He knew that an abstract theory, detached from real applications, would frighten off the reader. That is why he did collect an immense number of examples of all imaginable engineering areas and, at that time already, of biological and economical processes, in order to underline the common attributes. Most of the illustrations of the examples were drawn in excellent form and clearness by W. Oppelt himself – he was in this respect really adept! The tremendous worldwide success of the book is due to an appropriate limitation of the mathematical tools used.

The content of the last edition of the textbook is divided into eleven chapters (I–XI) .

Chapter I presents the introductory material like the problem and the method of attack. It is mainly based on a collection of physical systems and their block diagram models.

Chapter II supplies the background needed for the mathematical analysis of linear, time-continuous systems used in the book. Both time and sinusoidal responses are presented in order to characterize the transient behavior of control elements.

The next two chapters pay attention to quite a number of controlled processes (plants) and controller devices. Both the dynamic behavior and the design of engineering system examples are presented.

In Chapter III various controlled components are considered, arising mostly from industrial applications in the fields of mechanical, electrical, automotive, chemical or process engineering. All these processes are characterized by block diagrams and signal flow graphs.

Besides, a short introduction considers the fundamentals of random processes in automatic control.

The structure of controller devices is the topic of Chapter IV. A tremendous selection of mechanical, electrical, electronic, pneumatic, hydraulic and other designs are considered.

While the topics of the two previous chapters are treated independently of each other, now in Chapter V attention is paid to closed loop control systems, in which the functional components (plant and controller) are connected. The dynamic behavior and, in particular, the stability conditions of linear systems on one hand, and the synthesis methods on the other hand are considered by applying both time and frequency domain methods. Up to this point just simple SISO feedback systems are considered. Of course, the simple systems are used for the purpose of establishing definitions, a method of analysis, and the technique of computation.

Naturally, the systems are generally more complex such as interconnected multi-loop control or MIMO systems. Therefore in Chapter VI attention is now focused on complex control systems. For example, the advantages and disadvantages of feed-forward and cascade configurations are pointed out. Consequently, the control strategies are derived for a number of complex industrial systems.

In the preceding chapters, the analysis and design techniques discussed were mainly restricted to linear continuous-time feedback control systems. However, all practical systems are nonlinear to some extent and most of the commonly used tools and terminologies for linear systems are no longer valid. Nevertheless, there is a class of systems in which nonlinear elements are even intentionally introduced in order to improve system performance. The on-off or relay type servo-systems are common examples.

The next two chapters consider first systems with nonlinear characteristics and, subsequently, systems with a discontinuous nonlinearity. In the main part of Chapter VII the sinusoidal describing functions for simple isolated nonlinear elements, such as saturation, threshold and pre-load, are derived. The effect of nonlinear elements on the stability of closed loop control systems is studied afterwards.

Chapter VIII covers discontinuous control processes. At the beginning, attention is given to the dynamic behavior of feedback systems with switching and programmed controllers. Next the phase plane method and its application are studied. Then a short introduction to sampled-data control systems follows, and finally some typical examples how to apply time-optimal switching curves in the state space are presented.

The last three Chapters are dedicated to rather new developments. Chapter IX contains remarks on computer logic, on functional design of digital computers, on digital computer technology, and on digital control systems. Simulation devices and analog computer applications are addressed in Chapter X. A short introduction to adaptive control systems is the subject of the last Chapter XI.

The textbook was mainly addressed to advanced students and practicing engineers. The work is a classic which quickly became the standard in the field. A large number of figures and numerouss tables are characteristic for the book.

BÜCHER DER TECHNIK

Herausgeber: Dr.-Ing. Alfred Kuhlenkamp

NOTDRUCK

Dr.-Ing. Winfried Oppelt

Grundgesetze der Regelung

Mit 32 Bildern und 28 Tafeln

WOLFENBÜTTELER VERLAGSANSTALT G. m. b. H.
WOLFENBÜTTEL-HANNOVER
1947

BÜCHER DER TECHNIK

Herausgeber: Dr.-Ing. Alfred Kuhlenkamp

Dr.-Ing. Winfried Oppelt

Stetige Regelvorgänge

Wissenschaftliche Verlagsanstalt K. G. Hannover

in Gemeinschaft mit

Wolfenbütteler Verlagsanstalt G. m. b. H. Wolfenbüttel

1949

Winfried Oppelt

Kleines Handbuch technischer Regelvorgänge

Fünfte, neubearbeitete und erweiterte Auflage

1. Auflage 1954
2. neubearbeitete und erweiterte Auflage 1956
3. neubearbeitete und erweiterte Auflage 1960
4. neubearbeitete und erweiterte Auflage 1964
Verbesserter Nachdruck der 4. Auflage 1967
5. neubearbeitete und erweiterte Auflage 1972

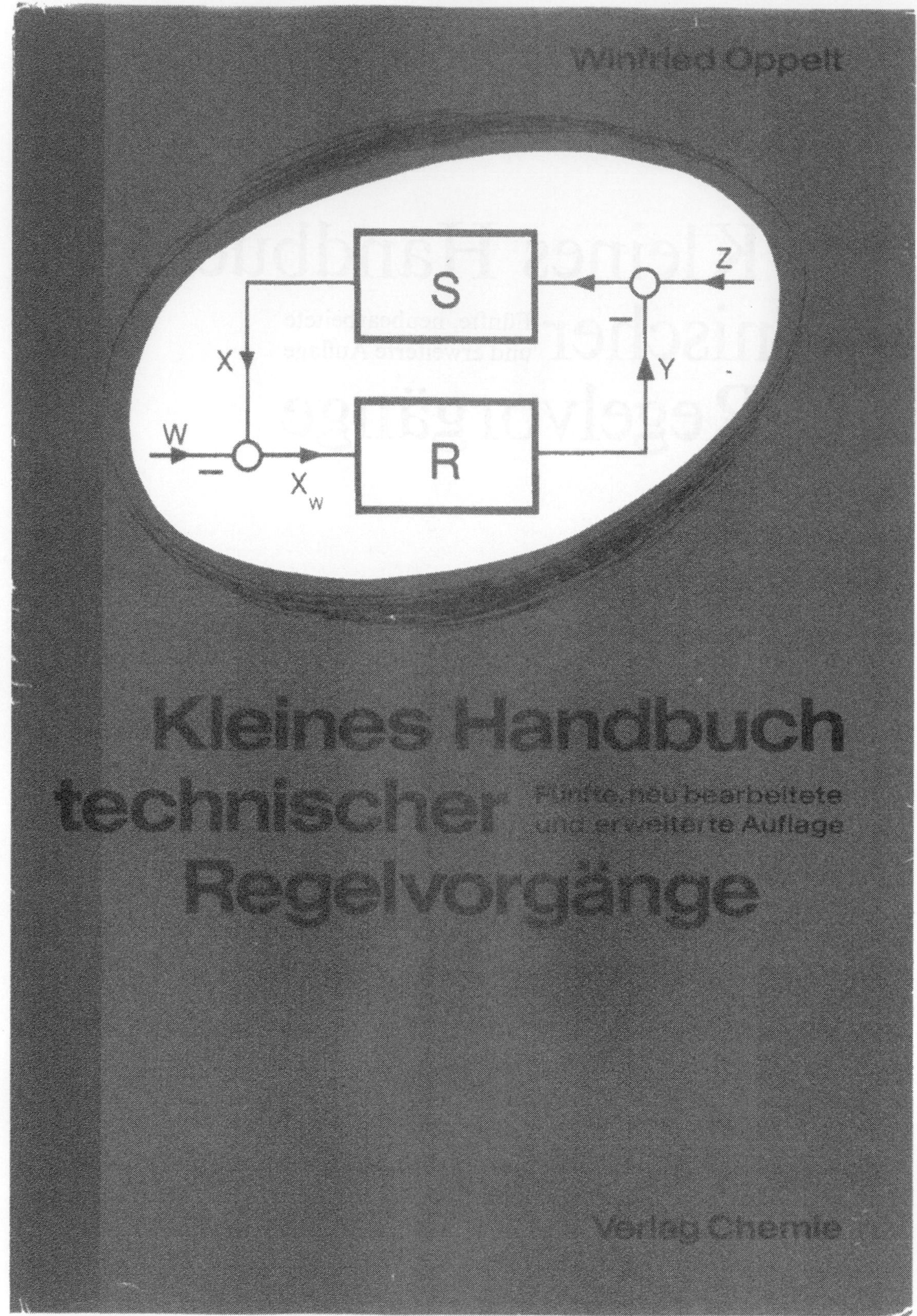
Winfried Oppelt
S
Z
X
Y
W
R
X_w
Kleines Handbuch technischer Regelvorgänge
Fünfte, neu bearbeitete und erweiterte Auflage
Verlag Chemie

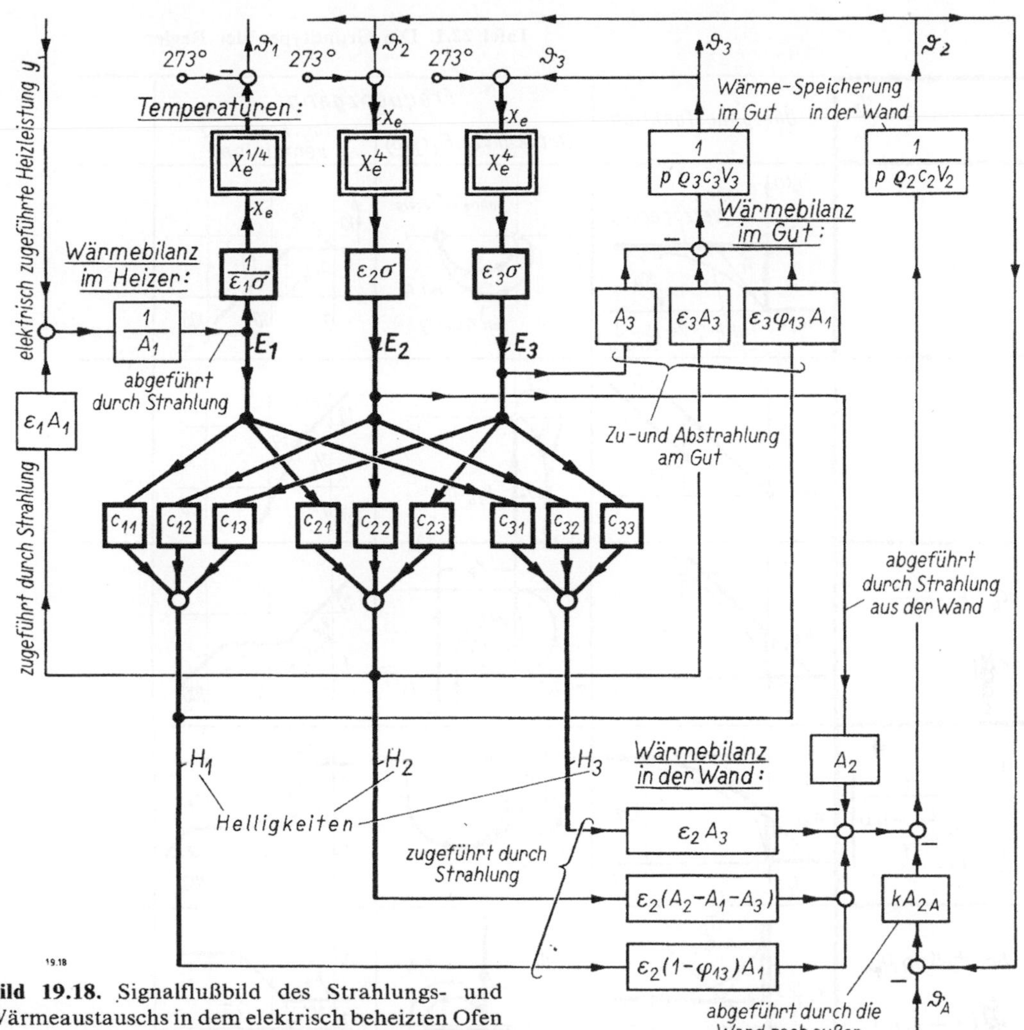

Bild 19.18. Signalflußbild des Strahlungs- und Wärmeaustauschs in dem elektrisch beheizten Ofen aus Bild 19.17. Mit dicken Linien hervorgehoben ist in der Bildmitte der Aufbau der Eigenstrahlung E_1, E_2 und E_3 sowie der Helligkeiten H_1, H_2 und H_3 von Heizer, Wand und Gut aus ihren Temperaturen ϑ_1, ϑ_2 und ϑ_3. Die Temperaturen selbst ergeben sich aus den Wärmebilanzen, die mit dünnen Linien gezeichnet sind.

Die „dicke" Wand. Bei den bisherigen Betrachtungen hatten wir angenommen, daß die Temperatur im Innern der Wand an allen Stellen dieselbe sei. Dies kann als Näherung für verhältnismäßig dünne Wände angesetzt werden. Die Wandtemperatur ergibt sich dann aus der Speicherwärme der Wand, die als Unterschied zwischen den zu- und abströmenden Wärmeflüssen aufgenommen werden muß. Bild 19.21 zeigt diese Zusammenhänge. Dort ist der an den Wandungen übergehende Wärmefluß angesetzt zu

$$Q_h = \alpha A(\vartheta_1 - \vartheta_{12}), \tag{19.14}$$

wobei α die Wärmeübergangszahl ist. Aus dem Signalflußbild ergibt sich durch Zusammenziehen ein Verzögerungsverhalten 1. Ordnung, Bild 19.21 rechts.

Tafel 22.1. Die Grundtypen der Regler

	F-Funktion $F_R = \frac{y_R}{x_w}$	Übergangsfunktion	Frequenzgang: Ortskurve $F_R(j\omega)$	Frequenzgang: log. Frequenzkennlinien
P	K_R			
I	$\frac{K_{IR}}{p}$			
PI	$K_R + \frac{K_{IR}}{p} = K_R\left(1 + \frac{1}{T_I p}\right)$			
PP	$K_{R1}\left[1 + \frac{1}{\frac{1}{\frac{K_{R2}}{K_{R1}} - 1} + T_I p}\right]$			
PD	$K_R + K_{DR} p = K_R\left(1 + T_D p\right)$			
PID	$K_R + \frac{K_{IR}}{p} + K_{DR} p = K_R\left(1 + \frac{1}{T_I p} + T_D p\right)$			
PPD	$K_{R1}\left[1 + \frac{1}{\frac{1}{\frac{K_{R2}}{K_{R1}} - 1} + T_I p} + T_D p\right]$			

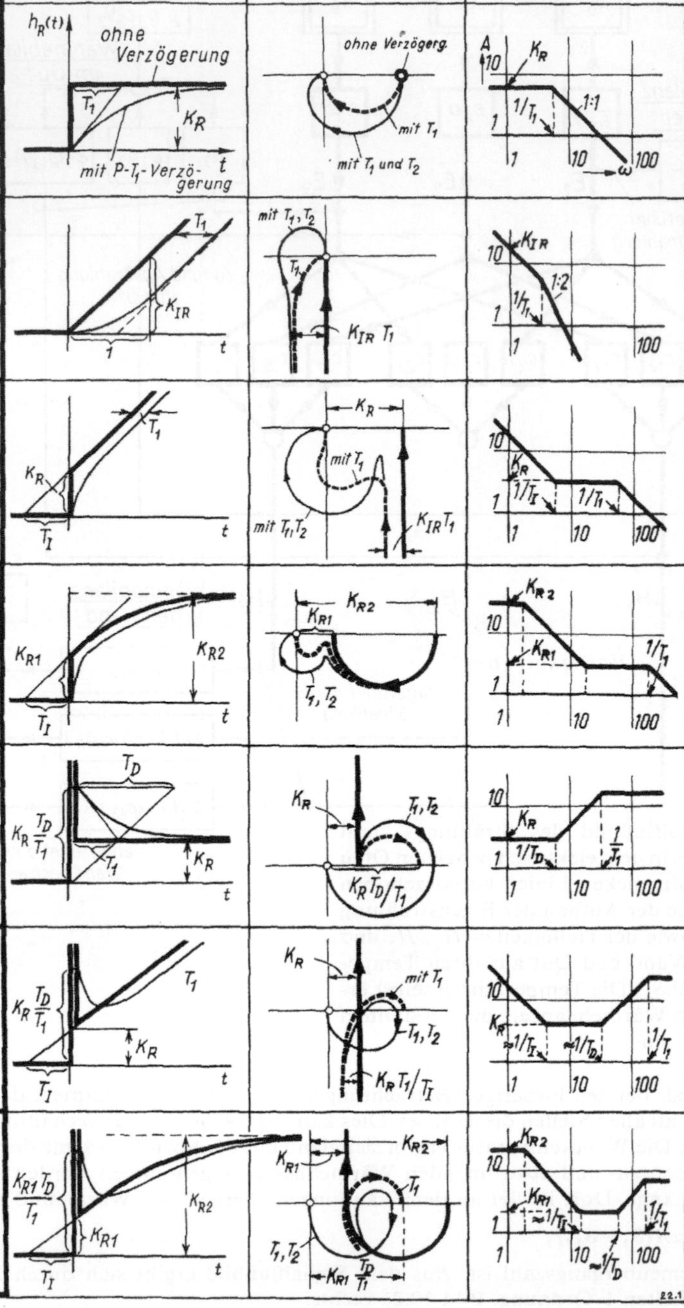

und ihre Darstellung.

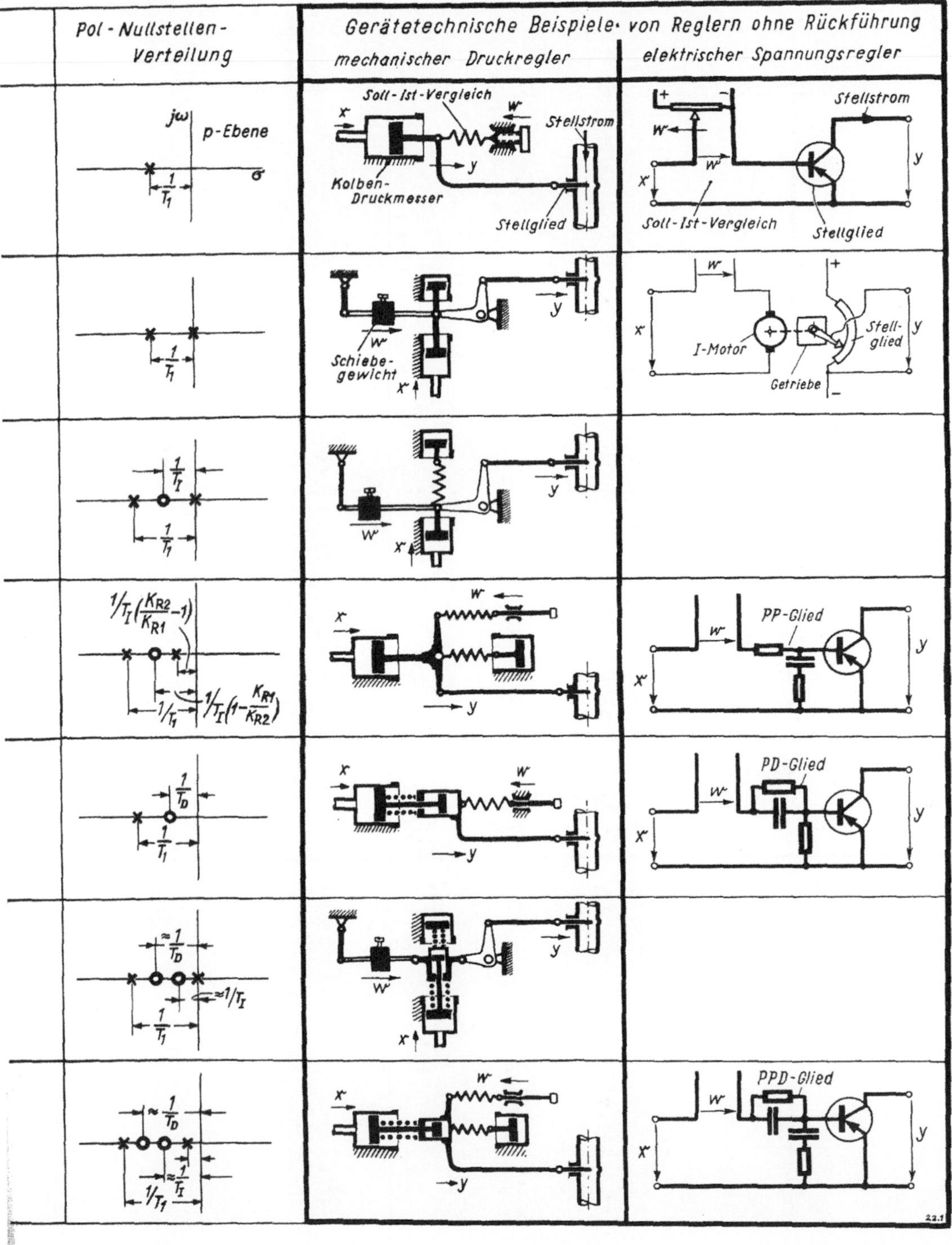

EARLY CONTROL TEXTBOOKS
in
HUNGARY

Automatika (Automation) (Lecture notes)

Andor Frigyes and István Nagy
Tankönyvkiadó, 1954, 1957

Szabályozások Dinamikája (Dynamics of Control Systems)

Frigyes Csáki
Akadémiai Kiadó, Budapest, 1966

Automatika (Automation)

Frigyes Csáki and Ruth Bars
Tankönyvkiadó, Budapest, 1969, 1971, 1974, 1983, 1986

Contributed by

Ruth Bars and István Nagy

Budapest University of Technology and Economics
Budapest, Hungary

Introduction

The Electrical Engineering Faculty (School) at the Technical University of Budapest was founded in 1949, with specialisations in power engineering and telecommunication. Professors coming from the Mechanical Engineering Faculty and from industry became responsible for creating the profiles of the new departments. Excellent experts worked on incorporating the most up-to-date knowledge into engineering education. From the very beginning, the departments combined three activities: education, research and industrial projects. Research and industrial experiences fertilised the material of engineering courses.

The instructors worked with great enthusiasm on creating the Hungarian engineering literature. A great number of university lecture notes and textbooks were published in the first years that later served generations of electrical engineers.

At the Department of Electrical Machines and Drives, a strong school was created guided by Professors Pál K. Kovács and István Rácz. Combining new concepts with traditional methods in discussing the operation and control of electrical drives, they created new approaches which - through their books and publications - became internationally appreciated and cited.

The subject of feedback control systems was taught at the Technical University of Budapest (TUB) in Hungary based on the so called German tradition using the time domain up to the end of the 1940's. The teachers involved in the subject were mechanical engineers. At the beginning of the 1950's, younger members of faculty became responsible for teaching feedback control systems. They had read a number of recently published English books such as

James, H.M.-Nichols, N.B.-Philips, R.S.: Theory of Servomechanisms. New York, McGraw-Hill Book Company, Inc., 1947.

Brown, G.S.-Campbell, D.P.: Principles of Servomechanisms. New York, John Wiley and Sons, Inc., 1948.

Chestnut, H.-Mayer, R.: Servomechanisms and Regulating System Design, New York, John Wiley and Sons.

The subject "Automation" appeared in the curriculum already in 1953, the lectures given by Professor Andor Frigyes. The first lecture notes were published in 1954 by Andor Frigyes and István Nagy (see below). The lecture notes were the advent in the University teaching of modern control theory in Hungary. The course of feedback control systems was based on the books listed. It is interesting to note that the Hungarian instructors did not have the English edition of the James-Nichols-Phillips book. Only the Russian translation of the book was available at that time and they studied this edition.

Two new Control Departments were established, arising from the Department of Electrical Machines and Drives. The Department of Automation (now Department of Automation and Applied Informatics) was established in 1961 under the direction of Professor Frigyes Csáki. The Department of Process Control (now Department of Control Engineering and Information Technology) was established in 1964, headed by Professor Andor Frigyes. Both departments offered basic and advanced control courses. The departments had a close link with the Research Institute of Automation (SzTAKI) of the Hungarian Academy of Sciences and with industry.

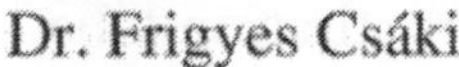

Dr. Frigyes Csáki

Dr. Andor Frigyes

Professor Frigyes Csáki had an impressively immense activity in control research and education. He was the founder of the Hungarian control literature. His books gave an overview and also a deep discussion of different areas of control theory. His most important scientific contribution was a systematic presentation and supplementary work over an extremely wide range of disciplines. A number of his books have been translated into English, German and Russian. These books served the control education and research community as basic reference sources and textbooks for a long time. Till now they are considered as a bible of control ideas. He recognized and called attention of his colleagues to upcoming new trends and ideas. His always up-to-date knowledge immediately affected the control curriculum and control research at the department. His books achieved a wide international reputation. Two of his books, one co-authored with Ruth Bars, are presented below.

Csáki wrote a book on state-space methods for control systems, which was used both in basic and advanced control courses. (Csáki Frigyes: Fejezetek a Szabályozástechnikából, Állapotegyenletek. Műszaki Könyvkiadó, Budapest, 1973.) This book has been translated into German and English (Die Zustandsraum-Methode in der Regelungstechnik, Akadémiai Kiadó, Budapest, 1973, State-space Methods for Control Systems, Akadémiai Kiadó, Budapest, 1977).

His main book, bible of Hungarian control engineers, was Modern Control Theories, Csáki Frigyes: Korszerű Szabályozáselmélet, Nemlineáris, Optimális és Adaptív Rendszerek, Akadémiai Kiadó, Budapest, 1970. It has been translated into English and Russian (Modern Control Theories, Nonlinear, Optimal and Adaptive Systems, Akadémiai Kiadó, Budapest, 1972, Sovremennaja Teorija Upravlenija, Nelinejnije, Optimalnije I Adaptivnije Sistemi, Izd. Mir, 1975,.). This book, reflecting immediately the newest ideas, attracted much attention in the international control community.

Automatika (Automation) (Lecture notes)

Andor Frigyes and István Nagy
Tankönyvkiadó, 1954, 1957 (402 pages)

In the Introduction, the concept, notion, classification and history of control were treated. The structure of control loop closed the chapter (26 pages).

The next chapter was a long one describing the most frequently applied components of the control loop. First the electric and then the hydraulic and pneumatic components were discussed. Among the sensors the resistive, inductive and capacitive ones, the control transformer, the phase discriminators and the speed sensors were described. After the voltage and current stabilizators the notebook covered the topics of vacuum-tube amplifiers, tiratron-tube, magnetic amplifiers, saturable reactors, dc generators and the amplydin. Among the actuators the dc and the ac servomotors were treated. The descriptions of hydraulic and pneumatic components concluded the chapter (122 pages).

Next in the second volume of the notebook the authors dealt with theoretical tools of the control systems such as: Laplace transformation, transfer function, frequency characteristic, Nyquist diagram, time response, weighting function, Faltung theorem (convolution), proportional, integrator and differenciator elements, block diagram technique and classification of control systems: 0, 1 and 2 type systems (92 pages).

The next chapter dealt with the stability problem. First it described the physical background. Routh-Hurwitz, Mihajlov and Nyquist stability criteria were included and even the proof of the Nyquist criterion was given (34 pages). A separate chapter described the Bode diagram technique, (24 pages). The quality measures such as overshoots, steady-state error, control time, M curve, phase margin etc. were inserted in the next chapter (31 pages). The problem and methods of design were mainly treated by approximate amplitude Bode diagrams, (24 pages).

The last chapter presented a number of applications in the field of feedback control systems. Some of the examples were: temperature control by pneumatic elements, speed control of steam turbine, excitation control of turbine generator by magnetic amplifier, speed control of a Ward-Leonard system, etc. (48 pages).

BUDAPESTI MŰSZAKI EGYETEM
VILLAMOSMÉRNÖKI KAR

Frigyes Andor
műegyetemi adjunktus
és
Nagy István
aspiráns

AUTOMATIKA

KÉZIRAT

BUDAPEST, 1957

diagramm a képzetes ill. a valós tengelyhez simulva a végtelenbe nyulik és a "Nyquist stabilitási kritérium bizonyitása" cimü fejezetben mondottakból beláthatóan, 1 tipus szabályozás esetében a jobb félsikban rajzolt végtelen sugaru félkörrel, 2 tipus szabályozás esetében pedig egy végtelen sugaru körrel záródik.

196/a.ábra.

196/b.ábra.

2 tipusu

196/c.ábra.

A fentiek alapján a 196.ábrákban mutatjuk meg, hogy a 184. ábra első sorában feltüntetett 0, 1 és 2 tipusu szabályozások jellegzetes amplitudó-fázis diagrammjai hogyan módosulnak, ha a teljes amplitudó-fázis diagrammot rajzoljuk fel.

A Nyquist stabilitási kritérium, abban az esetben, ha a fel nyitott kör átviteli függvényében a nevezőnek a jobb félsikon nincsen zérus helye /az átviteli függvénynek a jobb félsikon ni

55-41050 — 112 —

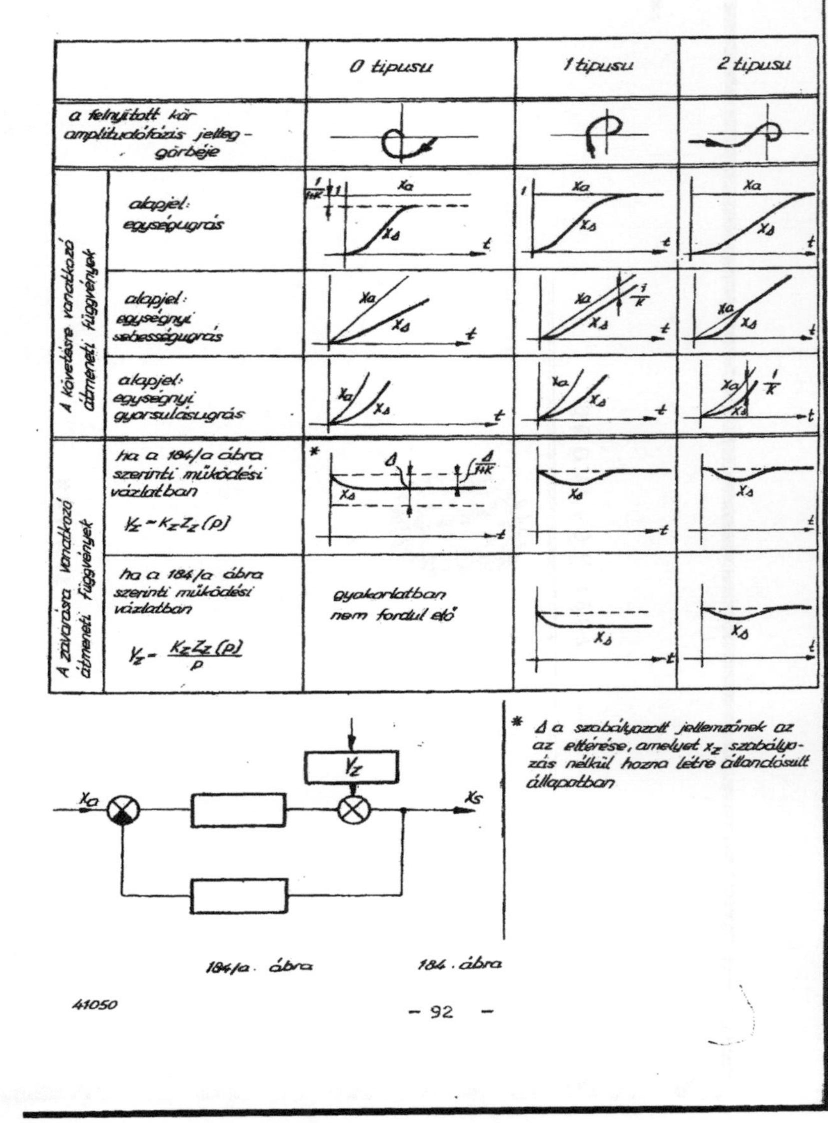

		0 tipusu	1 tipusu	2 tipusu
a felnyitott kör amplitudófázis jelleggörbéje				
A követésre vonatkozó átmeneti függvények	alapjel: egységugrás	$\frac{1}{1+K}$, x_a, x_Δ, t	1, x_a, x_Δ, t	x_a, x_Δ, t
	alapjel: egységnyi sebességugrás	x_a, x_Δ, t	x_a, x_Δ, $\frac{1}{K}$, t	x_a, x_Δ, t
	alapjel: egységnyi gyorsulásugrás	x_a, x_Δ, t	x_a, x_Δ, t	x_a, x_Δ, $\frac{1}{K}$, t
A zavarásra vonatkozó átmeneti függvények	ha a 184/a ábra szerinti működési vázlatban $Y_z = K_z Z_z(p)$	* Δ, $\frac{\Delta}{1+K}$, x_Δ, t	x_Δ, t	x_Δ, t
	ha a 184/a ábra szerinti működési vázlatban $Y_z = \frac{K_z Z_z(p)}{p}$	gyakorlatban nem fordul elő	x_Δ, t	x_Δ, t

* Δ a szabályozott jellemzőnek az az eltérése, amelyet x_z szabályozás nélkül hozna létre állandósult állapotban

184/a. ábra

184. ábra

41050 — 92 —

Az ilyen elemmel történő kompenzálás hatásának bemutatására induljunk ki ismét a 236.ábra "a"-val jelzett görbéjéből. A 253.ábrán ismét felrajzoltuk ezt a görbét és a PID kompenzáló elem "c"-vel jelölt jelleggörbéjét. A kompenzáló elem időállandóinak alkalmas felvételével a kompenzált szabályozás felnyitott körének logaritmikus amplitudó-körfrekvencia jelleggörbéjére a "b" görbét nyerjük. Látható, hogy ebben az esetben az eredetileg labilis kör stabilissá vált, ugyanakkor a statikus hiba / $\omega \longrightarrow 0$ mellett érvényes aszimptota/ nem változott, a vágási körfrekvencia pedig eléggé jobbra kerül, különösen ha az erősitési tényezőt K_1-gyel megnöveljük.

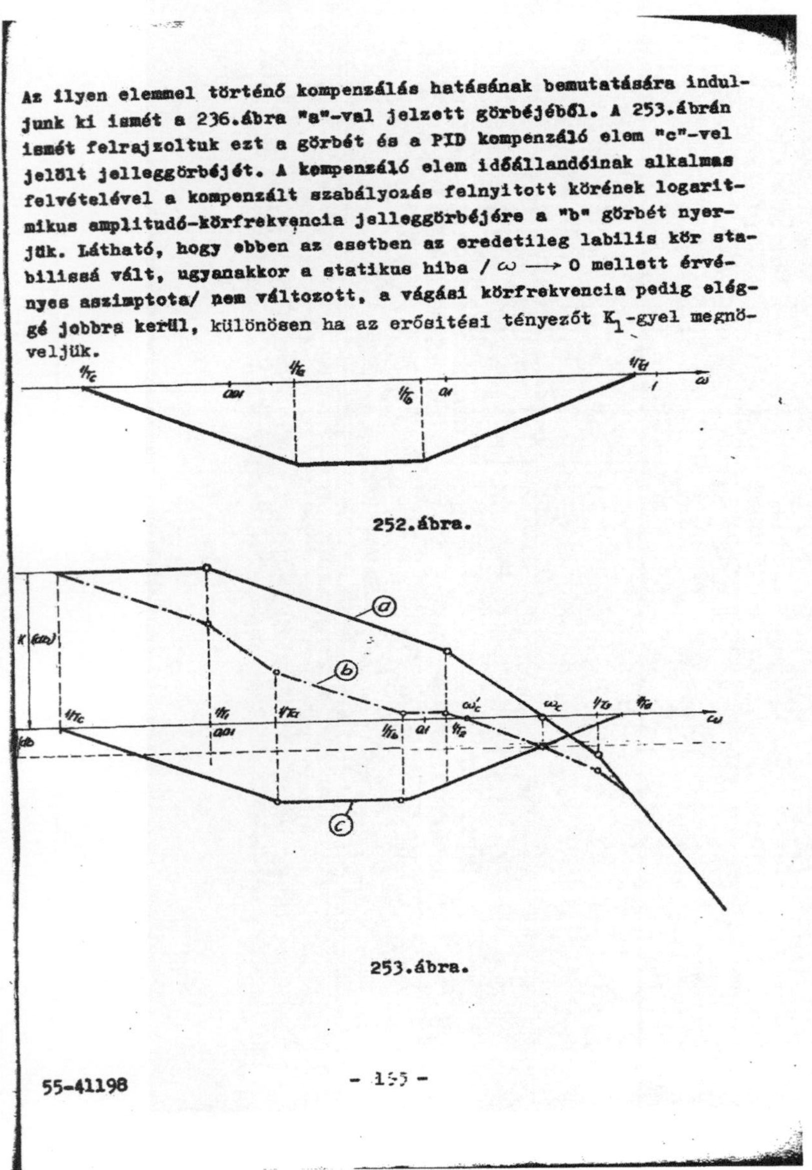

252.ábra.

253.ábra.

55-41198

- 155 -

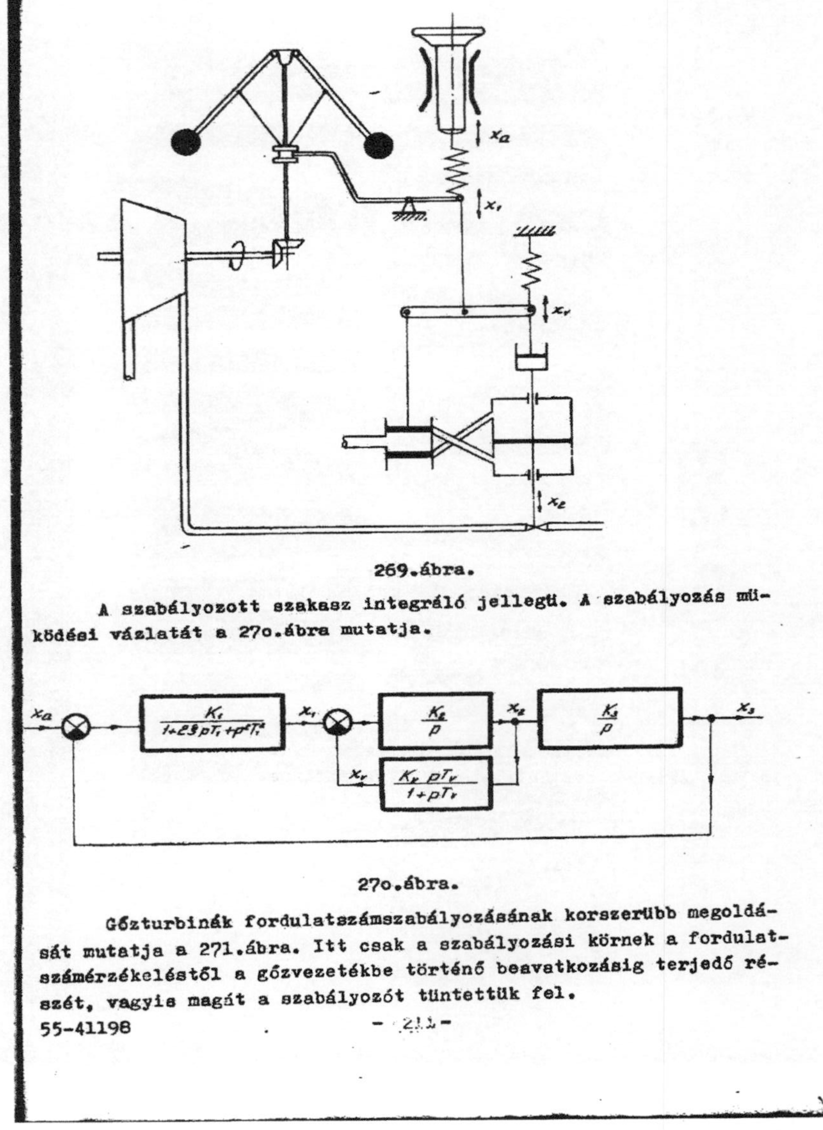

269.ábra.

A szabályozott szakasz integráló jellegü. A szabályozás müködési vázlatát a 27o.ábra mutatja.

27o.ábra.

Gőzturbinák fordulatszámszabályozásának korszerübb megoldását mutatja a 271.ábra. Itt csak a szabályozási körnek a fordulatszámérzékeléstől a gőzvezetékbe történő beavatkozásig terjedő részét, vagyis magát a szabályozót tüntettük fel.

55-41198

- 211 -

Csáki Frigyes

SZABÁLYOZÁSOK DINAMIKÁJA

Szabályozások Dinamikája (Dynamics of Control Systems)

Frigyes Csáki
Akadémiai Kiadó, Budapest, 1966

Csáki Frigyes: Szabályozások Dinamikája (Dynamics of Control Systems), is a comprehensive book. The first chapter summarizes linear control theory. The second part gives basics of optimization. The third part discusses in detail the analysis and design of stochastic control systems. The forth part is a detailed analysis of sampled-data control systems. This book was used as a textbook for advanced control courses between 1966 and 1980. It still has actualities.

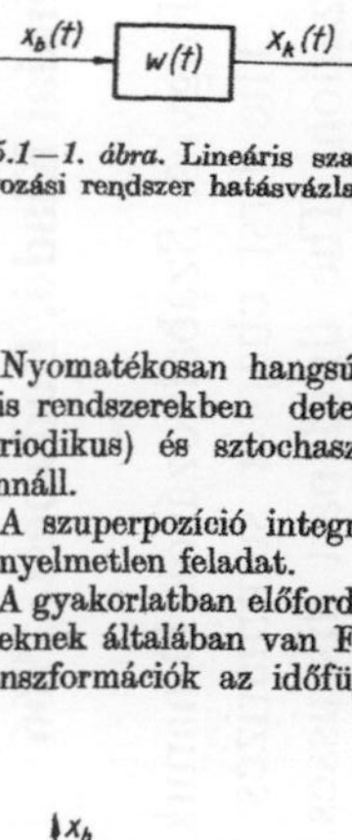

3.5.1—1. ábra. Lineáris szabályozási rendszer hatásvázlata

összetevők összege. Határátmenettel kapjuk az integrálösszefüggést.

Az előbbi (3.5.1–1) és (3.5.1–2) integrálösszefüggést gyakran a következő szimbolikus formában adják meg:

$$x_k(t) = w(t) * x_b(t) \tag{3.5.1-3}$$

Nyomatékosan hangsúlyozzuk, hogy mindkét integrálösszefüggés lineáris rendszerekben determinisztikus (szabályos, tranziens, aperiodikus, periodikus) és sztochasztikus (szabálytalan, véletlen) jelekre egyaránt fennáll.

A szuperpozíció integrál kiszámítása azonban általában fáradságos és kényelmetlen feladat.

A gyakorlatban előforduló, negatív időkre zérus értékű, determinisztikus jeleknek általában van FOURIER- vagy LAPLACE-transzformáltjuk. Ezek a transzformációk az időfüggvényekre vonatkozó szuperpozíció integrált a

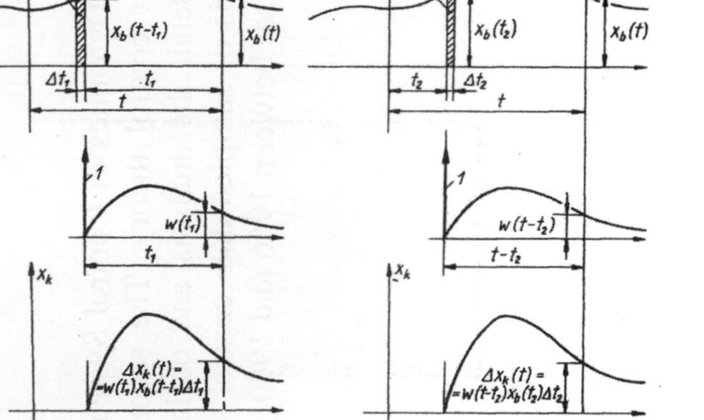

3.5.1—2. ábra. Az impulzusokra bontott bemenőjel hatására keletkező kimenőjel összetevők meghatározása. Első változat

3.5.1—3. ábra. Az impulzusokra bontott bemenőjel hatására keletkező kimenőjel összetevők meghatározása. Második változat

274

transzformált függvények közönséges szorzatába viszik át, ezzel azután a feladat lényegesen leegyszerűsödik:

$$X_k(s) = \int_0^{\infty} x_k(t)\, e^{-st}\, dt =$$

$$= \int_0^{\infty} \Big[\int_0^{\infty} w(t - t_2)\, x_b(t_2)\, dt_2 \Big] e^{-st}\, dt \tag{3.5.1-4}$$

$x_b(t)$ → $w(t)$ → $x_k(t) = w(t) * x_b(t)$, $x_k(t) = \int_{-\infty}^{\infty} w(t_1)\, x_b(t-t_1)\, dt_1$

$X_b(s)$ → $W(s)$ → $X_k(s) = W(s)X_b(s)$

3.5.1—4. ábra. A bemenőjel és a kimenőjel kapcsolata az idő és az operátor tartományban

Megjegyezzük, hogy a t_2 szerinti integrált most elegendő 0 alsóhatártól elkezdeni, mert az $x_b(t_2)$ bemenőjel $t_2 < 0$ időkre zérus. A szóban forgó integrál két egymástól független integrál szorzatára bontható:

$$X_k(s) = \int_0^{\infty} w(t_1)\, e^{-st_1}\, dt_1 \int_0^{\infty} x_b(t_2)\, e^{-st_2}\, dt_2$$

ahol $t - t_2 = t_1$ ill. $t - t_1 = t_2$. Végül

$$W(s) = \mathscr{L}[w(t)]$$

$$X_b(s) = \mathscr{L}[x_b(t)]$$

figyelembevételével (*3.5.1—4. ábra*):

$$X_k(s) = W(s)\, X_b(s) \tag{3.5.1-5}$$

Ezzel szemben a sztochasztikus jeleknek nincs sem LAPLACE- sem FOURIER-transzformáltjuk. A sztochasztikus jelek korrelációfüggvényeinek azonban már lehet kétoldalas LAPLACE- vagy FOURIER-transzformáltja, utóbbiak éppen a teljesítmény-sűrűségspektrumok.

Ezért először keressük az összefüggést a lineáris rendszerek bemenő- és kimenőjelének korrelációfüggvényei között.

3.5.2. KAPCSOLAT A KORRELÁCIÓFÜGGVÉNYEK KÖZÖTT

A kimenőjel értéke a $t + \tau$ időpontban:

$$x_k(t + \tau) = \int_{-\infty}^{\infty} w(t_3)\, x_b(t + \tau - t_3)\, dt_3 \tag{3.5.2-1}$$

Itt a t_3 időt a t_1, t_2 és t időtől való megkülönböztetés céljából vezettük be.

Vizsgáljuk meg most a bemenőjel és a kimenőjel keresztkorreláció-függvényét:

$$\varphi_{bk}(\tau) = \lim_{T \to \infty} \frac{1}{2T} \int_{-T}^{T} x_b(t)\, x_k(t + \tau)\, dt \tag{3.5.2-2}$$

18*

275

Automatika (Automation)

Frigyes Csaki and Ruth Bars
Tankonyvkiado, Budapest, 1969, 1971, 1974, 1983, 1986

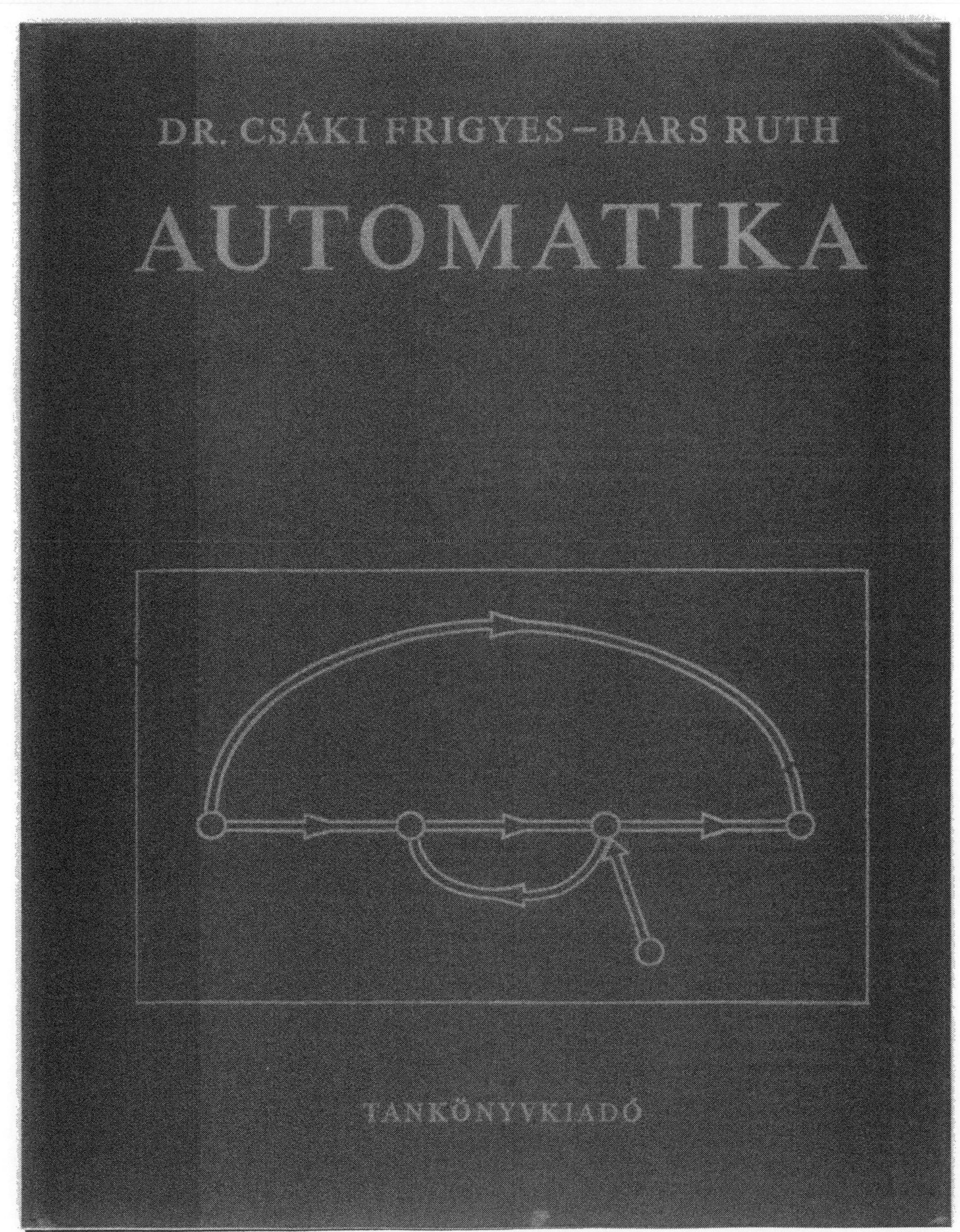

The textbook Csáki-Bars: Automatika was the first textbook for the basic control course. The book covered the following topics: historical introduction to control, basic concepts and structures, open- and closed-loop control, control examples, analysis in the time-, operator and frequency domain, characteristics of typical elements (lag, second order oscillating element, dead time, etc.), transfer functions, static and dynamic response, stability, PID compensation, models and control of electric machines, analog simulation, etc. Outlook, new trends: state-space methods, nonlinear control systems, optimal control, sampled-data control systems, computer control, stochastic control systems, adaptive control.

A fentiek miatt a holtidős szabályozás stabilizálása, illetve minőségjavítása igen fontos feladat.

A rendszer alkalmas módon választott integrálási időállandójú soros integráló taggal kompenzálható (10.3-4. ábra). Az eredő felnyitott kör NYQUIST-diagramjának meghatározását a 10.3-5. ábra mutatja. Állapítsuk meg az integráló tag kritikus integrálási időállandóját, amelyre a szabályozás éppen a stabilitás határára kerül.

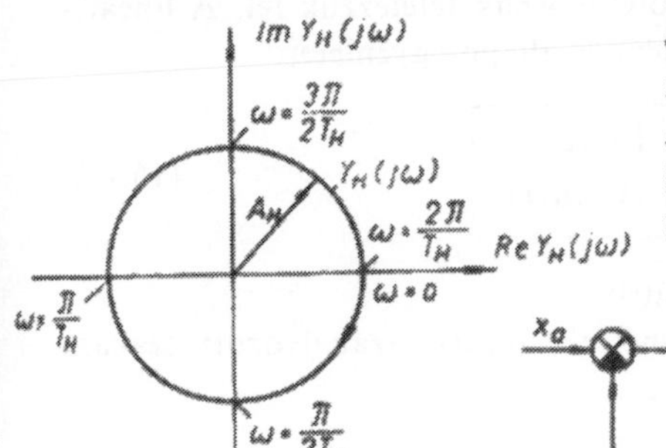

10.3-3. ábra. Holtidős szabályozás NYQUIST-diagramja

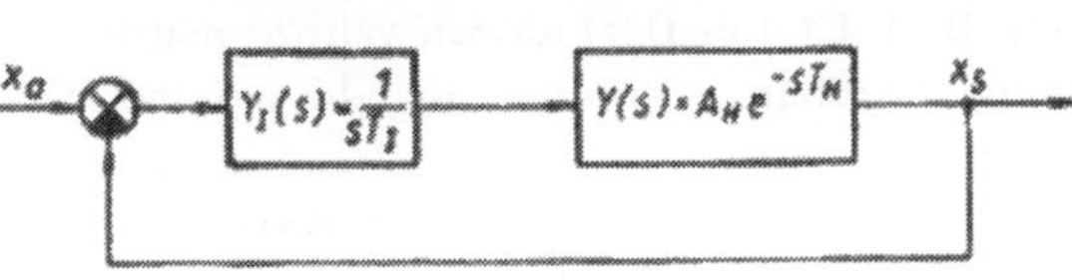

10.3-4. ábra. Holtidős szabályozás integráló taggal kompenzálható

Az eredő NYQUIST-diagram az $\omega_K = \frac{\pi}{2T_H}$ körfrekvencián metszi először a negatív valós tengelyt (mivel ezen a körfrekvencián mindkét tag fáziseltolása $-90°$).

Ha az abszolút érték 1-nél kisebb, a rendszer stabilis. Tehát:

$$\frac{1}{\omega_K T_I} A_H \leqq 1.$$

A rendszer a stabilitás határán van, ha a metszésponthoz tartozó vektor abszolút

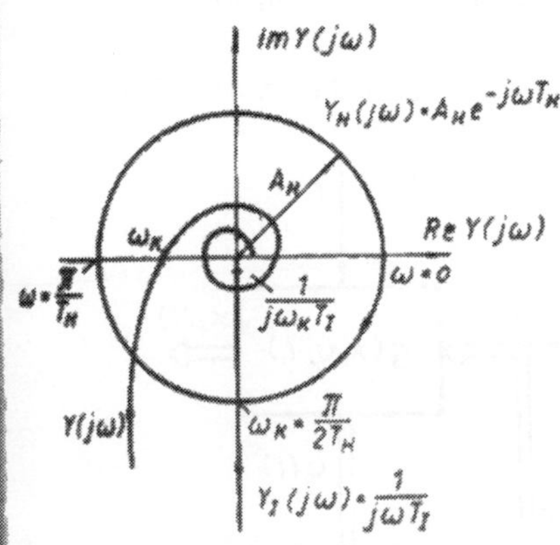

10.3-5. ábra. Integráló taggal kompenzált holtidős szabályozás NYQUIST-diagramjának meghatározása

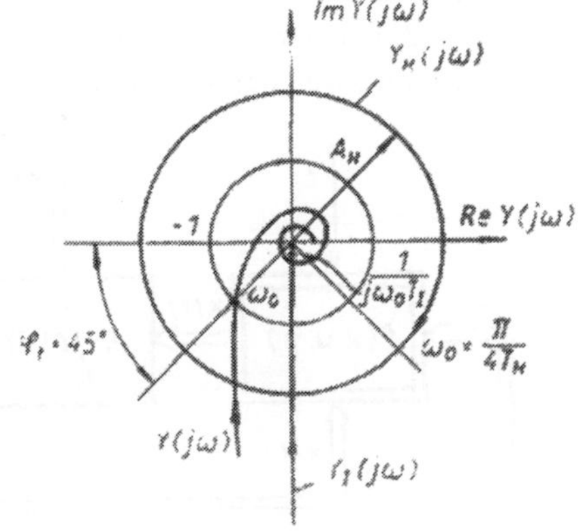

10.3-6. ábra. Az integrálási időállandó meghatározása $\varphi_t = 45°$ fázistöbblethez

403

Itt $\mathbf{x}$ az állapotvektor, $\mathbf{u}$ a szabályozott szakasz bemenőjeleiből (a beavatkozójelekből vagy a módosított jellemzőkből) alkotott vektor. $\mathbf{x}_s$ a szabályozott szakasz kimenőjeleiből, a szabályozott jellemzőkből alkotott vektor. (Így $\mathbf{u}(t)=\mathbf{x}_b(t)$ és $\mathbf{x}_s(t)=\mathbf{x}_k(t)$.) Az $\mathbf{f}$ és $\mathbf{g}$ vektorfüggvényeket folytonosnak tételezzük fel. A lineáris nemautonom rendszerek, illetve szakaszok általános állapotegyenletei:

$$\boxed{\begin{aligned}\dot{\mathbf{x}}(t) &= \mathbf{A}(t)\mathbf{x}(t)+\mathbf{B}(t)\mathbf{u}(t),\\ \mathbf{x}_s(t) &= \mathbf{C}(t)\mathbf{x}(t)+\mathbf{D}(t)\mathbf{u}(t).\end{aligned}} \qquad (16.1\text{-}2)$$

Itt $\mathbf{A}(t)$, $\mathbf{B}(t)$, $\mathbf{C}(t)$ és $\mathbf{D}(t)$ időben változó mátrix.

A leggyakrabban előforduló állandó paraméterű lineáris szabályozott szakasz egyenletei:

$$\boxed{\begin{aligned}\dot{\mathbf{x}}(t) &= \mathbf{A}\mathbf{x}(t)+\mathbf{B}\mathbf{u}(t),\\ \mathbf{x}_s(t) &= \mathbf{C}\mathbf{x}(t)+\mathbf{D}\mathbf{u}(t).\end{aligned}} \qquad (16.1\text{-}3)$$

vagy rövidített formában:

$$\boxed{\begin{aligned}\dot{\mathbf{x}} &= \mathbf{A}\mathbf{x}+\mathbf{B}\mathbf{u},\\ \mathbf{x}_s &= \mathbf{C}\mathbf{x}+\mathbf{D}\mathbf{u}.\end{aligned}} \qquad (16.1\text{-}4)$$

Itt $\mathbf{x}$ $n\times 1$-es, $\mathbf{u}$ $r\times 1$-es, $\mathbf{x}_s$ $q\times 1$-es oszlopvektor (oszlopmátrix), míg $\mathbf{A}$ $n\times n$, $\mathbf{B}$ $n\times r$, $\mathbf{C}$ $q\times n$ és $\mathbf{D}$ $q\times r$ méretű állandó mátrix.

Végül megemlítjük, hogy az egyváltozós, állandó paraméterű lineáris szabályozott szakasz egyenletei:

$$\boxed{\begin{aligned}\dot{\mathbf{x}} &= \mathbf{A}\mathbf{x}+\mathbf{b}u,\\ x_s &= \mathbf{c}^T\mathbf{x}+du,\end{aligned}} \qquad (16.1\text{-}5)$$

ahol $\mathbf{b}$ $n\times 1$ dimenziós oszlopvektor (oszlopmátrix), $\mathbf{c}^T$ $1\times n$ dimenziós sorvektor (sormátrix), d pedig skalár. Most u és x_s is skalár.

A nemlineáris rendszer hatásvázlatát a 16.1-1. ábra mutatja be. A 16.1-2. ábrán

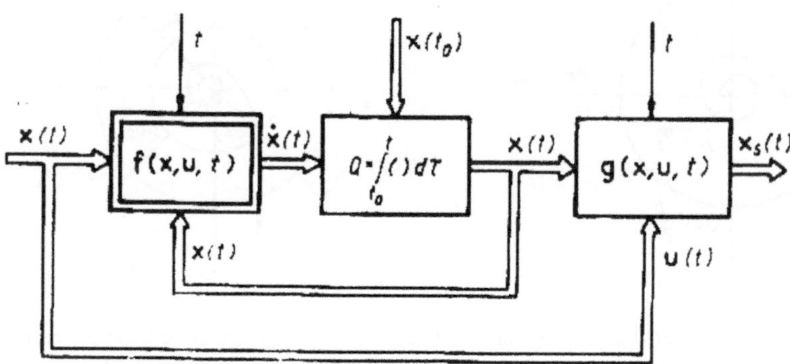

16.1-1. ábra. Nemlineáris folytonos-folyamatos működésű rendszer hatásvázlata

636

EARLY CONTROL TEXTBOOKS
in
ITALY

La Regolazione delle Turbine Idrauliche

Giuseppe Evangelisti
Nicola Zanichelli Editore, Bologna, 1947

Teoria della Regolazione Automatica

Giorgio Quazza
ANIPLA, 1962

Lezioni di Controlli Automatici - Teoria dei Sistemi Lineari e Stazionari

Antonio Lepschy, Antonio Ruberti
Edizioni Scientifiche SIDEREA, Roma, 1963

Elementi di servomeccanismi e controlli

Sergio Barabaschi and Renzo Tasselli
Zanichelli, Bologna, 1965

Contributed by

Alberto Isidori
Universita degli Studi di Roma "La Sapienza"

Sergio Bittanti
Politecnico di Milano

Claudio Bonivento and Giovanni Marro
University of Bologna

La Regolazione delle Turbine Idrauliche

Giuseppe Evangelisti
Nicola Zanichelli Editore, Bologna, 1947, 280 pages.

For a period about fifteen years, from the time of its appearance in 1947 until the early sixties, "La Regolazione delle Turbine Idrauliche", by Giuseppe Evandeglisti, was the only book written in Italian covering in a systematic manner problems and methods of automatic control. While the main purpose of the book was to address a variety of problems arising in the regulation of hydraulic power plants, the book covers in a rigorous and extensive form a number of topics that eventually became classical topics of any textbook in control, such as mathematical modelling, Routh-Hurwitz stability criterion, Laplace transform, steady state and transient analysis, etc.

Giuseppe Evangelisti (1903-1981) became Professor of Hydraulic Constructions at the University of Bologna in 1939. At that time, as a scientist and engineer, he had achieved a great international reputation in this field for his brilliant results on the theory of the water hammer phenomenon and on the design of hydraulic power plants. In the forties, new scientific curiosities brought him in touch with methods of analysis and design developed within the emerging areas of Automatic Control and Computer Engineering, which he specifically applied to the problem of the regulation of hydraulic turbines. Evangelisti promoted a lot of relevant pioneering initiatives in Italy and abroad, in these areas. He was one of the scientists that signed (1956) the famous Heidelberg resolution "in favour of an international union of Automatic Control", he founded (1957) the "Centro Calcoli e Servomeccanismi" at the University of Bologna (whose activities eventually merged into the Department of Electronics, Computer and Systems), he was a member of the first IFAC Council (1958) and very active in the scientific preparation of the first IFAC Congress in Moscow (1960). He taught regular courses of Automatic Control in the period 1960-1965 at the University of Bologna, for which he prepared a comprehensive set of lecture notes (1961).

"La Regolazione delle Turbine Idrauliche" covers material for the most part based on his personal research work during the difficult war period 1942-44. It is organized into nine chapters and covers the entire problem of the design of hydroelectric power generators. A first part of the book is devoted to the development of mathematical models of each individual subsystem (the turbines with their regulation devices, the supplied electric power network, the surge tank and the outflow pipe). The intermediate part of the book presents the mathematical tools for the analysis of stability and performance. Finally, the solution of the whole regulation problem is addressed in detail.

The first chapter reviews in detail all basic control schemes for the regulation of water turbines, whose components are then thoroughly analyzed and modelled in the second chapter, in the form of sets of first order differential equations. Among these schemes, one can recognize the classical proportional-integral-derivative feedback controllers realized by suitable connections of mechanical and oleo-dynamic devices. The next three chapters are devoted to what we would call today "automatic control theory". Chapter 3 covers the use of state-space representation, from general nonlinear models to the analysis of small oscillations via linear approximation, the Routh-Hurwitz stability criterion and Laplace transform. Chapter 4 covers the analysis of forced and transient response and Chapter 5 presents in detail approximation methods and algorithms for the numerical computation of solutions. The next three chapters describe how the general methods of analysis and design presented in Chapters 3-5 can be effectively used in a variety of problems dealing with the regulation of hydraulic turbines. Finally, the book is concluded by a ninth chapter dealing with a full nonlinear analysis of the regulation problem under large perturbations.

Contributed by

Claudio Bonivento and Giovanni Marro
University of Bologna
Italy

GIUSEPPE EVANGELISTI

ORDINARIO DI COSTRUZIONI IDRAULICHE NELL'UNIVERSITÀ DI BOLOGNA

LA REGOLAZIONE DELLE TURBINE IDRAULICHE

NICOLA ZANICHELLI EDITORE

BOLOGNA 1947

tagonista della molla *m*. Cessati i rapidi movimenti del servomotore, l'olio lentamente si travasa, e lo stantuffo si sposta, finchè la molla non ha

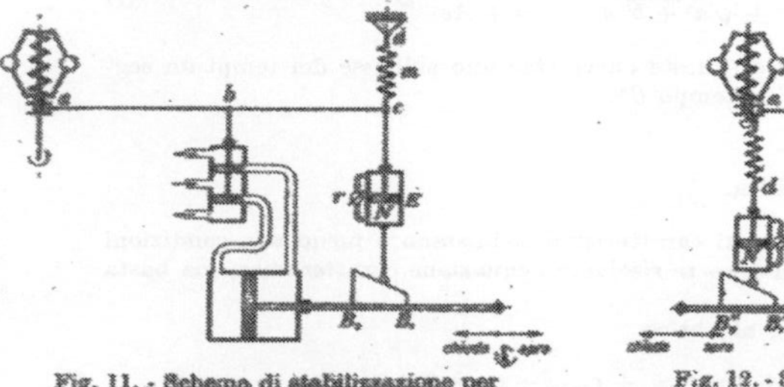

Fig. 11. - Schema di stabilizzazione per asservimento cedevole esterno.

Fig. 12. - Schema di stabilizzazione per asservimento cedevole interno.

ripreso la sua posizione d'equilibrio; il punto *c*, che fissa univocamente la velocità di regime, viene con ciò ricondotto alla sua posizione iniziale.

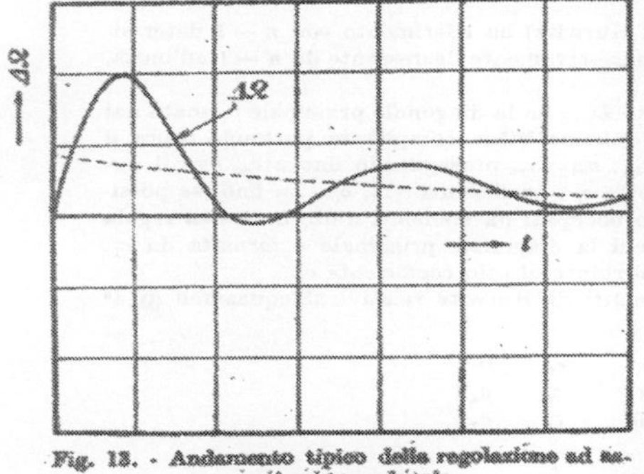

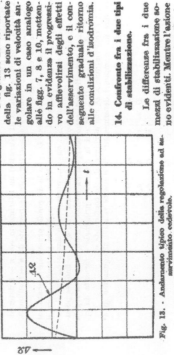

Fig. 13. - Andamento tipico della regolazione ad asservimento cedevole.

Nel grafico dimostrativo della fig. 13 sono riportate le variazioni di velocità angolare in un caso analogo alle figg. 7, 8 e 10, mettendo in evidenza il progressivo affievolirsi degli effetti dell'asservimento, e il conseguente graduale ritorno alle condizioni d'isodromia.

14. Confronto fra i due tipi di stabilizzazione.

Le differenze fra i due mezzi di stabilizzazione sono evidenti. Mentre l'azione accelerometrica si esplica in funzione dell'accelerazione istantanea del gruppo, qualunque sia la manovra che la stia provocando, quella dell'asservimento cedevole dipende dal tipo di manovra in atto, indipendentemente dalle condizioni di moto della macchina.

Ulteriore diversità risiede nel fatto che l'asservimento ha un'azione costantemente reattiva, nel senso che agisce sempre in senso contrario

CAPITOLO TERZO

LE EQUAZIONI DELLE PICCOLE OSCILLAZIONI

§ I – LE EQUAZIONI ALLE VARIAZIONI.

44. Richiami teorici.

Un semplice esame delle equazioni del moto dedotte nel capitolo precedente basta per dimostrare che manca ogni possibilità d'integrazione in termini finiti. Ciò non ostante, l'analisi ci offre il modo di giungere a risultati positivi, purchè ci si accontenti di indagare fenomeni confinati in un intorno sufficientemente ristretto dello stato di regime. Un tale mezzo è costituito da quel criterio delle piccole oscillazioni che per vastità di applicazioni e importanza di risultati occupa un posto di primo piano nella meccanica classica e tecnica. Pur senza entrare in dettagli, conviene render conto dei fondamenti teorici e delle modalità d'applicazione del procedimento (¹).

Si abbia un qualsiasi sistema meccanico in movimento, e sia il moto definito da n parametri $x_1, x_2 \ldots x_n$, funzioni del tempo t: per quanto è noto dalla meccanica razionale, le equazioni del moto si possono scrivere nella forma (normale)

$$
\begin{aligned}
\frac{dx_1}{dt} &= F_1(x_1 \ldots x_2 \ldots x_n, t) \\
&\cdots\cdots\cdots \\
\frac{dx_2}{dt} &= F_2(x_1 \ldots x_2 \ldots x_n, t) \qquad (a) \\
&\cdots\cdots\cdots \\
\frac{dx_n}{dt} &= F_n(x_1 \ldots x_2 \ldots x_n, t).
\end{aligned}
$$

Ammettiamo che le funzioni F siano dotate di derivate parziali del primo or-

(¹) Il metodo era già stato impiegato, in questioni di analisi pura, da G. DARBOUX, ma deve la sua introduzione nella meccanica applicata (specialmente in problemi di meccanica celeste) a H. POINCARÉ. Per maggiori ragguagli vedi É. GOURSAT: *Cours d'analyse mathématique*. Paris, Gauthier-Villars, 1915 (Tomo III, Cap. XXIII); T. LEVI CIVITA e U. AMALDI: *Lezioni di meccanica razionale*. Bologna, Zanichelli, 1926, vol. I, p. II; G. KRALL: *Meccanica tecnica delle vibrazioni*. Bologna, Zanichelli, 1940, vol. I, Cap. IV.

costante di tempo, di significato analogo alla (*f*). Alla loro volta periodo $T^\times$ e costante di tempo $\theta^\times$ definiscono il parametro numerico

$$\varepsilon = \frac{T^\times}{2\theta^\times} = \pi \left|\frac{r}{s}\right| \qquad (h)$$

che prende il nome di decremento logaritmico $\left(\text{relativo al semiperiodo } \frac{T^\times}{2}\right)$.

Le ragioni di queste denominazioni sono più che note dalla fisica matematica elementare. In un termine aperiodico (*b*) risulta

$$\left[\frac{d}{dt}\Delta x\right]_{t=0} = -\frac{1}{\theta}\,[\Delta x]_{t=0}\,, \qquad (i)$$

e quindi la tangente al diagramma del moto nell'istante iniziale stacca sull'asse dei tempi un segmento uguale precisamente alla costante di tempo θ (fig. 37). Nel moto oscillatorio smorzato (fig. 38) le (*c*) e (*h*) ci dicono che le ordinate distanziate di un semiperiodo $\frac{T^\times}{2}$ sono, in valore assoluto, in progressione geometrica di ragione $e^{-\varepsilon}$ (6); è pure

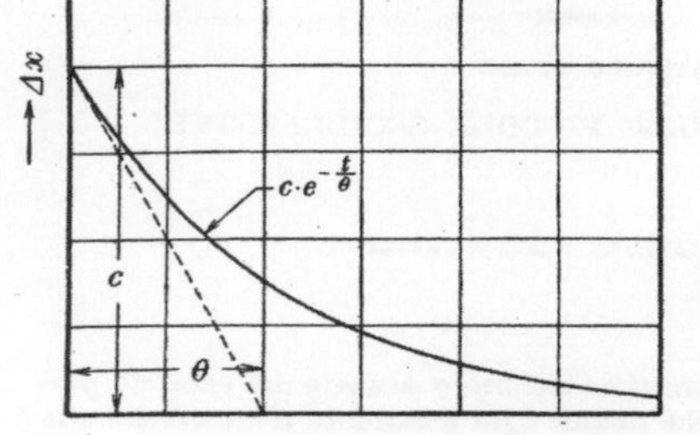

Fig. 37. - Andamento tipo di una componente di moto aperiodica.

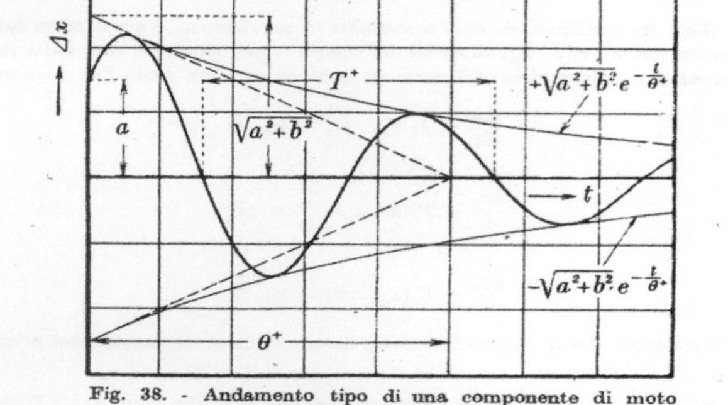

Fig. 38. - Andamento tipo di una componente di moto oscillatoria smorzata.

(6) Questa proprietà del decremento logaritmico riesce utile per il tracciamento materiale delle curve (*c*): tracciato infatti il primo semiperiodo, se ne ottiene il secondo cambiando segno alle ordinate e alterandole nel rapporto fisso $e^{-\varepsilon}$, e così via.

noto che il diagramma del moto è tangente alle due esponenziali (dette curve d'estinzione, e anch'esse segnate in fig. 38)

$$\Delta x = \pm\sqrt{a^2+b^2}\,e^{-\frac{t}{\theta^\times}} = \pm A e^{-\frac{t}{\theta^\times}}\,, \qquad (j)$$

e che le tangenti all'origine di queste curve staccano sull'asse dei tempi un segmento uguale alla costante di tempo $\theta^\times$.

50. Le condizioni di stabilità.

Per decidere se gli esponenti caratteristici soddisfano o meno alle condizioni richieste per la stabilità non occorre risolvere l'equazione caratteristica, ma basta conoscerne i coefficienti.

Invero, data l'equazione algebrica

$$c_0 x^n + c_1 x^{n-1} \ldots + c_{n-1}x + c_n = 0\,,$$

affinchè le radici siano o reali negative o complesse a parte reale negativa occorre e basta che siano contemporaneamente soddisfatte le due seguenti condizioni:

a) tutti i coefficienti devono avere ugual segno;

b) assunto positivo il segno dei coefficienti, devono risultare positivi tutti i determinanti di Hurwitz.

La seconda condizione (detta di Hurwitz) ha riferimento con $n-1$ determinanti D_{n-1}, $D_{n-2} \ldots D_1$, di ordine progressivamente decrescente da $n-1$ all'unità, ottenuti nel modo che segue.

Il determinante di ordine massimo D_{n-1} ha la diagonale principale formata dai coefficienti c_1, $c_2 \ldots c_\nu \ldots c_{n-1}$; ogni colonna viene completata portando sopra il generico elemento c_ν i coefficienti $c_{\nu+1}$, $c_{\nu+2} \ldots$, proseguendo fino a c_n, ove il numero di posti liberi lo consenta, e sotto c_ν i coefficienti $c_{\nu-1}$, $c_{\nu-2} \ldots$ fino (se possibile) a c_0: i posti rimasti vuoti sono occupati da elementi nulli. La stessa regola vale per il determinante D_{n-2}, in cui la diagonale principale è formata da c_1, $c_2 \ldots c_{n-2}$, e così via, fino a D_1 che si riduce al solo coefficiente c_1.

Per fare un esempio, i determinanti di Hurwitz relativi all'equazione di 4° grado valgono

$$D_3 = \begin{vmatrix} c_1 & c_3 & 0 \\ c_0 & c_2 & c_4 \\ 0 & c_1 & c_3 \end{vmatrix}$$

$$D_2 = \begin{vmatrix} c_1 & c_3 \\ c_0 & c_2 \end{vmatrix}$$

$$D_1 = c_1\,.$$

Se il grado dell'equazione non è eccessivamente alto gli sviluppi dei determinanti portano a relazioni abbastanza semplici. Eliminate le inequazioni superflue — cioè i determinanti necessariamente positivi se risultano positivi quelli di or-

Teoria della Regolazione Automatica

Giorgio Quazza
ANIPLA, 1962 (263 pages)

Born in Mosso, a small town not far from Biella, Italy, in 1924, Giorgio Guazza began his studies in engineering in the Politecnico of Turin in November 1941. He was an outstanding student; however, the dramatic war events of those days led him to join the Italian Resistance Movement. While continuing to study, he was already engaged in the Resistance in the mountains of the Italian Alps. It was there, on December 4, 1944, that he was taken prisoner during a roundup. After a period in a prison camp in Bolzano, he was sent to the concentration camp of Mauthausen, where he had to remain until May 1945 when the camp was liberated. Weakened by the internment, he was then down to his last physical resources. After a period of rehabilitation, he resumed his studies in the summer of 1945, and during the period from September 1945 to November 1946 he was able to pass all the remaining 25 exams. He presented his thesis on the stability of large power transmission lines in December 1946. He then spent a period of time in a Turin company working at the control of electrical machines, and then moved to the USA for three years (1950 – 1953), working first at MIT and then at the Brooklyn Polytechnic Institute. His studies focused on the emerging theory of servomechanisms, and for this he obtained his Doctorate degree in 1953.

Upon his return to Italy, he worked again as a control system designer in industry, until he joined the Italian Electrical Power Agency (ENEL) in Milan in 1964. Here he created the Automatic Control Research Center (*Centro Ricerca di Automatica* - CRA) in 1967, the purpose of which was to cope with control problems in the Italian electricity generation system. His research activity ranged from servomechanisms and analogue computing to speed governors for thermal and hydro–electric power systems. Soon, he became an internationally acknowledged expert in on-line power system control.

To put those years into perspective, let us mention some events taking place in Italy. In April 1956, an international meeting was organized at the Museum of Science and Technology in Milan, the *Convegno Internazionale sui Problemi dell'Automatismo*, held under the aegis of the Italian National Research Council (CNR). As chronicled in the newspapers of those days, the enthusiasm and expectations for control and automation were so remarkable, not only among specialists but also in the common people, that the conference had more than 1000 participants. Giorgio Quazza (then 32) attended the presentations with tremendous interest, and this experience had a deep impact on his engineering evolution. Also in 1956, the first exhibition on control and automation named the *Biennial Italian Automation and Instrumentation* (BIAS) was held. Last but not least, the *Associazione Nazionale per l'Automazione* (ANIPLA) was founded (again in 1956). This association has been very active, establishing a significant link between academia and industry, through the organization of conferences on various application areas and many tutorials for control engineers in industry.

At the end of the '50's and early '60's, ANIPLA invited Giorgio to give courses for the associates and it was the publisher of the first volume written by Giorgio, *Teoria della Regolazione Automatica* (1962), a comprehensive text of 263 pages on most aspects of basic control methods. The summary of the volume is as follows. 1: Analogies between electrical, mechanical, thermal and pneumatic phenomena; 2: Control systems in open loop and closed loop; 3: Performance of control systems; 4: Closed loop control systems; 5: Determining the static and dynamic characteristics of a control loop; 6: Improving the time response of a system with partial feedback. 7: Systems with partial feedback and many controlled variables; 8: Non-linear systems and discontinuous control. Appendix: Complex functions, Laplace Transforms, Stability criteria, Relationship between the phase diagram and the attenuation diagram, Root locus method, Correlation between transient response and frequency response, Distributed parameter systems, Sampled data systems.

Meanwhile, at the Technical University of Milan (Politecnico di Milano) a group of young and active researchers in control was growing up and invited Giorgio to give some specialization courses for engineers. The volume *Introduzione alla Ottimizzazione dei Processi*, published in 1964 by Tamburini Editore, Milan, was prepared for such courses. This volume (119 pages) presents first the role of optimization in control, by describing methods such as Kuhn and Tucker, Arrow and Wolfowitz, and then turns to calculus of variation, Pontryagin maximum principle, and dynamic programming. The copyright page of each copy of this book carries the hand-written signature of the author.

COPYRIGHT © 1964 - TAMBURINI EDITORE s. p. a.
Tutte le copie debbono portare la firma dell'Autore

The involvement of Quazza in the Politecnico culminated in 1971, when he was called to teach a regular course, *Process Control*, for students in their fifth year in Electronic Engineering. For this course, he prepared his third book, with the same title, (CLUP, 1973). It covers many aspects of process control, and has been used for decades by many students.

While directing the CRA of ENEL and teaching at the Politecnico di Milano, Giorgio Quazza was also very active in IFAC, where he had come to be appreciated by everybody as a genuine gentleman. He first served as Vice-Chairman of the Technical Committee on Systems Engineering (1969 – 1972), then as Chairman of the Technical Committee on Applications (1972 – 1975) and in 1975 was elected to the Executive Council. In all these tasks he was appreciated not only for his efficiency, but also for his humor and presence. But, to the regret of all, a mountaineering accident in 1978 robbed IFAC and the control engineering community of one of its most illustrious and hardworking members. His corpse was found in a deep crevice after days of continuous search on August 11th. Giorgio was planning to cross the glacier Ventina of Mount Rose, from Ayas to Zermatt, in the company of a close friend. The friend could not reach him, so he decided to make the crossing alone. In Zermatt he had an appointment with Professor John Coales, former president of IFAC, for a trip together in the mountains.

Giorgio Quazza will be remembered as one of the towering figures of control engineering of the past century, and a man of many qualities, the most eminent being that he was a person upon whom anyone could place complete reliance in time of trouble, as written in his obituary (Automatica, Vol.15, 1979).

Contributed by

Sergio Bittanti
Politechnico di Milano, Italy

POLITECNICO DI MILANO

CORSO DI CULTURA DI TECNICHE DELL'AUTOMAZIONE

TEORIA DELLA REGOLAZIONE AUTOMATICA

A CURA DI

GIORGIO QUAZZA

ANIPLA
1962

$$\begin{array}{l|l} s^3+14s^2+48s+80 & s^2+4s+8 \\ \underline{4s^3+\ 8s} & \overline{s+10} \\ 10s^2+40s+80 & \\ \underline{10s^2+40s+80} & \\ \quad = \quad = & \end{array}$$

l'ulteriore radice $s=-10$.
Si sono così determinate tutte le radici di $P(s)$.

) Intersezioni di curve componenti.

Per via grafica, le radici di $P(s)$ sono naturalmente quelle in cui il grafico di $P(s)$ attraversa l'asse delle ascisse. Ma se, p. es., $P(s)$ è di grado 5°, le radici di $P(s)$ sono anche le intersezioni delle curve:

$$s^3(s^2+a_4s+a_3) \quad e \quad -(a_2s^2+a_1s+a_0)$$

talora semplificabili convenientemente. Per via analitica, può, p. es., convenire identificarle attraverso la:

$$s^3 = -\frac{a_2s^2+a_1s+a_0}{s^2+a_4s+a_3}$$

o con altra disposizione suggerita da criteri di opportunità di calcolo.

4. Criterio di Nyquist. (1)

a) Sia data una curva chiusa C nel piano complesso s. La funzione $w = f(s)$, com'è noto, stabilisce una corrispondenza tra il piano $s = \sigma + j\omega$ e il piano $w = u + iv$, definita appunto dalla funzione $f(s)$, che di C offre una rappresentazione conforme W nel piano w, anch'essa curva chiusa. La funzione:

$$\log w = \log |w| + j(\varphi + 2k\pi)$$

è (v. Appendice I, 3) una funzione polidroma, che ha per punto di diramazione l'origine, cioè dove w si annulla.
Dopo ogni giro compiuto da w lungo la curva W attorno all'origine, log w ritorna al suo valore iniziale, mentre la parte immaginaria di log w cresce di 2π. Si immagini di percorrere la curva C nel piano s in senso orario; in corrispondenza W, trasformata di C, eseguirà un certo numero N di giri in senso orario intorno all'origine nel piano w. La differenza tra il valore assunto da log w dopo che la curva chiusa C, e quindi W, sia stata interamente percorsa (e perciò si sia ritornati al punto iniziale) e il valore che log w aveva nel punto di partenza, è perciò: $-j2\pi$ N.
Di conseguenza:

$$\oint_{\curvearrowleft W} d\log w = -\oint_{\curvearrowright W} d\log w = \log w_1 - \log w_2 = -j2\pi N \qquad (14)$$

che è relazione valida qualunque sia il contorno chiuso $W = f(C)$, anche se ha tratti all'infinito. D'altra parte, si supponga che $f(s)$ abbia nell'insieme A delimitato da C solo zeri e poli, ma nessuna singolarità essenziale,

(1) Nyquist, Regeneration Theory, Bell System Tech. Journal, 1932.
Bode, Network Analysis and Feedback Amplifier Design, Van Nostrand, 1945.

- 44 -

A titolo di esempio, si vedano nella Fig. A 18 le curve W corrispondenti alle funzioni:

$$G(s) = \frac{1}{1+sT} \; ; \; G(s) = \frac{1}{s(1+sT)} \; ;$$

$$G(s) = \frac{1}{1+2\xi sT+s^2T^2} \; ;$$

$$G(s) = \frac{K}{s(1+2\xi sT+s^2T^2)} \; ; \; G(s) = K\frac{1+sT_2}{s^2(1+sT_1)^2}$$

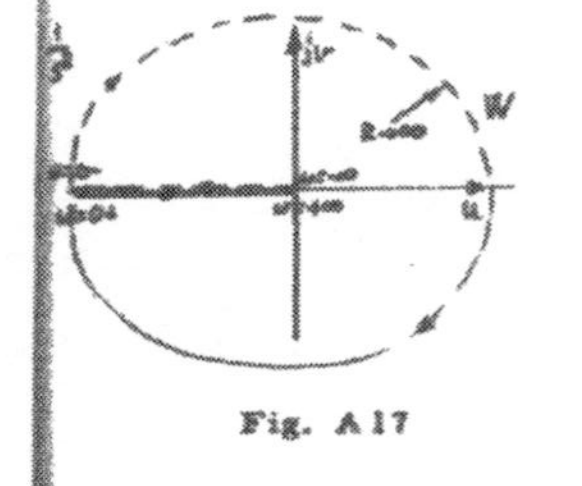

Fig. A 17

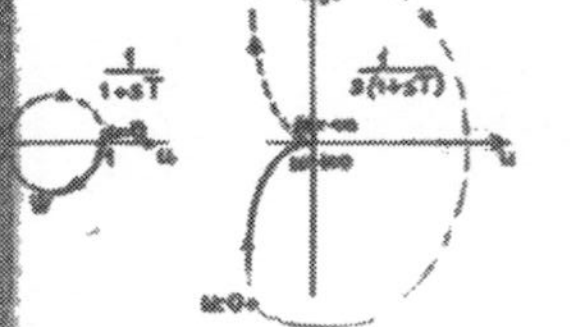
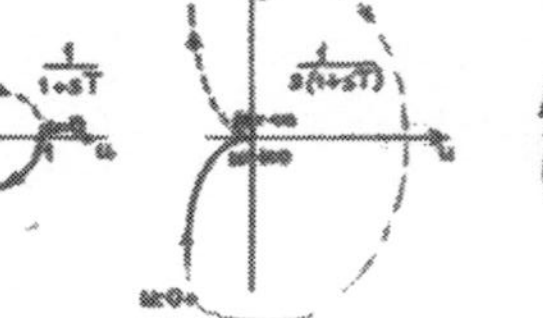
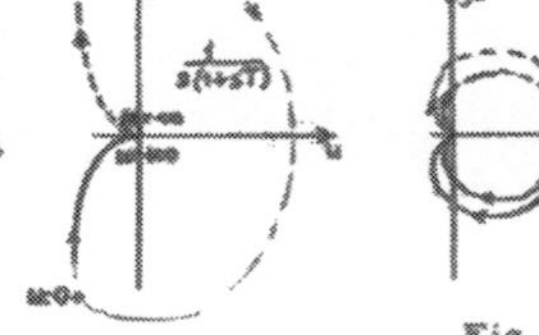
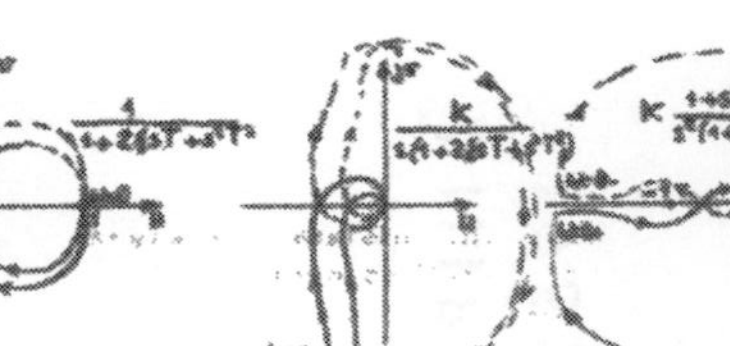
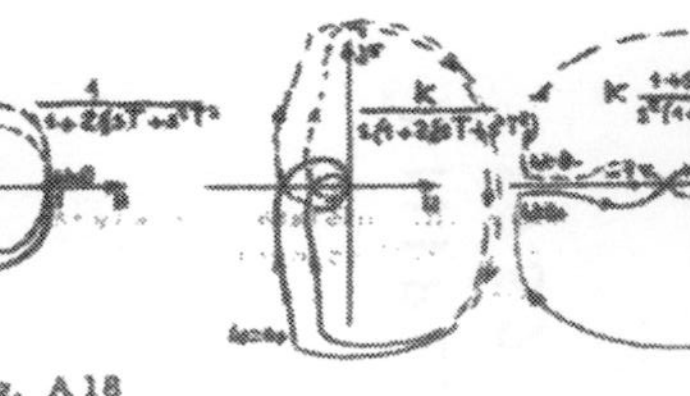

Fig. A 18

c) La funzione di trasferimento:

$$H(s) = \frac{G(s)}{1+G(s)}$$

dell'anello chiuso, a reazione unitaria, descrive (v. 1) un sistema stabile se il numero Z di zeri di $1+G(s)$ nel semipiano destro è uguale a zero. D'altra parte, $1+G(s)$ non può avere ovviamente altri poli che quelli di $G(s)$: se $G(s)$ è nota, è noto il numero di poli P di $1+G(s)$ nel semipiano destro. Si aggiunga che $H(s)$ non può avere di per sè altri poli che gli zeri di $1+G(s)$, giacchè i poli del numeratore sono cancellati dai poli del denominatore: in altre parole, se $G(s) = \frac{N(s)}{D(s)}$ è espresso come quoziente di polinomi, è:

$$H(s) = \frac{N(s)}{D(s)+N(s)}$$

cioè i poli di $G(s)$, cioè le radici di $D(s)$, non appaiono più come poli di $H(s)$.
Il "diagramma di Nyquist", cioè il luogo W trasformata di C a mezzo della funzione $w = G(s)$, può essere scelto a rappresentare la funzione $1+G(s)$ spostando l'origine delle coordinate del piano w nel punto $-1+j0$. Perciò, parlare di giri del luogo $1+G(s)$ attorno all'origine è equivalente a parlare di giri di $G(s)$ attorno al punto $(-1+j0)$. Di conseguenza, il teorema su enunciato si può esprimere anche come segue:
il numero di giri N in senso orario che il luogo $G(s)$ (1) trasformata del con

(1) Data la forma di C, e con l'intesa di tener nel dovuto conto il semicerchietto attorno all'origine, il contorno W può essere chiamato semplicemente $G(j\omega)$.

-47 -

APPENDICE V

Metodo dei luoghi delle radici

1. Definizione di luogo delle radici.

Sia:

$$G(s) = KG_1(s) = K \frac{(s - z_1)(s - z_2)\ldots.(s - z_m)}{(s - p_1)(s - p_2)\ldots.(s - p_n)}$$

la funzione di trasferimento dell'anello aperto, e si chiamino ψ_i gli argomenti dei "vettori" $(s - z_i)$ tracciati dagli zeri z_i al punto s, φ_i gli argomenti dei "vettori" $(s - p_i)$ valutati a partire dall'asse reale positivo.

I poli della funzione di trasferimento dell'anello chiuso, cioè le radici dell'equazione caratteristica:

$$1 + G(s) = 0$$

sono i punti del piano complesso s che soddisfano alle due equazioni:

$$\text{ang } G(s) = (\psi_1 + \psi_2 + \ldots. + \psi_m) - (\varphi_1 + \varphi_2 + \ldots + \varphi_n) = 180^\circ + k\ 360^\circ \quad (1)$$

$$|G(s)| = 1 \quad \text{cioè} \quad \frac{|sp_1| \cdot |sp_2| \cdot \quad |sp_n|}{|sz_1| \cdot |sz_2| \cdot \quad |sz_m|} = K \quad (2)$$

k denotando un numero intero positivo o negativo o nullo, e gli $|sp_i|$, $|sz_i|$ essendo le lunghezze dei vettori condotti dai poli o dagli zeri a s.

Nell'insieme dei punti s, definito dall'equazione (1), le radici dell'equazione caratteristica sono quelle,e solo quelle, che soddisfano all'equazione (2). Al variare del "guadagno" K dell'anello, le radici cambiano di posizione, pur continuando a soddisfare l'equazione (1), cioè a appartenere all'insieme definito da (1). Se si faccia variare K da zero a infinito, le radici si spostano nel piano s, descrivendo l'intiero insieme suddetto, che si chiama perciò luogo delle radici, eventualmente formato di segmenti o parti separate (luoghi delle radici).

Poichè per ogni data G(s) è relativamente agevole tracciare graficamente i luoghi delle radici, sulla base della (1), e determinare con la (2) la posizione delle radici corrispondente a un determinato valore di K, nonchè descrivere gli effetti di circuiti stabilizzatori, si intuisce la possibilità di valersi, sia per l'analisi che per la sintesi dei sistemi di regolazione, di un metodo fondato sull'uso dei luoghi delle radici. (1) Benchè esso sia di impiego meno facile e rapido dell'analisi frequenziale in coordinate logaritmiche, presenta tuttavia il vantaggio di fornire molto più direttamente gli elementi essenziali per

(1) W. R. Evans, Graphical Analysis of Control Systems, Trans. AIEE, 1948.
W. R. Evans, Control System Dynamics, Mc Graw Hill, 1954.
Thaler and Brown, Servomechanism Analysis, Mc Graw Hill, 1953.
J. G. Truxal, Automatic Feedback Control System Synthesis, Mc Graw Hill, 1955.

- 59 - (Pag. 58 in bianco)

valutare la risposta transitoria e consente, in particolare, sintesi di sistemi definiti in base a caratteristiche di risposta al gradino, come tempo di salita, sovraelongazione, smorzamento, tempo di assestamento, da ottenersi.

2. Qualche esempio di luoghi delle radici.

a) Si consideri la:

$$G(s) = \frac{K}{s(s + a)}$$

I luoghi corrispondenti sono tracciati in Fig. A 27 e appaiono costituiti dal segmento $\overline{-a0}$ e dal suo asse (la freccia indica il senso di K crescente). Invero, preso un punto Q su detto segmento, l'angolo di $\overline{-aQ}$ è 0, l'angolo $\overline{0Q}$ è 180°, perciò vale la (1). Così vale la (1) per un punto R, perchè $\varphi_2 = \beta$, essendo il triangolo isoscele: perciò $(\varphi_1 + \varphi_2) = 180°$.
Una radice complessa come R è naturalmente accompagnata dalla sua coniugata R', e nella risposta transitoria al gradino vi corrisponde un'oscillazione smorzata di costante di tempo

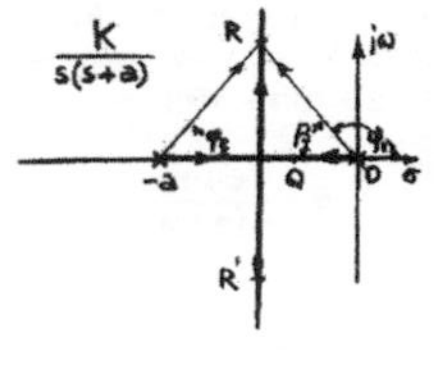

$$1/\xi\omega_n = \frac{1}{a/2}$$

Fig. A 27

sempre la stessa comunque vari il guadagno K (se $K > a^2/4$). Al crescere di K, definito da (2) come il prodotto delle distanze della radice da 0 e da (-a), R si muove verso l'alto, cioè diminuisce lo smorzamento dell'oscillazione smorzata, dato da:

$$\text{cotg}\,\beta = \frac{a}{2\,\omega_R} = \frac{\xi}{\sqrt{1 - \xi^2}}, \text{ ovvero } \beta = \text{arc cos}\,\xi$$

In generale, l'angolo β che la congiungente una radice con l'origine forma con l'asse reale, definisce il coefficiente di smorzamento dell'oscillazione smorzata relativa.
Radici sull'asse reale, come Q, individuano un guadagno K così basso $(< \frac{a^2}{4})$ che la risposta al gradino è semplicemente esponenziale.

b) La:

$$G(s) = \frac{K}{s(s + a)(s + b)}$$

definisce il luogo di Fig. A 28.
Esso entra nel semipiano destro, il che indica che, se K cresce al di là di un certo limite, l'anello chiuso può diventare instabile, a differenza del caso a). Si riconosce che la intersezione J del luogo con l'asse immaginario deve corrispondere a $\varphi_2 + \varphi_3 = 90°$, cioè i triangoli rettangoli P_3OJ e JOP_2 sono simili, quindi:

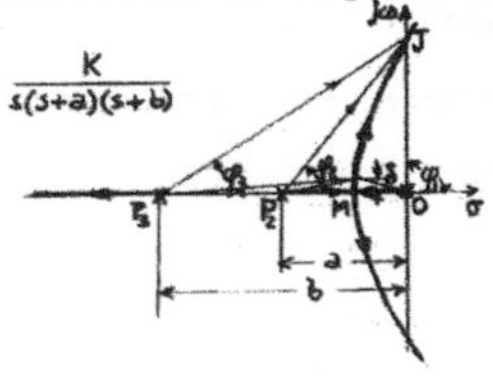

$$\tan\varphi_3 = \frac{JO}{OP_3} = \frac{OP_2}{JO} \quad \text{cioè} \quad \tan\varphi_3 = \sqrt{\frac{OP_2}{OP_3}}$$

che definisce φ_3 cioè J. Così l'intersezione delle sezioni complesse dei luoghi con l'asse reale è determinata dall'ovvia uguaglianza di piccoli

Fig. A 28

- 60 -

angoli:

$$\frac{\delta}{MP_2} + \frac{\delta}{MP_3} = \frac{\delta}{MO}$$

cioè

$$\frac{1}{MO} = \frac{1}{MP_2} + \frac{1}{MP_3}$$

c) La:

$$G(s) = K \frac{(s+b)}{s(s+a)}$$

definisce i luoghi di Fig. A 29, in cui l'effetto dello zero (-b) è indicato dalla rotazione dei luoghi verso il semipiano sinistro (l'instabilità non può mai verificarsi). Si noti che i poli sono indicati con il segno x, gli zeri con o.

Fig. A 29

3. Regole per la costruzione dei luoghi.

Ricordando la (2), le radici per K → O devono essere i poli di G(s), per K → ∞ gli zeri di G(s): perciò, descritti per K crescente, i luoghi partono dai poli e terminano agli zeri di G(s) (eventualmente agli zeri a s → ∞). Perciò anche il numero di luoghi deve essere uguale al numero n di poli di G(s) (o al numero m di zeri, se fosse m > n), ciascuno contato col proprio ordine di molteplicità.

Inoltre:

gli asintoti per s → ∞ sono definiti da $\frac{s^n}{s^m} = -K$, ovvero $\arg s^{n-m} = 180^\circ + k\,360^\circ$ cioè:

$$\arg s = \frac{1}{n-m}(180^\circ + k\,360^\circ)$$

ed invero solo (n-m) luoghi vanno a ∞: gli altri, come osservato sopra, terminano a zeri finiti di G(s);

gli asintoti suddetti tagliano tutti l'asse reale nel punto:

$$s_1 = \frac{\sum p_i - \sum z_i}{n-m}$$

le sezioni complesse sono coniugate (i polinomi sono a coefficienti reali); sono luoghi i tratti di asse reale alla sinistra di un numero dispari di zeri e poli reali di G(s);

le intersezioni con l'asse immaginario si possono trovare in generale (ma v. il caso particolare 2 b)) con la prova di Routh, v. Appendice III, 2 b) e 3 e), con che si determina pure il guadagno K che annulla la penultima riga della matrice di Routh;

- 61 -

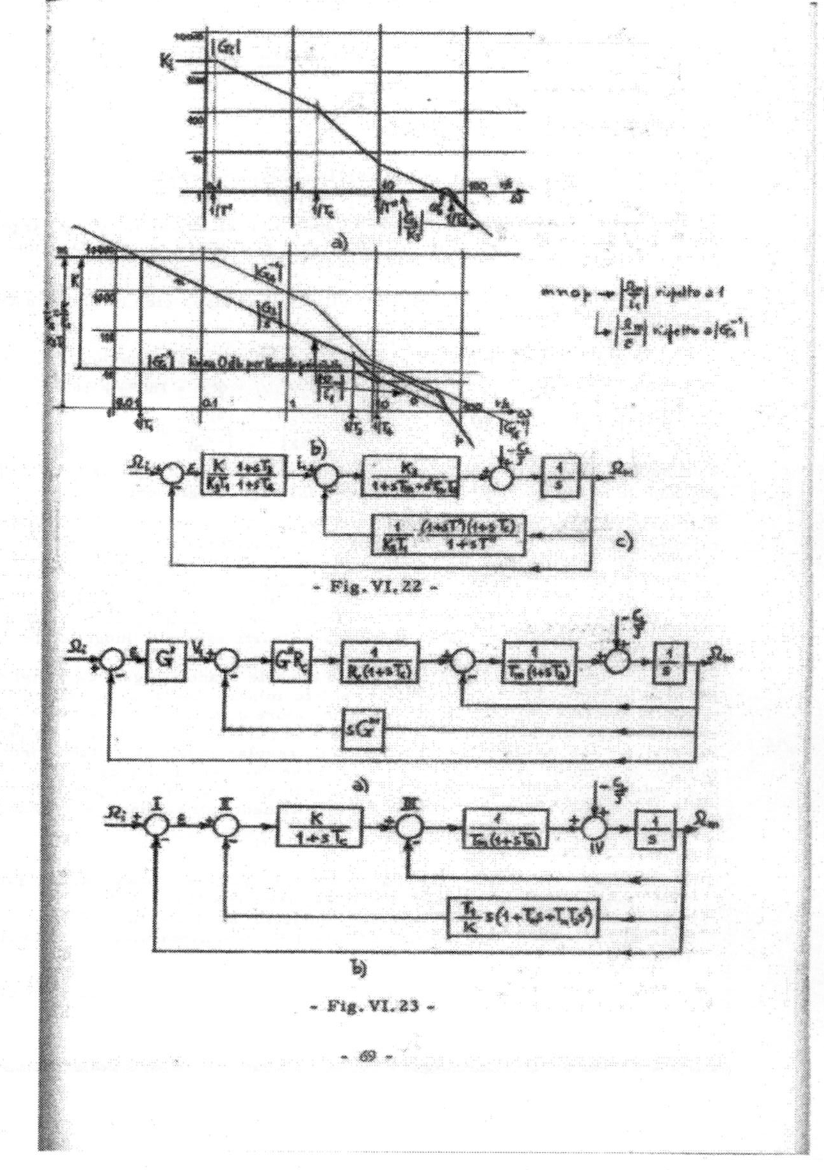

- Fig. VI. 22 -

- Fig. VI. 23 -

- 69 -

Lezioni di Controlli Automatici - Teoria dei Sistemi Lineari e Stazionari

(Lectures on Automatic Control - Linear and Time-Invariant Systems)

Antonio Lepschy, Antonio Ruberti
Edizioni Scientifiche SIDEREA, Roma, 1963

The first edition of the book "Lezioni di Controlli Automatici - Teoria dei Sistemi Lineari e Stazionari " (*Lectures on Automatic Control - Linear and Time-Invariant Systems*) by Antonio Lepschy and Antonio Ruberti was published in 1963 (in two volumes); the second edition (in one comprehensive volume) appeared in 1967. The late Antonio Ruberti (who after having been Rector of the University of Rome was appointed Minister for Higher Education and Scientific Research, and then European Commissioner for Scientific Research in Bruxelles) was at that time a professor of Automatic Control with the School of Engineering of the University of Rome. Antonio Lepschy was a professor of Automatic Control at the University of Trieste. The book has been used as a textbook for automatic control courses not only at the Universities of Rome and Trieste but also in many other Italian Universities.

The book is divided into three parts. The first one has an introductory style and presents the notion of control and a classification of control actions. Two appendixes supply a short history of automatic control and the terminology of control systems and components. The second chapter discusses the role of control in the field of automation and in the context of cybernetics.

The second part of the volume deals with the theory of linear time-invariant systems described by input-output relations. First the temporal behaviour of a linear time-invariant system is analysed by discussing the properties of the associated differential equation and by evaluating the free and the forced response of the system. Next the analysis in terms of Laplace transform is developed and the analytic and graphical representations of the transfer functions are discussed.

The third part is devoted to the analysis and synthesis of feedback control systems. Block diagrams, signal flow graphs and their algebra are introduced, emphasizing their use in modelling and analysis of feedback loops. A chapter is devoted to stability analysis: the Routh and Hurwitz criteria and the Nyquist criterion are introduced and discussed in detail; the meaning of stability margins for feedback systems is discussed. The steady-state and transient response to canonical inputs (in particular to the unit step) and the concepts of system promptness and precision are discussed, introducing the notions of loop type, rise time, overshoot, settling time etc. as well as indexes like ISE, ITAE etc.

Next, the problem of assessing the closed-loop behaviour starting from the analytic (poles and zeros) and/or graphical properties of the open-loop transferences is addressed. The root locus method and the use of the Nichols chart (and of other similar loci of the $G(j\omega)$ plane) are presented.

The last chapters of the book are devoted to the problem of controller synthesis. The standard PID controllers and the relevant tuning formulas are presented; the lag-, lead- and lag-lead actions are discussed and the procedures for computing their parameters are illustrated. A chapter was devoted also to the so called *direct synthesis* methods. Some notions about control system synthesis with two degrees of freedom are also presented.

Soon after the first edition of the book was published, the authors edited a textbook on control system components (A. Lepschy, A. Ruberti (eds): *Componenti dei Sistemi di Controllo*; vol. I "*Strumentazione*", vol.II "*Amplificatori e Motori*", Edizioni Scientifiche Siderea, Roma, 1964). Later Antonio Ruberti and Alberto Isidori published a book on system theory discussing the state variables approach (A. Ruberti, A. Isidori, *Teoria dei Sistemi*, Boringhieri, Torino, 1979).

Contributed by

Alberto Isidori
Universita degli Studi di Roma "La Sapienza"

Prof. Ing. Antonio LEPSCHY
Inc. di Controlli Automatici
Università di Trieste

Prof. Ing. Antonio RUBERTI
Inc. di Controlli Automatici
Università di Roma

LEZIONI DI CONTROLLI AUTOMATICI

TEORIA DEI SISTEMI LINEARI NORMALI

VOLUME I

Edizioni Scientifiche SIDEREA

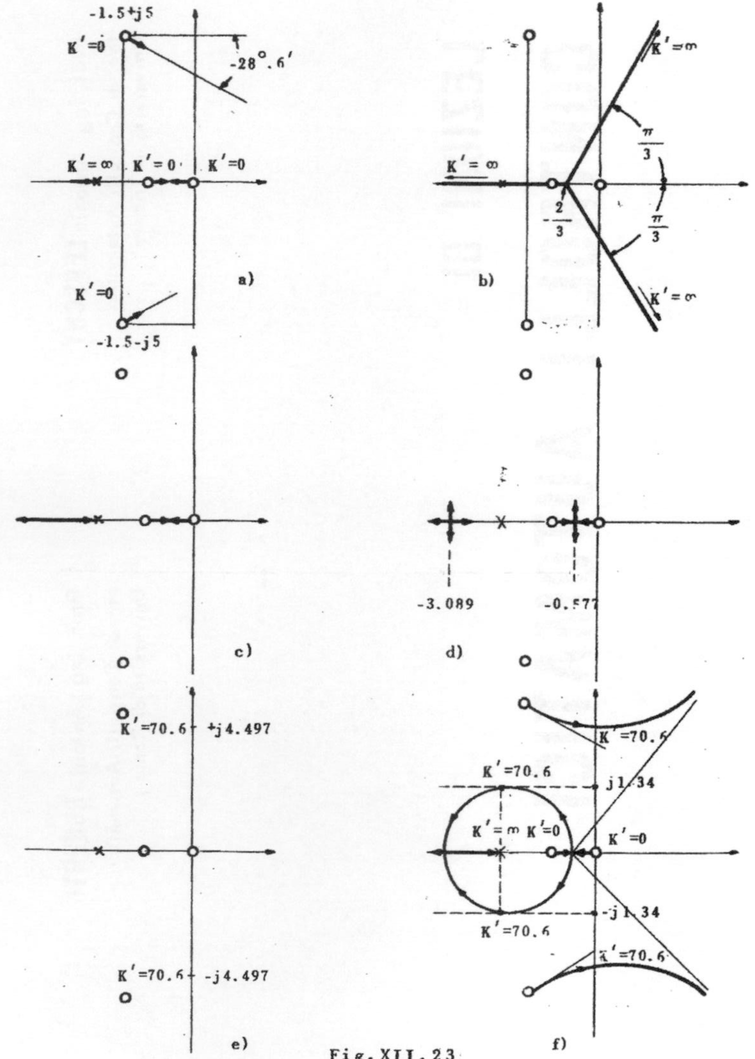

Fig.XII.23

cui corrisponde la segunete tabella di Routh:

(4)	1	30,25	2 K'
(3)	4	27,25+K'	
(2)	93,75 − K'	8 K'	
(1)	$\dfrac{2554,6875 + 34,5\,K' - K'^2}{93,75 - K'}$		
(0)	8 K'		

L'elemento della riga (1) si annulla per K'=70,656 e per K' = =-36,156; questo secondo valore non ha interesse in quanto si sta tracciando il luogo positivo. Le radici immaginarie corrispondenti al valore positivo di K' che interessa sono $\pm$ j 4,497 (fig.XII.23e).

In base a questi elementi si puo' disegnare il luogo, che assume, cosi', l'andamento qualitativo presentato in fig.XII.23f.

Dopo aver aggiustato tale andamento si tratta, anche in questo caso, di graduare il luogo; il valore di K' corrispondente alla intersezione del luogo con l'asse immaginario e' gia' stato determinato (K' = =70,656); mettendo a sistema la (XII.44) e la (XII.45) si possono trovare gli altri due punti del luogo corrispondenti allo stesso valore di K' che risultano $r_3 = -2 + j1,34$ e $r_4 = -2 - j1,34$. Nel proseguire la graduazione del luogo si potranno, anzitutto, trovare mediante la(XII.30) i valori di K' corrispondenti ai due punti doppi reali, dopo di che applicando ancora la (XII.44) e la (XII.45) si trovera', per ciascuno di essi, l'altra coppia di radici corrispondenti allo stesso K'. Si ritiene opportuno mettere in evidenza che per i tratti complessi del luogo e quindi, in particolare, per gli interi rami uscenti dai due poli coniugati ed intersecanti l'asse immaginario e' possibile determinare il valore di K' corrispondente a ciascun punto di uno solo dei rami, attribuirlo per simmetria anche al punto coniugato e calcolare, infine, le altre due radici che corrispondono allo stesso valore di K' con la (XII.44) e con la (XII.45). E' pero' possibile procedere anche sistematicamente alla graduazione del luogo applicando separatamente per ogni ramo la (XII.30).

→ *Regoli per il tracciamento del luogo.* Il tracciamento del luogo delle radici comporta la considerazione della (XII.28) e cioe' la misura e la somma algebrica di angoli; la sua graduazione implica invece l'applicazione della (XII.30) o della (XII.31) e cioe' la misura di lunghezze e la loro moltiplicazione e divisione.

Per eseguire queste operazioni si puo' ricorrere ad opportuni accorgimenti ed in particolare all'impiego di regoli, il piu' noto dei quali e' quello proposto da parte di

$$1 + j\,\omega\frac{\tau_i}{m_i} \over 1 + j\,\omega\,\tau_i \qquad (m_i > 1) \tag{XIII.33}$$

In questo caso la G_1 assume la forma:

$$G_1 = K_1 \frac{1 + j\,\omega\dfrac{\tau_i}{m_i}}{1 + j\,\omega\,\tau_i} \tag{XIII.34}$$

Si supponga allora che si sia scelto K_1 in relazione alle esigenze del comportamento a regime e che la funzione G_2, a titolo di esemplificazione, abbia un polo nell'origine ed un polo reale negativo. Il diagramma di Bode (asintotico per i moduli) della funzione K_1G_2 ha dunque l'andamento riportato in fig.XIII.20; esso attraversa l'asse del-

torno al polo (fig.X.8a) corrisponde un percorso (cfr.fig. X.8b) che lascia àlla propria sinistra il punto improprio cosi' come quello infinitesimo lasciava alla propria sinistra il polo; si puo' anche dire, piu' rapidamente, che la chiusura all'infinito va fatta in senso orario, se il percorso uncinato ha lasciato il polo alla sinistra. La rotazione corrispondente a questo percorso e' di tante volte π quante volte ne indica la molteplicita' del polo considerato.

Il caso corrispondente ad una coppia di poli coniugati (semplici) sull'asse immaginario (di ordinata $\pm j\omega_0$) e' esemplificato in fig.X.9 (1).

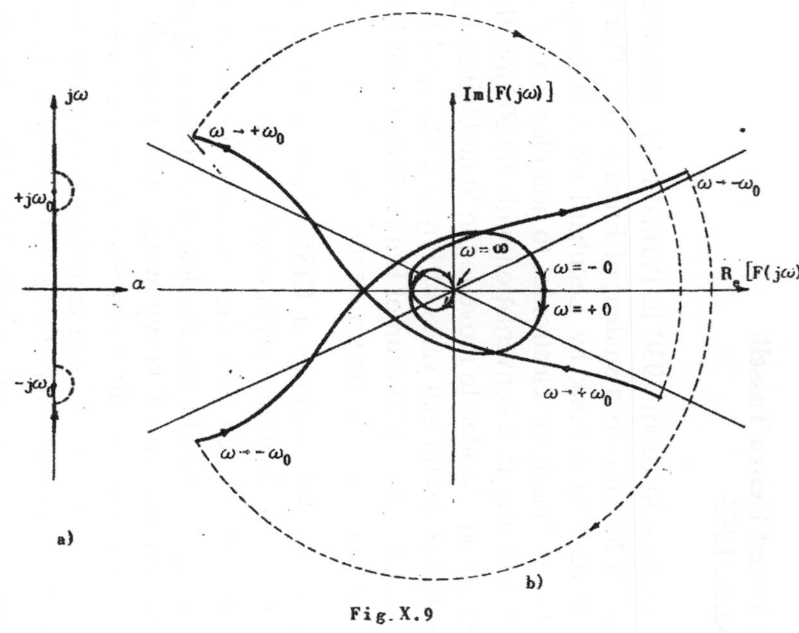

Fig. X.9

(1) - E' chiaro che scegliendo percorsi che lasciano alla propria sinistra i poli di F con parte reale nulla tali poli non devono essere considerati nel computo di P_p che figura nella (X.15). Se i percorsi uncinati fossero stati

Elementi di servomeccanismi e controlli

Sergio Barabaschi and Renzo Tasselli
Zanichelli, Bologna, 1965

In 1948, Sergio Barabaschi (born 1930 in Parma, Italy) enrolled in the Technical University of Milan (Politecnico di Milano) as a student of Engineering. During his studies, around 1950, he had the chance of reading a book by Leonhard on *Regelungstechnik*. It was this book that triggered his interest in "applying electronics to non-electronics", as he still refers to it today. He investigated this possibility in a conversation with the Rector, the geologist Gino Cassinis, and, following his suggestion, decided to move to the State University of Milan, where he completed his preparation with two additional years of Physics. His goal to write his thesis on control was achieved, in 1952, under the supervision of Emilio Gatti (born 1922 in Turin, Italy, now an IEEE Life Fellow). The subject was the control of the delay of certain trains of impulses.

In January 1953, he started working at CISE (*Centro Informazioni Studi ed Esperienze*), an advanced private research institution established in 1946 in Milan. The objective of CISE was to explore the possibilities offered by nuclear science for the generation of electric energy. The specific task of young Barabaschi was the design of the hydrogen temperature control of a small pilot plant for the production of heavy water. The project's leader was Mario Silvestri, a professor of technical physics in the *Politecnico di Milano*, who had a special gift for the depth and broadness of his technical and non technical knowledge. Barabaschi then moved on to the development of a dynamic simulator of a nuclear reactor under the coordination of Emilio Gatti, with whom, in June 1955, he visited Caltech in the US to present their result.

In April 1956, he took part in the *Convegno Internazionale sui Problemi dell'Atomatismo,* held in Milan, where he met many scientists including Leonhard, the author of *Regelungstechnik.*

In those times, there was increasing interest in nuclear reactors as a new source of energy, thus a center, the *Nuclear Research Center* (CRN - *Centro Ricerche Nucleari*), was created in *Ispra*, a town approximately 50 km north of Milan. Barabaschi moved to that Center where the primary goal was to set up the first Italian nuclear reactor, called ISPRA-1. In 1956, he spent a period at Allis Chalmers in Buffalo, N.Y., US, to prepare the control specifications for the new reactor.

On his return to CRN, he worked for some three years on the development of the *Laboratory of Servomechanism and Control* of that Centre, where more than 30 technicians and 20 engineers were employed. Among these engineers, there was Renzo Tasselli (born 1934), who joined in 1957. The task of that laboratory was twofold. One goal was to provide tools for the design of reactor control systems, and the other the development of precise motion control systems tools to operate the robotic arms that were to be used for handling radioactive material. Barabaschi made a major contribution to the development of the robot Mascot-1 (*Manipolatore Automatico Servocomandato e Transistorizzato*).

On March 25, 1957, the *Treaty Establishing the European Community* was signed in Rome. On the same occasion, the *Treaty Establishing the European Atomic Energy Community* (EURATOM) was also signed. It was then that Italy proposed the *Ispra Center* as a location for EURATOM, moving two laboratories, including the *Laboratory of Servomechanism and Control*, to another location. The *Laboratory* directed by Barabaschi moved to the *Centro Nucleare della Casaccia* (some 20 km north of Rome) in 1963. Here Barabaschi remained until

1975, when *Ansaldo*, a major Italian company located in Genoa, asked for his assistance in the design of nuclear reactors. Sergio Barabaschi then began commuting between Rome and Genoa. It was during this years that Renzo Tasselli passed away in a tragic car accident in 1978.

The years from 1980 to 1993 marked a new period of life for Barabaschi. He joined the *Ansaldo Company* in Genoa full time, where he was called to work on automation and control, the passion of his life. In 1994, however, the Minister of University and Research, Salvini, asked for his assistance in Rome as Under-Secretary at that Ministry. From this the step to European commitments was short; from 1996 to 1998, he served as President of the *European Industrial Research and Management Association*, located in Paris, and subsequently as President of the *European Science and Technology Assembly* in Brussels until the year 2000. Finally he acted as Vice President of the *European Council of Applied Science and Engineering* (operating in Paris, at *Académie de France*) until 2004.

The story of the book *Elementi di servomeccanismi e controlli* is as follows. Together with Tasselli, Barabaschi started working on the book in 1961 (during that period both authors were working at CRN in *Ispra*). The text was completed some three years later when Barabaschi and Tasselli were at *La Casaccia* and eventually published in 1965 by Zanichelli – Bologna.

The idea was to write a text on control in the easiest way possible, with strong emphasis on instrumentation, actuators and transducers, in order to supply an effective aid for the training of technicians. The volume consists of 747 pages and contains a huge number of photographs, block diagrams and drawings of various type, especially figures of transient behavior, Nyquist, Nichols and Bode plots, and so on. This required a incredible amount of work since most of these drawings had to be hand-made by the authors.

The book is organized in 23 Chapters plus 4 Appendices as follows. After an introductory chapter on automatic control, some basics of differential equations are introduced in elementary terms, with a number of electro-mechanical examples. Then first-order and second-order systems are dealt with in Chapters 3 and 4 respectively. Transfer functions are the subject of Chapter 5, while feedback is dealt with in the subsequent chapter. The steady state behavior is then studied, followed by stability analysis. The specification requirements and the main compensation methods are treated in Chapter 10, and then attention turns to "Nonlinear System Theory". This is indeed the title of Chapter 11, where various topics are treated including the concept of describing function. The remaining 12 Chapters are devoted to the components of control systems, transducers (of velocity, acceleration, force, temperature, pressure, level, flow), amplifiers (based on electronic tubes and on transistors), magnetic amplifiers, motors, hydraulic and pneumatic systems as well as mechanical components. There are also 4 appendices (Fourier analysis, Laplace transforms, Proofs of root locus rules and Nyquist criterion).

This book was a great success selling more than 30.000 copies in the subsequent ten years.

Contributed by

Sergio Bittanti
Politecnico di Milano

4 Collana di testi di elettronica diretta dal Prof. Alessandro Alberigi Quaranta

Sergio Barabaschi Renzo Tasselli

Elementi di servomeccanismi e controlli

Zanichelli Bologna

Z

4.5 Funzione di trasmissione di un sistema del secondo ordine

La funzione di trasmissione dell'accelerometro in esame si ricava immediatamente dalla (4.4.1) e vale:

$$F(j\omega) = \frac{1}{(j\omega)^2 + 2\zeta\omega_n(j\omega) + \omega_n^2} = \frac{1/\omega_n^2}{\left(j\dfrac{\omega}{\omega_n}\right)^2 + 2\zeta j\left(\dfrac{\omega}{\omega_n}\right) + 1}. \qquad (4.5.1)$$

Se consideriamo la grandezza $(j\omega)$ come una variabile, si vede che il denominatore della funzione di trasmissione (4.5.1), uguagliato a zero, coincide con l'equazione caratteristica dell'accelerometro.

Pertanto, ricordando che con P_1 e P_2 sono state indicate le radici di tale equazione, la funzione di trasmissione (4.5.1) può anche essere scritta nella forma:

$$F(j\omega) = \frac{1/\omega_n^2}{\left(1 - \dfrac{j\omega}{P_1}\right)\left(1 - \dfrac{j\omega}{P_2}\right)}. \qquad (4.5.2)$$

Vediamo adesso quale è la rappresentazione grafica della funzione di trasmissione dell'accelerometro.

4.5.1 Rappresentazione di Bode

Occorre distinguere due casi a seconda che ζ sia maggiore o minore di 1. Nella prima evenienza, come sappiamo, le radici P_1 e P_2 sono reali e negative; indicando con T_1 e T_2 due costanti di tempo (intrinsecamente positive) di valore uguale all'inverso del reciproco di P_1 e P_2, l'espressione (4.5.2) si può scrivere nella forma:

$$\frac{F(j\omega)}{1/\omega_n^2} = \frac{1}{1 + j\omega T_1} \cdot \frac{1}{1 + j\omega T_2}, \qquad (4.5.1.1)$$

$$T_1 = -\frac{1}{P_1}; \quad T_2 = -\frac{1}{P_2}.$$

Si vede quindi che la funzione di trasmissione in studio si ottiene come prodotto delle funzioni di trasmissione di due particolari sistemi di primo grado.

Poiché la rappresentazione del Bode di questi ultimi ci è nota, sapremo quindi ottenere anche la rappresentazione dell'equivalente sistema

Fig. 4.5.1.1 La figura b. riporta il diagramma di Bode delle ampiezze per un sistema del secondo ordine (con guadagno statico unitario) caratterizzato da due radici di valore $-\frac{1}{T_1}$ e $-\frac{1}{T_2}$.

Esso si ottiene come somma dei diagrammi relativi a due sistemi del primo ordine, caratterizzati dalle costanti di tempo T_1 e T_2, diagrammi riportati, a tratto e a tratto e punto rispettivamente, nella figura a.

del secondo ordine. In particolare, si noti che il prodotto di due grandezze complesse (quali sono per una determinata frequenza le funzioni di trasmissione), si ottiene moltiplicando le ampiezze e sommando le fasi. Nel caso però che le ampiezze siano espresse in decibel, cioè in una unità logaritmica, il loro valore andrà sommato per ottenere l'ampiezza del termine prodotto.

Le figure 4.5.1.1 e 4.5.1.2 riportano le rappresentazioni del Bode, approssimate secondo le regole indicate nel paragrafo 3.6.1, sia singolarmente per i due termini a fattore dell'espressione (4.5.1.1), sia per la funzione di trasmissione globale (a meno, quest'ultima del guadagno statico $1/\omega_n^2$).

Si noti che per frequenze sufficientemente elevate, l'amplificazione del sistema decresce con una pendenza di 40 db/decade, mentre lo sfasamento tende al valore di $-180°$. Per pulsazioni abbastanza inferiori a $\frac{1}{T_1}$, l'attenuazione vale 0 db e lo sfasamento 0°; il comportamento del sistema per pulsazioni intermedie è caratterizzato da due « frequenze d'angolo » in corrispondenza ai valori $\omega_{c1} = \frac{1}{T_1}$ ed $\omega_{c2} = \frac{1}{T_2}$ nella curva delle attenuazioni.

Nel caso limite di $\zeta = 1$ le due radici P_1 e P_2 sono sempre reali negative, ma coincidenti; da quanto sopra detto si ricava immediatamente che la curva approssimata delle attenuazioni, riportata in figura

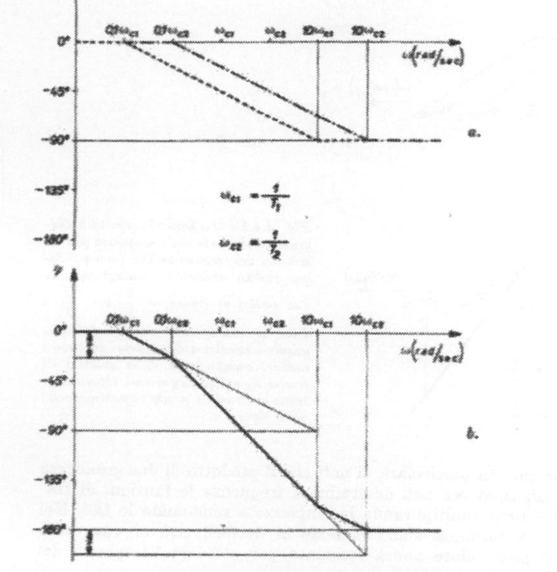

Fig. 4.5.1.2 La figura b. mostra il diagramma di Bode delle fasi per un sistema del secondo ordine caratterizzato da due radici reali di valore $-\frac{1}{T_1}$ e $-\frac{1}{T_2}$. Esso si ottiene come somma degli analoghi diagrammi relativi a due sistemi del primo ordine con costanti di tempo T_1 e T_2 rispettivamente (figura a.).

4.5.1.3, presenta una pendenza di 40 db/decade per pulsazioni superiori a $\omega_c = \frac{1}{T_1} = \frac{1}{T_2} = \omega_n$ ed un valore di 0 db per valori inferiori; si noti il corrispondente andamento degli sfasamenti.

Si consideri infine il caso di $\zeta < 1$; in corrispondenza le due radici P_1 e P_2 dell'equazione caratteristica divengono complesse, impedendo così l'applicazione delle regole che consentono una rappresentazione approssimata dei diagrammi del Bode.

Si usa quindi ricorrere, in questo caso, a delle famiglie di curve calcolate una volta per tutte, e riportate nelle figure 4.5.1.4 (diagramma delle attenuazioni) e 4.5.1.5 (diagramma degli sfasamenti); ogni curva è caratterizzata da un particolare valore del coefficiente di smorzamento

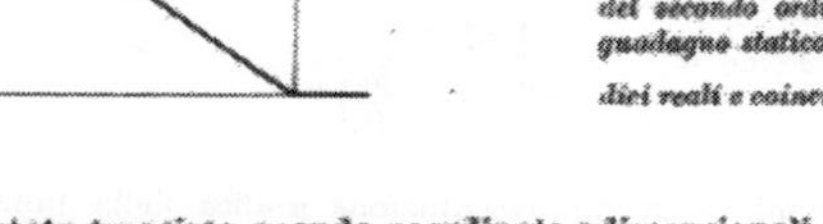

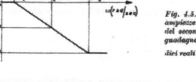

Fig. 4.5.1.3 Diagrammi di Bode delle ampiezze e delle fasi per un sistema del secondo ordine caratterizzato da un guadagno statico unitario e con due radici reali e coincidenti di valore $-\frac{1}{\omega_c}$.

ζ ed è stata tracciata usando coordinate adimensionali che ne permettono un impiego universale.

Esaminando le curve dell'attenuazione, si può notare la presenza di un fenomeno che non avevamo finora incontrato: la risonanza. Tale fenomeno consiste nella esaltazione della risposta in corrispondenza di una determinata frequenza detta appunto frequenza di risonanza del sistema.

Nel caso che l'entrata presenti una frequenza di tale valore (o di valore sufficientemente vicino) l'uscita assume ampiezze maggiori di quelle presenti nel caso statico (o nel caso di frequenze di eccitazione particolarmente basse).

Si può dimostrare che la presenza della risonanza è legata alla presenza di radici complesse nell'equazione caratteristica del sistema in esame.

La pulsazione ω_R in corrispondenza della quale si ha la massima esaltazione dell'uscita, la pulsazione di risonanza appunto, si può ottenere in funzione del coefficiente di smorzamento mediante l'impiego del diagramma di figura 4.5.1.6, o analiticamente dalla relazione approssimata:

(4.5.1.2) $$\omega_R \simeq \omega_n \sqrt{1 - 2\zeta^2}\,.$$

La pulsazione di risonanza presenta quindi il suo valore massimo per $\zeta = 0$ nel qual caso coincidono i valori: frequenza di risonanza, frequenza naturale, frequenza naturale non smorzata. Si noterà che tanto minore è lo smorzamento e tanto più accentuato è il picco di risonanza.

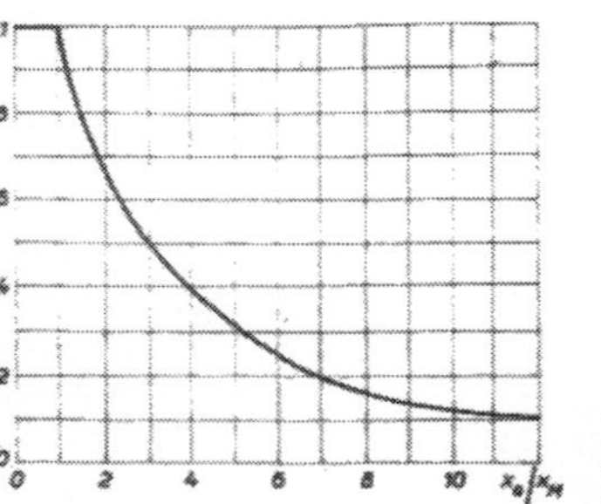

Fig. 11.5.11 Andamento della funzione descrittiva $B(x_0)$ del sistema rappresentato dalla caratteristica di fig. 11.5.9.

sariamente esso si satura quando il segnale di ingresso supera un certo valore.

Consideriamo infine il caso di un sistema del tipo precedente nel quale però sia presente oltre alla saturazione anche una zona morta $\pm x_d$; la caratteristica relativa è quella riportata in fig. 11.5.12.

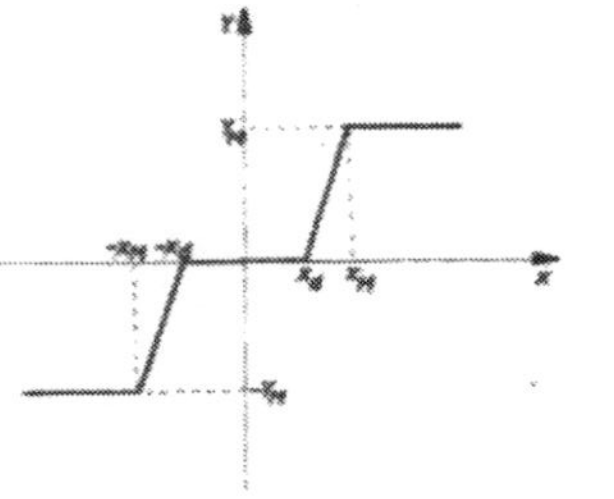

Fig. 11.5.12 Caratteristica uscita (y)-entrata (x), di un sistema con zona morta e saturazione.

La funzione descrittiva $B(x_0)$ dipende in questo caso da tre parametri x_M, x_d, K, visto che il massimo valore dell'uscita è dato dall'espressione

$$Y_M = K(x_M - x_d)\,.$$

La funzione $B(x_0)/K$ in funzione di x_0/x_M, è riportata per diversi valori di x_M/x_d nella fig. 11.5.13.

Fig. 11.5.13 Andamento della funzione descrittiva $B(x_0)$ per il sistema rappresentato dalla caratteristica di fig. 11.5.12. Con K si è indicato il guadagno corrispondente al tratto lineare della caratteristica citata.

Il caso x_M/x_d infinito con $x_d \neq 0$ è quello di un sistema lineare con sola zona morta.

Finalmente il caso limite $x_M = x_d$ è quello di un sistema on-off con zona morta $\pm x_d$.

11.6 Stabilità di un sistema non lineare

In questo paragrafo studieremo come si possa calcolare la stabilità di un sistema controreazionato che contenga un blocco non lineare *N.L.* facendo riferimento a quanto esposto nei paragrafi precedenti.

Si consideri il sistema di fig. 11.3.4 e si assuma, per il momento, che il blocco non lineare *N.L.* sia sostituito con un blocco lineare dove il guadagno K possa essere variato (vedi fig. 11.6.1).

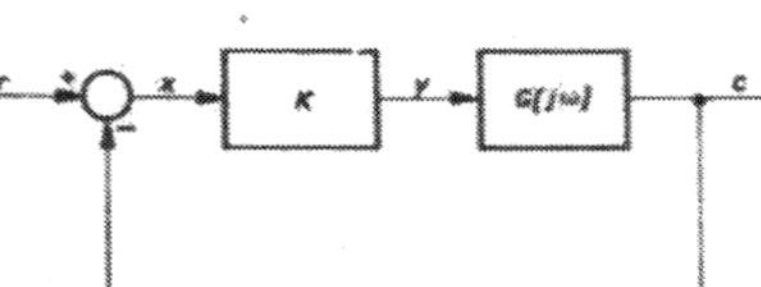

Fig. 11.6.1 Schema di un anello di regolazione con guadagno K variabile.

EARLY CONTROL TEXTBOOKS
in
JAPAN

Jidouseigyo no Riron to Jissai (Control Theory and Practice)

Author: Takeshi Samukawa
Date of Publication: September 1948

Jidouseigyo Nyumon (Introduction to Control Theory)

Author: Keisuke Izawa
Date of Publication: May 1954

Jidouseigyo Rironi (Automatic Control Theory)

Author: Yasundo Takahashi
Date of Publication: October 1954

Contributed by

Katsuhisa Furuta

Tokyo Denki University
Japan

Jidouseigyo no Riron to Jissai (Control Theory and Practice)

Takeshi Samukawa
JSME(Japanese Society of Mechanical Engineers), September 1948
Tokyo Japan

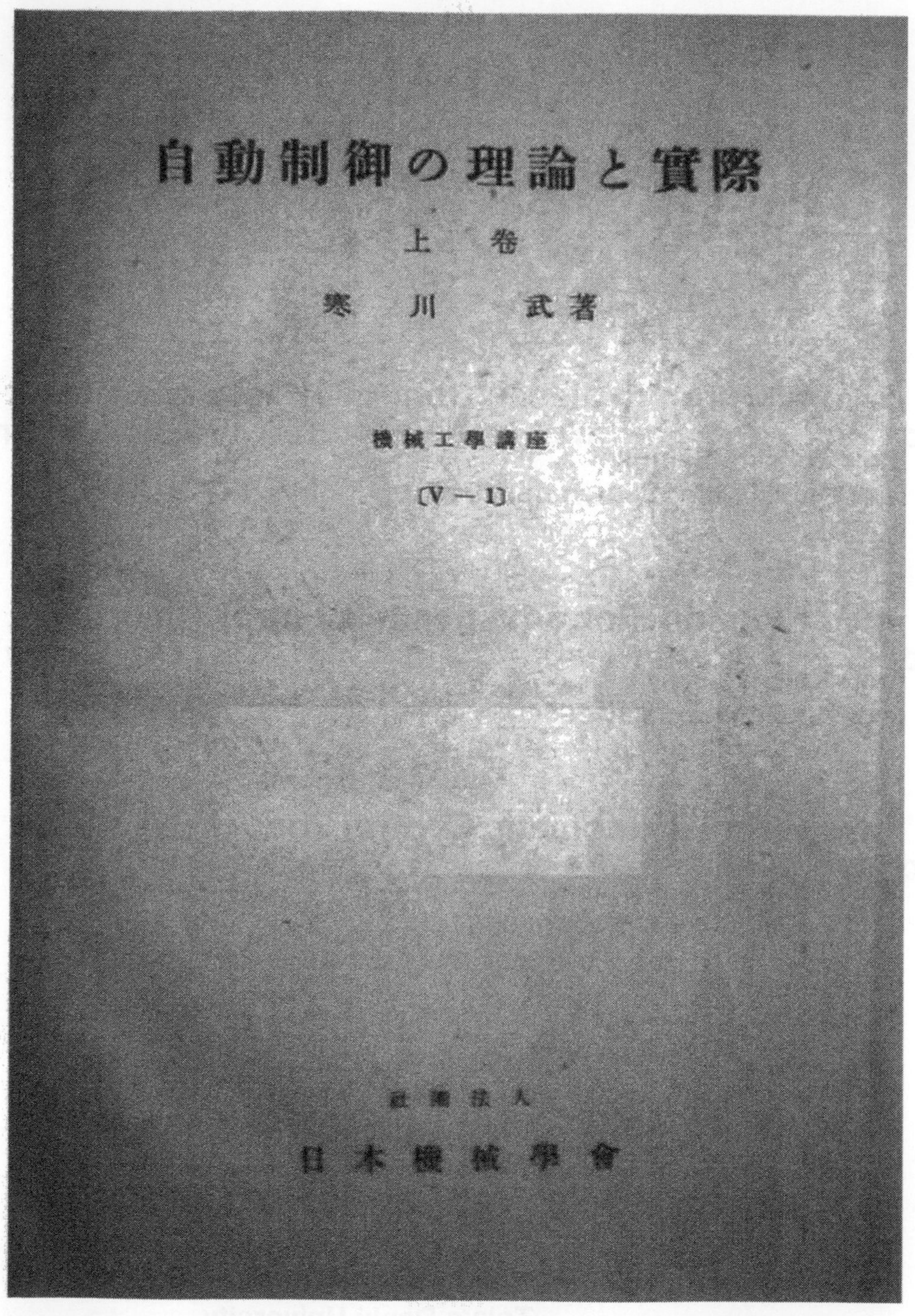

The author was a control engineer who worked for an Askania Japan Co. from 1935 to 1944. His health condition was not good in the time of World War II and he became a lecturer of Tokyo Institute of Technology by the recommendation of Professor Shigeo Sasaki. He engaged in writing his thesis to submit as the Doctoral Dissertation to Tokyo Institute of Technology, but unfortunately passed away in December 1945. The degree was not awarded. The contents were so new at that time for the Japanese engineers that the Japan Society of Mechanical Engineers (JSME) published the thesis as a book. The preface was written by Prof. S. Sasaki. The original handwritten dissertation itself was a treasure for his family, therefore it was hand transcribed by his friend for publication. Since most of the research was done while he worked for a German related company and in the time of World War II, the references were several German papers, and a few American/English papers. Unfortunately, he was not known due to his short life but he was one of the researchers in control engineering in the pre-history of control engineering in Japan.

The book consists of two volumes. The first volume had the following four chapters in 202 pages.

Chapter 1, Introduction

Chapter 2, Structure and Classification of Control Systems
(Static and astatic, Set point and servo control, Feedback and feed forward)

Chapter 3, Elements of Control Systems
(Electric and mechanical sensors, Actuator valves, controller)

Chapter 4, Control Theory
(Control methods, Stability degree, Control error and inverse response, Remarks on control methods, Dead time, Dead band, Friction and stiction, On-off control, Theory and practice.) Most of the mathematical models of plants are represented by differential equations. In this section, Hurwitz stability condition and Nyquist stability criterion are described. Dead time and nonlinear elements like dead zone, hysteresis and others were also considered.

The second volume has the following two chapters in 210 pages.

Chapter 5, Various Control Problems
(Control of rotational velocity, Control of pressure, Pressure control theory, Control of flow rate and pressure difference, Servo control I (Flow rate and pressure), Servo control II (Position and angle), Level control, Temperature control, Moisture control, Composition control, Directional control of vehicle, Trajectory control, Electric control)

Chapter 6, Evaluation of Control Theory through Experiments and Remarks on Previous Control Theory (Artus, Tolle, Oppelt, Stein-Wunsch)

The book was owned by the late Professor K.Izawa and its manuscript has been found.

Typical pages of the book are shown below. The first gives controlling function. The second shows the transient response and the third one is explaining the dead zone.

管の内部に収容される熱電對を用いる工業用温度計に比較すると後れが遙かに少い．また汽罐の水位の検出に熱膨脹管をいわゆるゲージガラスの代りに用い，この管の伸縮を利用して制御を行うことがある（コープス式水位制御）．この方法によると，水位の下降時の検出の後れが水位上昇時に比し遙かに少く，それが汽罐運轉に都合がよいとして好んで採用されている．

かくのごとく検出部の作動の後れの除去に多くの努力が拂われるのであるが，制御設定値を決定する設定部も検出部の一部を構成する關係上同様の趣旨をもつて構成される．たとえば設定部はばねまたは重錘を利用することが多いのであるが，重錘の使用はばねと異なり，ヒステレシスや經年變化を伴なわないので，計器特に0點式計器では好まれるのであるが，制御の場合には検出部の固有振動數が重錘を用いると低下する上に検出部が無定位的に作用する傾向を増し制御が不安定となるのでばねによつて設定するごとくする．

最近の傾向として，機械量を電氣的に検出することが多くなつて來ている．この場合，距離の制限がなくなり非常に便利ではあるが，單に在來の検出法を用い，その検出量を電氣的に變換するようなことをすると，機械量と電氣量との變換を行う部分の作用の後れが徒に追加されるおそれがあるから，電化に當つては制御作用をそのような後れのために悪化させぬように留意が必要である．

3・2　管制部の種類と構造

管制部は検出部の微弱な作動に應じて強力な操作を管制支配する部分であつて，構成それ自體としては單なる増幅装置に外ならぬことは Kniehahn の指摘する [31] 通りであるが，制御作用を決定する最も偶要な役割をもつ．

構造上具備すべき必要條件を擧げると下記のごとくである．

（1）摩擦および慣性に阻害されること少く，検出部に輕快に應動すること．

（2）補助動力によつて反力を生じ，そのために検出作用が亂されるようなことのないこと．

（3）管制特性は連續的で，なるべく廣い範圍に亘つて直線的であること．すなわち検出量の大小に應じて操作作用も變化すること．

（4）十分な微速に至るまで圓滑な操作が可能で，できれば微速のときほど操作力が強大であること．

管制部の構造は検出部のそれのごとく制御目的によつて一々相違することはなく，その種類は大體において限られている．しかも制御装置中最も特徴のある部分であるから，以下各型式について詳述することにする．

管制部は使用する補助動力の種類によつて，純機械式・壓力流體式・電氣式の3種に大別す

ることができる．

なお以下においては，検出法が電氣的手段によるものであつても，管制部が油壓利用の壓力流體式に屬するものであるときにはこれを飽くまでも油壓式として取扱うことにする．たとえば近時調速機の電化に關してしばしば論ぜられているが，その多くは検出法の電化を對象としており [30]，われわれからみれば，これは依然油壓式の範疇に入れて論ずるものとする．

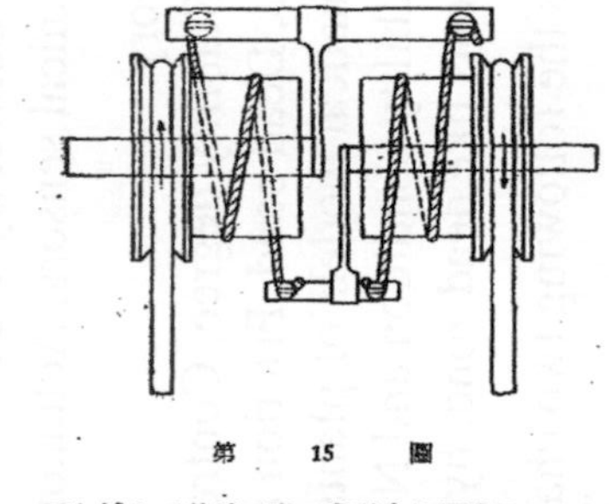

第　15　圖

3・2・1　純機械式

純機械式の管制部にはクラッチ機構を應用した不連續特性をもつものと，摩擦驅動を利用した連續特性をもつものとがある．第15圖は後者の例であつて，この機構は現在主としてトルク增力機として非饋還型の自動追尾装置および計算機械の一部等に利用される．

このトルク增力機の作動原理は，第16圖に示すごとく，船舶等の卷上機の卷胴（絞盤）にロープを卷きつけロープの一端を弱い人力で加減して他端の強い牽引力を調節するのと全く同じ原理に基ずいており，この構成が順逆の兩方面の回轉用に2組採用されている．

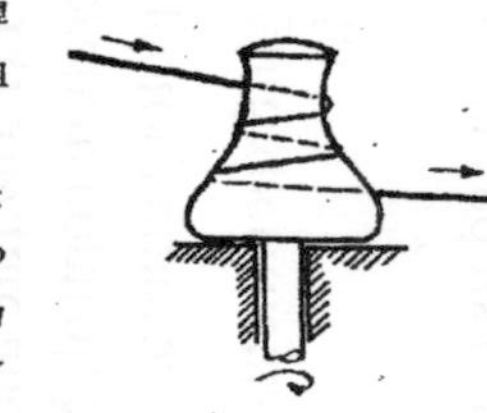

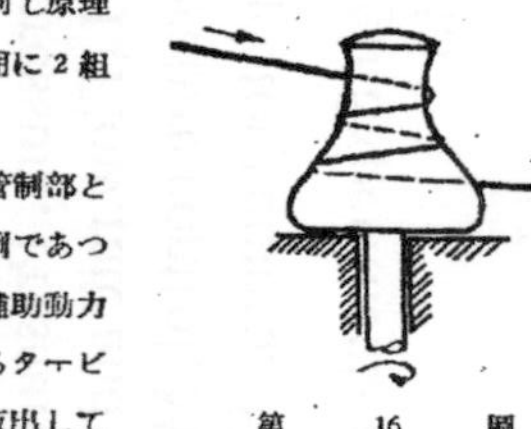

第　16　圖

第17圖のハルトマンの調速装置は摩擦接手を管制部として利用せる例であつて，この場合補助動力は被制御系たるタービン側より直接取出している點と差動齒車を應用して操作を行つている點が注目される．

第　17　圖

これらの機械的管制部は現在のところでは自動制御用としてはほとんど使用されていない．その原因として，動力の遠方傳達の不便・發熱に伴なう磨耗と摩擦係數の變化・必ず多少の不動帶（4・6節參照）を與える必要のあること等を擧げることができる．

3・2・2　壓力流體式

案内弁・噴射管・絞管制弁・

係數 b_1 の大きい系では變化が緩かに始まつてその後定常値に落ちつくのであるから，その意味でこの場合 b_1 は一般の後れの觀念と一致する．

係數 b_0 は定常値に落ちつき得るや否やを決定し，b_0 の大きい系では (c) 式より明かなごとく操作を大きく行つてもその割に變化 φ の定常値は僅かしか上下しない．すなわちこのように係數 b_0 は系が自身で平衡せしめるごとく作用するから，この種の被制御系は固有平衡性 (Selbstregelung, inherent self-regulation) をもつと稱せられる．

$\varphi \cdot \mu \cdot \eta$ を%で表わすときには，係數 b_1 は前節と同様に時間のディメンションをもつこととなり始動時間 T_a を表わし，係數 b_0 は無次元數となり“固有平衡率”(Ausgleichsgrad) と呼ばれることがある．

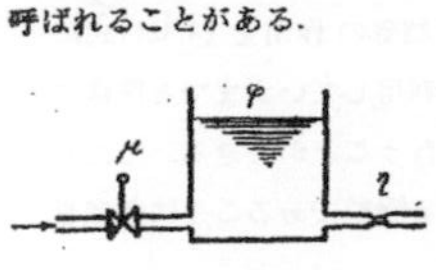

この系は電氣回路でいえば，C―R 回路における電壓に對する電流の關係に相當する．貯水槽による等價系を作れば第86圖のごとくこの場合も單一蓄積系を形成するが，この場合溢出堰が存在するために固有平衡性が生ずる．電氣工學方面では係數 b_1 を時定數と呼んでいる．

第 86 圖

この“壓力型”の被制御系が制御問題中最も多く現われる．Oppelt の例に倣つて，こゝでも壓力制御の被制御系によつて代表させることにする．ガス壓力制御の場合，操作 μ は加減弁の開閉によつて與えられるが，擾亂 η は減壓制御の場合ならばその加減弁前方の壓力變動やガス使用量の變化によつて與えられる．

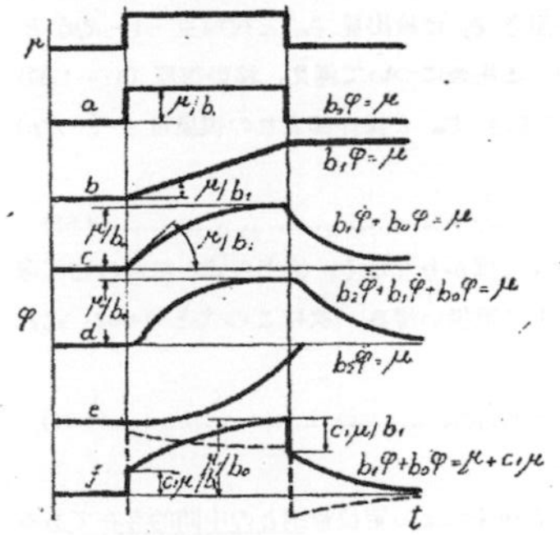

また溫度制御や回轉數制御（調速）の場合にもこの壓力型の系が多く現われて來る．さらに電氣量の制御，たとえば電壓の制御の場合の被制御系もこの型と見なし得ることが多い．

係數 b_1 および b_0 の決定は實驗によつて行えばよいが，係數 b_1 は過渡的な變化現象に關連して決定しなければならないので壓力制御のように b_1 の極めて小さい場合には，第5章で明かにする理論的計算より求める方がよい．これに對し溫度制御の場合のごとく係數 b_1 が大きく，しかも計算によつては正確に求め難い場合には下記の實驗による方法によつて求めるのが正確である．係數 b_0 は定常狀態によつて求めるものであるから，實驗的

第 87 圖

に定めることは容易である．

まず係數 b_0 は (c) 式より明かなごとく，

$$b_0 = \bar{\mu}/\varphi_{t=\infty} \quad \cdots\cdots (d)$$

によつて決定できる．次に蓄積能に對する係數 b_1 は變化曲線より次のごとくして求められる．

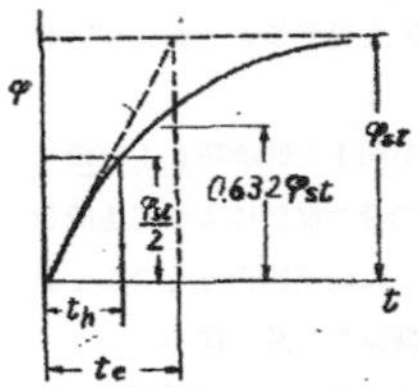

その第1の方法は第88圖に示すごとく，定常變化 $\varphi_{t=\infty}/2$ を起すに要した時間 t_h を求めて，

$$b_1 = 1{\cdot}44 \cdot b_0 \cdot t_h \quad \cdots\cdots (e)$$

なる式より決定する．時間 t_h は半減時間 (Halbfallzeit) とも呼ばれる．

第2の方法は定常變化 $\varphi_{t=\infty}$ の 63·2% の變化を行うに要した時間 t_e を求めることによつて計算する方法であつて，

$$b_1 = b_0 \cdot t_e \quad \cdots\cdots (f)$$

第 88 圖

なる關係式を利用する．

d. “溫度型”の被制御系

$$b_2\varphi'' + b_1\varphi' + b_0\varphi = \mu - k_0\eta \quad \cdots\cdots (4.21)$$

$$b_2\varphi'' + b_1\varphi' + b_0\varphi = \mu - (k_0\eta + k_1\eta') \quad \cdots\cdots (4.22)$$

この系に對して單位函數的な操作 $\mu = \bar{\mu} \cdot 1$ を與えると，變化 φ は第87圖の d 曲線のごとく減衰振動的にまたは非周期制動的に行われる．その意味でこの系は振動型の檢出部と類似し，係數 b 間の相互の關係如何によつて減衰振動的（判別式 $b_1^2 - 4b_0b_2 < 0$）にもなり，また非周期的（判別式 $b_1^2 - b_0b_2 > 0$）ともなる．操作をもとに戻せば變化も 0 に戻るが，これは固有平衡性の係數 b_0 が存在するためである．

この系は電氣回路で云えば $L-R-C$ 回路に相當し，また貯水槽による等價系を作れば第89圖に示すごとき二連蓄積系となる．

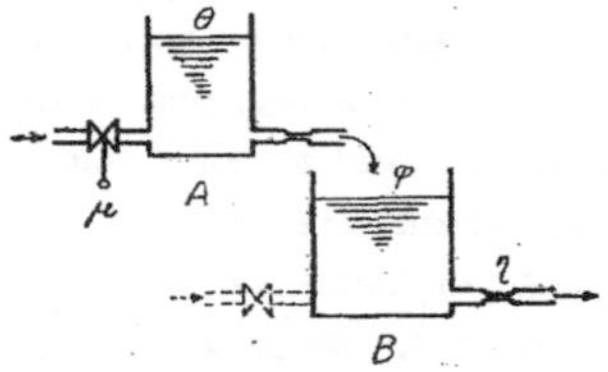

圖のごとく二連蓄積系として構成されているときには，單位函數的な操作に對して生ずる變化は非周期的であつて減衰振動的であり得ないことが次のごとく證明される．

［證明］ 二連蓄積系のときには圖より明かなごとく壓力型の系が2個直列に配置されて構成される．從つて，

第 89 圖

第1の“壓力型”の系 A $\quad b_{1A}\theta' + b_{0A}\theta = \mu \quad \cdots\cdots (a)$

第2の“壓力型”の系 B $\quad b_{1B}\varphi' + b_{0B}\varphi = \theta - \eta \quad \cdots\cdots (b)$

の結合であつて，A の系に生じた變化 θ がさらに B の系に影響して問題にしている變化 φ

4·6 不動帯の影響

以下においては管制部の不動帯が制御に對して如何なる影響を與えるかを考察する．

現實の管制部では多少の差はあるが不動帯をもつているのが普通である．特に，繼電式管制部は管制特性が不連續であるため 0 點を中心としてかなり大きい不動帯をもつている．また案内弁のごとき連續特性をもつ管制部でも正の重合（オーバラップ）を與えるとそれに相應せる不動帯が發生する．

從來より通常の制御の場合には，不動帯を與えると制御の安定性が向上し亂調防止上有効であることがしばしば經驗されている．そのために，亂調の防止を不動帯の賦與によつて達成せんと試みることも稀でなく，時には繼電式の制御裝置の製作者のなかには不動帯をもつことをもつてその裝置の利點の一つとして擧げているような我田引水的な議論さへ見受ける．

これに反し，原動機の調速の場合等では不動帯のために制御の安定度がかえつて低下し，亂調を誘起することさえあるのが經驗されている．

從つて以下において不動帯の制御におよぼす影響を理論的に考察して，上記の矛盾の起る機構を明かにし，さらに不動帯の取捨に對する指針を樹立する．

4·6·1 制御方程式　最初に，それ自體としては無定位的に作動するごとく構成されている管制部が不動帯をもつ場合について考える．復原その他の手段によつて無定位式のものが定位管制部として作動する場合は，無定位管制體に對する考察を擴張して論じ得る．なお本來の定位式管制部に不動帯を與えることは全く意味のないことであるから考察から除外する．

無定式の管制部では操作に慣性の影響が表われて來ない限り，檢出量 ξ に對して操作速度 μ が定まり，もし不動帯がなければ，第118圖の點線 AB に示すごとく $\xi-\mu$ 特性は直線をもつて表わし得る．しかるに不動帯が 0 點を中心として X なる幅をもつときには，$\xi-\mu$ 特性は實線 $CDEF$ となるものと考えることができる．ただしこの場合もいままでと同様に特性曲線は飽和しないものとしてその影響は考慮に入れないことにする．

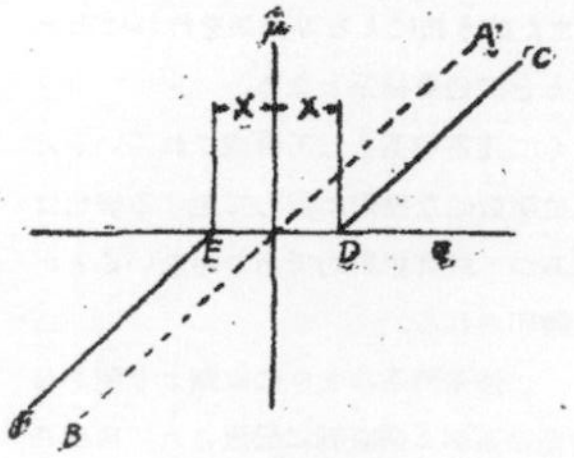

第 118 圖

いま管制部の動きを ξ_p，操作部の動きを μ とすれば，不動帯のない場合の $\xi-\mu$ 特性すなわち第 118 圖の AB は (4.10) 式，すなわち，

$$-\mu=K_p\xi_p \qquad (a)$$

によつて表われる．この式においては鋭敏度に對する係數 K_p は ξ に無關係な常數である．

次に不動帯のある場合の $\xi-\mu$ 特性すなわち第 118 圖の $CDEF$ を同様の形の式によつて表わすことにする．すなわち，

$$-\mu=xK_p\xi_p \qquad (4.82)$$

この場合には係數 K_p は從前通りやはり常數であるが，係數 x は ξ_p によつて變る變數であつて，

$$\left.\begin{array}{lll} -X<\xi_p<+X & \text{に對し} & x=0, \\ \xi_p>+X & \text{に對し} & x=1-(X/\xi_p), \\ -X>\xi_p & \text{に對し} & x=1+(X/\xi_p) \end{array}\right\} \qquad (4.83)$$

によつて表わされ，x は ξ_p に對して第 119 圖に示すごとく變化する．

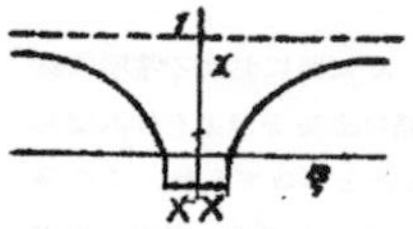

第 119 圖

このように不動帯をもつ無定位管制部の作用を (4.82) 式によつて表わすことにすれば，これを利用していままでと同様に制御方程式を誘導して理論的考察を行うことができる．ただし，係數 x は常係數でなく ξ_p に關する變數であることに留意を要する．

まず無定位管制の場合には，檢出部の動き ξ_m が管制部の動き ξ_p に相當するから，

$$\xi_p=\xi_m \qquad (4.84)$$

と置いて論ずればよい．

剛性復原による定位管制の場合には，管制部の動き ξ_p は檢出量 ξ_m と復原量 $-(-R\mu)$ との和となる．たとえば，管制部として案内弁を用いる場合について考え，移動復原 (3·4·1 節) でポートの開きを ξ_p とすると，ξ_p は檢出部の動き ξ_m と操作部よりの復原量 $-(-R\mu)$ との和である．すなわち，

$$\xi_p=\xi_m-(-R\mu)=\xi_m+R\mu \qquad (4.85)$$

が成立する．この式は移動復原の場合に對し成立するばかりでなく，檢出の式と組合わせて考えると設定復原に對しても同様に成立するものとして取扱い得る．次にこの式と (4.82) 式とを組合わせて，

$$-(\mu+xRK_p\mu)=xK_p\xi_m \qquad (4.86)$$

なる剛性復原の場合に對する管制操作の式を得る．

彈性復原に對しての考察は，これが無定位管制と復原による定位管制との中間的存在であるとして，こゝでは論ずることを省略するが，以上と全く同様の方法で不動帯をもつ場合に對する作用式を誘導することができる．

不動帯の制御におよぼす影響を論ずるに當り議論を簡單にするために，以下においては檢出

Jidouseigyo Nyumon (Introduction to Control Theory)

Keisuke Izawa
Corona Publishing Co., May 1954
Tokyo Japan

The author was said to be the first control researcher receiving the Doctor of Engineering Degree in control engineering. He was born in Gunma Prefecture in 1925 and graduated from Department of Measurement, The University of Tokyo, in 1948. He wrote the book "Introduction of Automatic Control" in 1954 at the age of 29. After serving as a lecturer at the University of Tokyo, he became an associate professor at Tokyo Institute of Technology in 1954 and full professor in 1962. He was an associate professor at Purdue University. He attended the first meeting of IFAC in Heidelberg, and attended the IFAC Congresses until he passed away by an automobile accident in March 1970. He was a very energetic and very active researcher in control theory. His research interest was bang-bang type control and he showed that the sampled data system with a bang-bang type element has the aperiodic like chaotic response.

The book was revised 4 times and has been the best sold book in control theory in Japan. It contains 6 chapters, addition and appendix. The uniqueness of the book is the introduction of the slide rule to calculate the gain and phase of linear systems, which was commercially sold by the Hemmi Slide Rule Campany. The following introduction of the book is based on the last version.

The first chapter describes the basic idea of the control and its brief history. Chapter 2 presents the fundamental ideas of control theory like Laplace Transform, Block Diagram, Frequency Response, Loop Transfer Function and Closed Loop Transfer Function, Slide Rules for computation of the gain and phase of the frequency characteristics, the calculation of the frequency characteristics from the transient response and the transient response from the frequency characteristics. Chapter 3 gives the stability and its condition, Routh condition, Hurwitz condition, Leonhard condition, Root Locus and Logarithmic Root Locus method. Chapter 4 gives control performance and characteristics such as parameters of steady state error, transient response, frequency characteristics, and r.m.s condition. He studied the gain and phase margins of the loop transfer functions for past examples and from their results their choices for servo mechanisms and process control are given. Chapter 5 presents the synthesis problem of control systems. The objective and procedure of control system synthesis are described first, then tuning of the gain, addition of compensation, cascade control system, and series and feedback compensation are written. The feedback compensation is shown to compensate the time delay, which describes the basic principle of the Smith Predictor. The synthesis problem is studied using a process model as an example, and cascade control is applied. Chapter 6 deals with discontinuous control including the bang-bang type and discrete-time control. Examples of discontinuous systems are given. Then the discrete-time control system is explained including Z-transform. The additional materials include the Routh condition and its calculation, the dynamics of gear train, how to choose P, PI, PID and I controller for the given process model, and correlation function.

OHM文庫
自動制御入門
(改訂新版)
工学博士
伊沢計介著
オーム社

Thus the book covers almost all control approaches for single input single output systems before the introduction of state space. The book was the most popular control textbook for a long time, until the author passed away, and it was translated into English (Introduction to Automatic Control, Elsevier, 1963).

The book had been a best seller in engineering books, and was especially useful for process control engineers. It is now out of print but had been read by many control engineers for a long time.

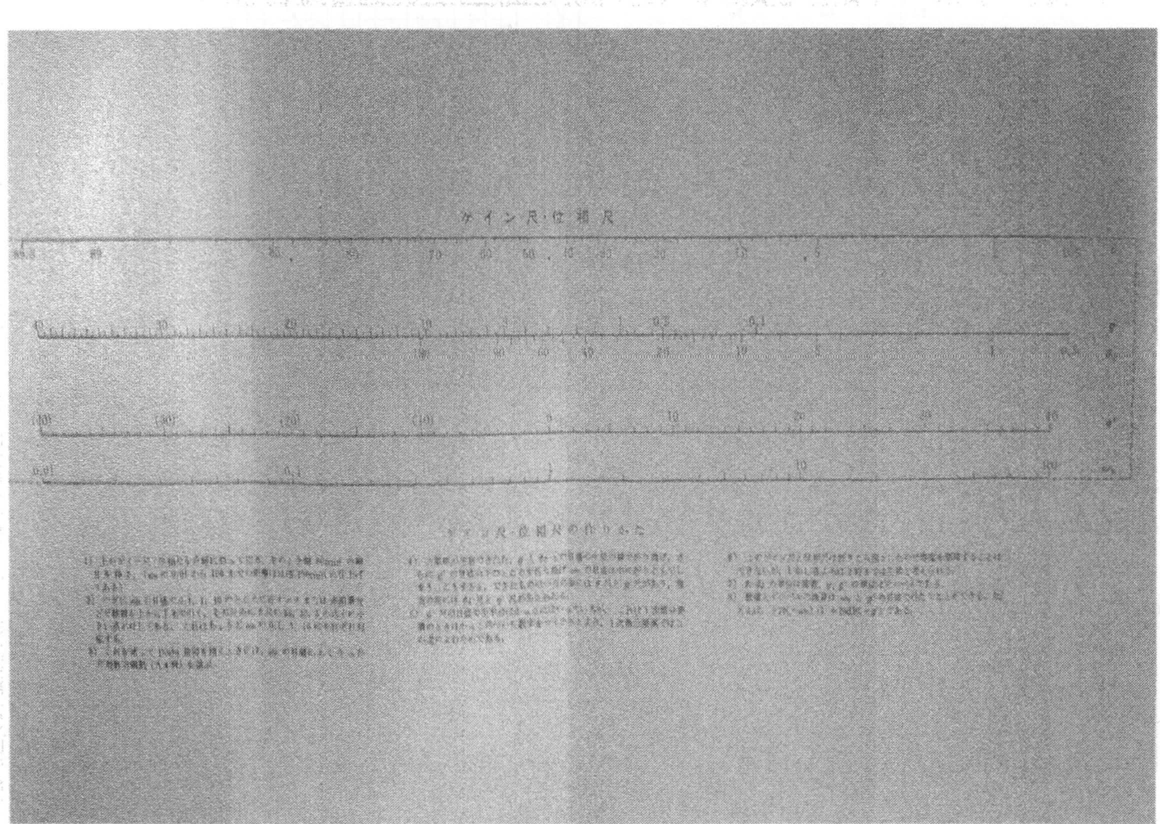

Slide rule explanation in the book

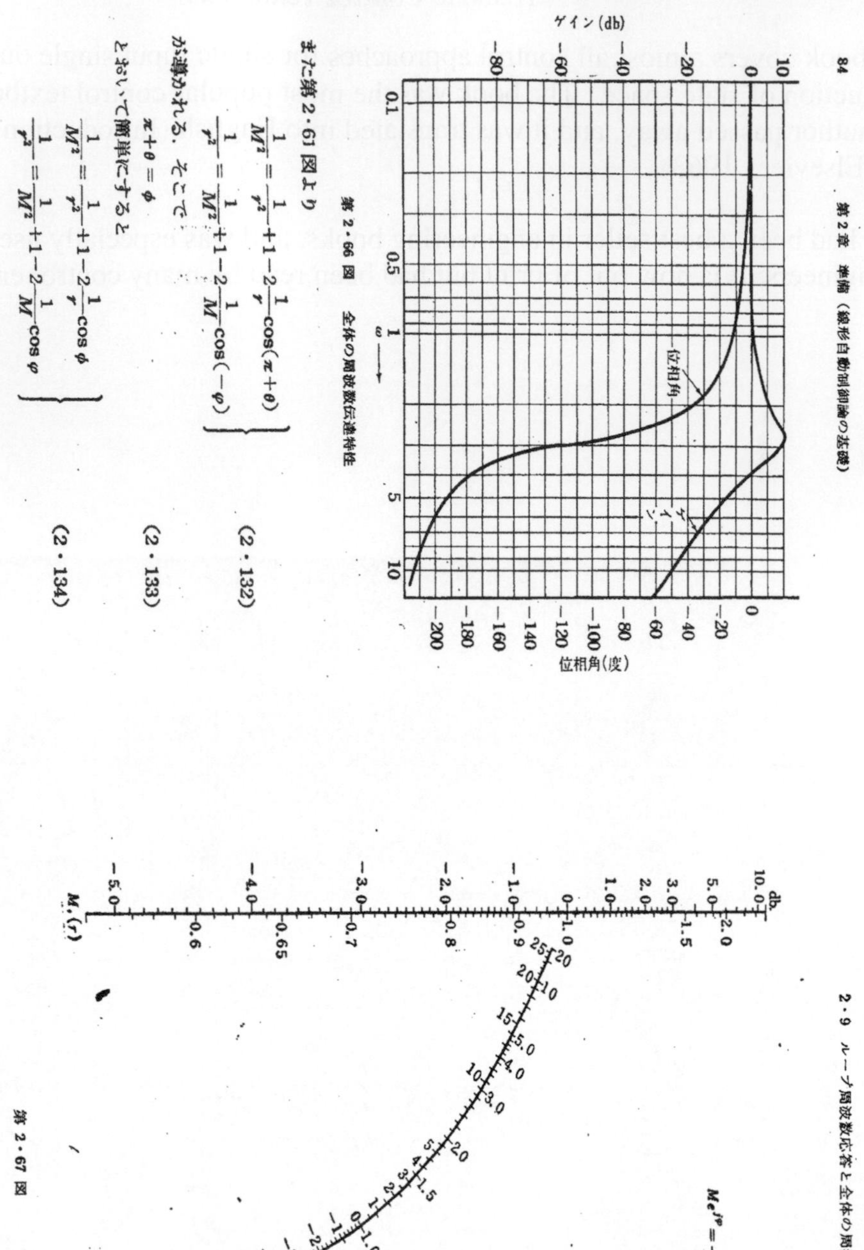

84　第2章　準備（線形自動制御論の基礎）

第2・66図　全体の周波数伝達特性

また第2・61図より

$$\left.\begin{aligned}\frac{1}{M^2}&=\frac{1}{r^2}+1-2\frac{1}{r}\cos(\pi+\theta)\\ \frac{1}{r^2}&=\frac{1}{M^2}+1-2\frac{1}{M}\cos(-\varphi)\end{aligned}\right\}\quad(2\cdot132)$$

が導かれる．そこで

$$\pi+\theta=\phi\quad(2\cdot133)$$

とおいて簡単にすると

$$\left.\begin{aligned}\frac{1}{M^2}&=\frac{1}{r^2}+1-2\frac{1}{r}\cos\phi\\ \frac{1}{r^2}&=\frac{1}{M^2}+1-2\frac{1}{M}\cos\varphi\end{aligned}\right\}\quad(2\cdot134)$$

2・9　ループ周波数応答と全体の周波数応答　85

$$Me^{j\varphi}=\frac{re^{j\theta}}{1+re^{j\theta}}$$

第2・67図

A typical page in Chapter 2 is the calculation diagram to compute from the loop transfer function to the closed loop transfer function.

y が $+A$ あるいは $-A$ という一定値をとるような制御装置による温度制御系について考える．第6·20図はこの系をブロック線図に示したものである．サンプリング機構がなければ，この不連続制御系は空間的な不連続性のみをもつたものであり，第6·14図に示した不連続制御系で動作すきまのないものとなる．

第6·20図　落下枠式オンオフ制御系

順序としてまずはじめにサンプリング機構（落下枠）がなくて常時連続的に（間歇的でなく）動作信号を監視し，その正負によりオンオフ操作が行われる場合の考察を位相面において行う．この場合には（6·25）～（6·29）と同様にして無次元時間 $\tau=t/T$ を用いると

$$z'+z=\begin{cases}-AK: & z(\tau-\lambda)>0\\ +AK: & z(\tau-\lambda)<0\end{cases} \qquad (6\cdot45)$$

$$\lambda\equiv L/T \qquad (6\cdot46)$$

を得る．これを z'，z を両座標とする位相面に描くと第6·21図のようになる．ここに得られた特性線は第6·16図のものと同じであるが，動作すきまがない（$h=0$）ので第6·16図のA点，D点はこの場合には z' 軸上に乗つている．それでもやはり第6·16図のときと同じくこの図でもみられるとおり，定つた振幅の持続振動をくりかえす．

さてつぎにサンプリング機構がつけ加わつた場合を考えよう．このときはオンオフの制御操作は落下枠がおりた瞬間の動作信号

の符号の正負によつてのみなされる．そしてその他の時刻での動作信号に対してはなんら制御動作を起さない．たとえば状況点P（第6·21図）が右側の特性線上を左上へ移行していくときに，サ

第6·21図

ンプリング機構はP点の縦座標 z を無次元時間で Tr/T ごとに検出している．もしもP点がA点へ到達した直後にサンプリングが行われたならば，状況点Pはそれよりむだ時間 λ だけ経過してB点に達するとただちにC点へとびうつる．しかしもしもP点がA点へ到達する直前にサンプリングが行われたならば，それから後にP点がA点を越えて左上へ移動していても制御操作の切り換えは行われない．そしてつぎのサンプリングの機会（A点から T_r/T だけたつた時刻）にはじめて $z>0$ であることが検出され，制御操作の切換えが行われる．したがつて状況点はその時刻からさらにむだ時間 λ を経過してからもう1つの特性線上へとびうつる．

第6·22図は以上のことをまとめたものである．すなわち状況点がA点からG点およびD点からH点まで移行するのに要する

A typical page of Chapter 6 describes sampled-data control with relay in the closed loop.

Jidouseigyo Rironi (Automatic Control Theory)

Yasundo Takahashi
Iwanami Syoten, October 1954
Tokyo Japan

Professor Yasundo Takahashi is considered as one of the pioneers of control engineering in Japan, together with Professor Yoshikazu Sawaragi. He was a professor of control engineering of The University of Tokyo, then became a professor t the University of California at Berkeley in 1958. After retirement from UC Berkeley in 1979, he worked for Toyohashi University of Technology till 1982. He was a member of the IFAC Japan NMO, National Committee of Automatic Control, Science Council of Japan, in 1957. He received the Oldenburger Medal in 1978. He passed away in October 1996. His first book on automatic control was published in 1948 by Kagakugizyutsusya Publishing Co. The company was located at Kanazawa City in Toyama Prefecture. The contents of the book were unique in the sense that a template for drawing the Bode Diagram was given. Later he published a book on computation of automatic control problems using Personal Computer, with the same company. But the book was not read widely so here we shall introduce the book published by Iwanami Syoten, one of the biggest publishers in Japan.

The book is 240 pages including the appendices. Chapter 1 is the introduction giving examples and the basic idea of automatic control. In the example, the servomechanism of the lathe is discussed. Chapter 2 is the representation of dynamics of linear systems. Transfer function and frequency characteristics including Bode diagram are described. A brief explanation of Laplace transform is given in the appendices. Chapter 3 gives examples of lower order systems. Chapter 4 handles the connection of systems using block diagrams. Series and parallel connection of the systems and the idea of P, I, D control are included. Chapter 5 presents higher order systems. The distributed parameter system like heat transfer is described. Dead time is also mentioned. Chapter 6 presents the characteristics of control systems. Steady state control errors and stability are analyzed. Hurwitz condition is given. Various PID tuning methods from the time response including Ziegler-Nichols are discussed. Chapter 7 deals with the frequency characteristics of single loop systems. The computation of the closed loop frequency characteristics from loop frequency characteristics is explained. The Ziegler-Nichols PID tuning based on frequency characteristics is presented. Describing function is also discussed. Chapter 8 discusses multi-loop systems including two controlled variables using the example of the pressure and level control of a boiler. Chapter 9 gives the computation of statistical characteristics like auto and cross-correlation functions. Chapter 10 deals with sampled-data control systems. Z-transform and frequency characteristics are given. Appendices are Laplace transform and pulse transfer function.

The author mentioned that the contents were from his lectures at undergraduate and graduate courses. The level is quite high for those days and the book can be used even currently if state space considerations are added. After he retired from the University of California, he published a revision of the book with Iwanami Shoten.

Professor Yasundo Takahashi was a mentor of control researchers in Japan just after World War II. His group studying automatic control became the Society of Instrument and Control Engineers.

Typical pages in Chapter 6 describe examples of process control and servomechanism, and the tuning criterion of PID controller.

EARLY CONTROL TEXTBOOKS

in

KOREA

Automatic Control Analysis and Control Devices

H.Y. Cheon, B.S. Hong and C.B. Park
1976

Linear Control System Engineering

J.S. Kim
1988

Contributed by

Oh-Kyu Kwon

Inha University,
Inchon, Korea

Introduction

Hangeul, the Korean alphabet, was invented in 1443 during the reign of King Sejong of the Joseon Dynasty. It is composed of 10 vowels and 14 consonants. The 'Hunminjeongeum,' a historical document which provides instructions to educate people using Hangeul, is registered with UNESCO. UNESCO awards a "King Sejong Literacy Prize" every year in memory of the inventor of Hangeul.

When the first control textbooks came out in English (e.g. *Theory of Servomechanisms*, by H. M. James, N. B. Nichols and R.E. Phillips, published in 1947), Korea was under difficult times just after gaining national independence from Japan. It took about 30 years after the first English control book was published that the first Korean book on automatic control was written in Hangeul. Until then, almost all control books in Korea were translated versions of books originally written in English.

Contributed by

Oh-Kyu Kwon
School of Electrical Engineering
Inha University,
Inchon, Korea

Automatic Control Analysis and Control Devices

H.Y. Cheon, B.S. Hong and C.B. Park
1976

The oldest Korean control textbook was written in 1976 by H.Y. Cheon, B.S. Hong and C.B. Park, which is shown in Figure 1, 2, and 3 and its title can be translated into English as *Automatic Control Analysis and Control Devices*. It dealt with the classical control techniques including PID control. But it is hard to say that this is a fully Korean book since it used Korean only for postpositional words and all the terminologies and most of contents were written in Chinese characters, as shown in the figures.

Contents : *Automatic Control Analysis and Control Devices* (H.S. Cheon, 1976)

自動制御解析과 制御機器

(改正 增補版)

工學博士 千 熙 英

洪奉植・朴椿培

共 著

淸 文 閣

目 次

表 16.1(계속)

制御動作	目標値의 變化에 대한 偏差의 傳達函數 $G_{rz}(j\omega)$	積分性없는 對象에 대한 制御應答의 最終値			
		階段入力에 대한 應答 $\{r(t)=1\}$		定速度入力에 대한 應答 $\{r(t)=t\}$	
		制御量	殘留偏差	制御量	殘留偏差
比例	$\dfrac{1}{1+K_pkG(j\omega)}$	$\dfrac{K_pk}{1+K_pk}$	$\dfrac{1}{1+K_pk} \fallingdotseq \dfrac{1}{K_pk}$	∞	∞
比例 + 積分	$\dfrac{1}{1+K_p\left(1+\dfrac{1}{T_Ij\omega k}\right)G(j\omega)}$	1	0	∞	$\dfrac{T_I}{K_pk}$
比例 + 積分 + 微分	$\dfrac{1}{1+K_p\left(1+\dfrac{1}{T_Ij\omega}+T_Dj\omega\right)kG(j\omega)}$	1	0	∞	$\dfrac{T_I}{K_pk}$

크게 하는 것이 制御의 精密度를 높이는데 바람직하다.

그러나 K_p를 크게 하면 比例動作範圍가 좁아지는 것에 注意하여야 한다. K_p가 너무 크면 경우에 따라 不安定하게 되는 경우가 있다.

(b) **比例積分制御**(PI 動作; proportional and integral action, reset control)

이 動作을 하는 調節部에서 1次要素를 制御對象으로 하는 制御系는 階段變化에 대하여 殘留偏差가 없는 것이 長點이다. 따라서 實際로 많이 使用된다.

調節部動作을 數式으로 表示하면

$$
\left.
\begin{aligned}
y(t) &= K_pz(t)+k_i\int z(t)\,dt \\
&= K_p\left[z+\frac{1}{T_I}\int z(t)\,dt\right] \\
G_{zy}(s) &= K_p\left[1+\frac{1}{sT_I}\right]
\end{aligned}
\right\}
$$

여기서 T_I를 積分時間, $1/T_I$를 리세트率(reset rate)라 한다. 위 式에서 아는바와 같이 操作量 y와 偏差 z과의 定常値는 一義的으로 對應하지 않는다. 若干의 偏差가 있어도 第2項의 積分이 効果를 나타내어 z는 0으로 收斂한다. 즉 積分動作은 偏差가 생긴 경우 比例帶의 位置를 移動시켜 結局 z를 0까지 가져가는 役割을 한다고 본다. 一般的으로 對象의 傳達遲延이나 遲延時間이 커질수록 比例帶를 增加시킬 必要가 있지만 너무커지면 應答의 振動週期가 커져 應答이 늦어진다.

(c) **比例積分微分動作**(PID 動作, 3項動作; proportional integral and

Linear Control System Engineering

J.S. Kim
1988

In the strict sense of the word, the first Korean control textbook was published in 1988 by J.S. Kim when the 24th Olympic Game was held in Korea. Figures 4-6 show the book cover and the Table of Contents of Kim's book. Its title is to be translated in English as *Linear Control System Engineering*, and an English translation of the Table of Contents is included below. This book has an important meaning in that it is fully written in Korean and uses Korean terms in control engineering. It covers modern control theory and multivariable systems as well as the classical subjects. It has been widely used as a textbook and a reference in Korean literature and provided a motive to publish a Korean dictionary of terminology in electrical and control engineering.

Contents : *Linear Control System Engineering* (J.S. Kim, 1988)

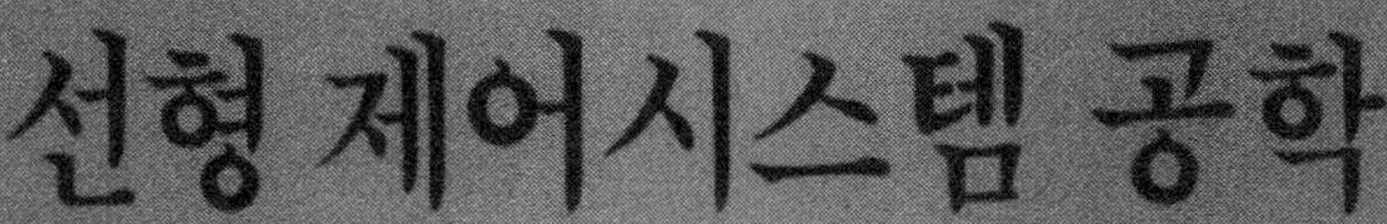

김 종 식

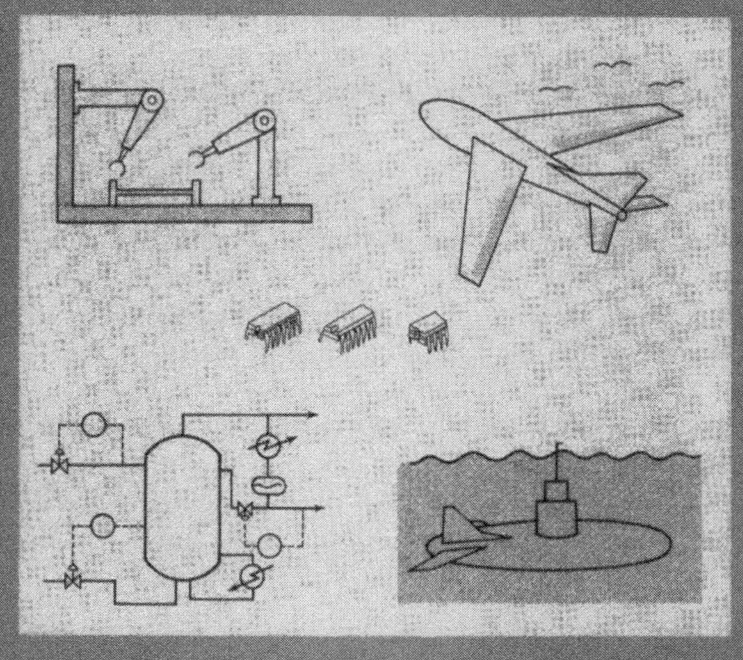

淸 文 閣

차　　례

robustness) 문제를 고려하지 않은 설계법이므로 소프트웨어적으로 설계된 제어시스템이 이론적으로 안정할지라도 실제 제작된 하드웨어적인 제어시스템은 불안정할 수 있다.

안정도-강인성 문제는 1970년대 후반 Safonov에 의하여 그 중요성이 인식되기 시작하였고, 1980년대 초 Doyle과 Stein은 이 안정도-강인성 문제를 고려할 수 있을 뿐만 아니라 복잡한 다변수 제어시스템을 조직적으로 설계할 수 있는 시간 및 주파수역 설계법인 LQG/LTR(Linear Quadratic Gaussian control with Loop Transfer Recovery) 방법을 개발하였다. 이 방법은 제어시스템 설계과정에서 모델링 오차의 영향을 고려한 보다 실제적인 공학적 최적 제어이론이다.

위와 같은 제어이론의 역사적 배경을 통하여 알 수 있듯이 제어시스템의 설계방법은 다양하다. 주어진 시스템의 복잡성, 그리고 요구되는 제어시스템의 정밀도 등에 따라 주파수역 설계법, 시간역 설계법, 그리고 시간 및 주파수역 설계법 중 적절한 방법을 선정할 수 있는 능력도 가져야 할 것이다. 막연히 복잡한 문제까지 다룰 수 있고 정확하고 조직적인 설계법만이 좋은 것은 아니다. 시스템의 특성, 설계사양, 그리고 경제성 등을 고려하여 적절한 설계법을 이용하는 것이 바람직하다.

1.2 제어 용어의 정의

제어시스템을 이해하기 위해서는 우선 제어에 관한 용어를 정확히 파악하는 것이 중요하므로, 자주 사용되고 있는 기본용어들을 간단히 설명하기로 한다.

그림 1.1은 일반적인 제어시스템의 블럭선도(block diagram)이다.

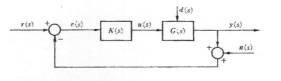

그림 1.1 일반적인 제어시스템

여기서 $G(s)$: 플랜트(plant)
제어를 하고자 하는 시스템(system)

$K(s)$: 보상기(compensator)
오차신호에 따라 제어입력을 만들어 주는 요소

$r(s)$: 기준입력(reference input)
목표값 혹은 요구값에 해당되는 입력

$d(s)$: 외란(disturbance)

외부로부터 제어변수를 교란시키는 입력

$n(s)$: 센서잡음(sensor noise)
센서를 통하여 들어오는 잡음 입력

$y(s)$: 출력(output)
관심있는 시스템의 변수

$e(s)$: 오차신호(error signal)
기준입력과 측정된 출력의 차이로 생기는 신호

$u(s)$: 제어입력(control input)
보상기에서 생성된 플랜트를 조작하기 위한 입력

1) 시스템(system)

어떤 주어진 목적을 달성하기 위하여 상호작용을 하는 여러 개의 요소가 모여 하나의 복합체를 이루고 있는 실체를 시스템이라고 한다. 제어공학에서는 제어할 시스템, 즉 플랜트를 먼저 모델링하고 그 시스템의 동특성을 알아야만 제어기법을 적용할 수 있다. 시스템의 한계가 정해지면 그 시스템 주위 외곽을 환경(environment)이라고 한다. 주위 환경으로부터 시스템에 영향을 미치는 물리량을 입력(input)이라고 하고, 그 입력에 의하여 플랜트의 상태가 변하게 된다. 그리고 시스템의 상태를 나타내는 양(quantity)을 상태변수라고 하고 그 중에서 관심있는 시스템의 변수를 출력(output)이라고 한다.

2) 시스템 동특성(system dynamics)

일반적으로 동역학이라 함은 어떤 시스템에 힘이 작용하여 그 물체가 운동을 하는 현상을 취급하는 역학으로 정의하지만, 제어공학에서는 넓은 뜻으로 시스템이 주어진 입력에 대하여 시간에 따라 변화하는 현상을 말하며, 이것은 시스템의 동적인 특성을 나타내므로 시스템 동특성이라 한다.

3) 개루프(open-loop) 제어시스템

플랜트의 출력이 제어입력을 생성하는 보상기에 아무런 영향을 주지 않는 제어시스템을 말한다. 즉, 출력이 측정되지도 않고 피드백(feedback)되지도 않는 제어시스템을 말한다.

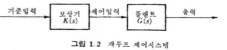

그림 1.2 개루프 제어시스템

4) 폐루프(closed-loop) 제어시스템

주로 피드백 제어시스템이라고 하며, 어떤 플랜트를 목적에 맞도록 제어하기 위하여 플랜트의 출력을 피드백하여 기준입력과 비교하여 그 차이가 없어질 때까지 계속 제어할 수 있는 제어시스템을 말한다.

EARLY CONTROL TEXTBOOKS
in
THE NETHERLANDS

Elementaire Theorie van de Automatische Procesregeling
(Elementary Theory of Automatic ProcessControl)

J. Stigter
Kluwer, Deventer, The Netherlands, 1962.

Regeltechniek en Automatisering in de Procesindustrie
(Control and Automation in Process Industry)

P.M.E.M. van der Grinten,
Prisma-Boeken, Utrecht/Antwerpen, 1968.

Regeltechniek

J.C. Cool, F.J. Schijff, T.J. Viersma
Agon Elsevier, Amsterdam/Brussel, 1969.

Contributed by

Paul Van den Hof

Delft University of Technology
Delft, The Netherlands

Elementaire Theorie van de Automatische Procesregeling
(Elementary Theory of Automatic ProcessControl)

J. Stigter
Kluwer, Deventer, The Netherlands, 1962, 110pp.

The writer of this book is an engineering advisor and lecturer at a polytechnic school in Dordrecht, The Netherlands. He had a background in Mechanical Engineering, presumably from the Technical High School in Delft (currently Delft University of Technology).

The book discusses the elementary concepts of open-loop and closed-loop control, block diagrams and closed-loop transfers, first for static systems, and then for first-order and second-order dynamical systems. The frequency response function is introduced and Bode diagrams are analyzed, as well as polar (Nyquist) diagrams and Nichols diagrams. Stability and control performance (gain and phase margin) are explained using the Nyquist diagram, while the effects of P, I and D actions are explained. At the end of the book some nonlinear phenomena are discussed: actuator limitations, dead zone, hysteresis, friction. Typical examples include mixing systems and level control systems.

ELEMENTAIRE THEORIE VAN DE AUTOMATISCHE PROCESREGELING

Ir. J. STIGTER w.i.

Technisch adviseur en leraar aan de H.T.S. te Dordrecht

N.V. UITGEVERSMAATSCHAPPIJ Æ. E. KLUWER

Deventer - Antwerpen

INHOUD

Teneinde het probleem beter voor berekening toegankelijk te maken stelt men wel in een bepaalde evenwichtstoestand de waarde van K constant. In het geval van Kp moet men er dan rekening mee houden dat dit slechts in een klein gebied geldt: De waarde van Q kan immers zeer sterk variëren. Ondanks het feit, dat op grond van de werking van de regelaar H niet sterk verandert, en dus bij grote Q ook een grote h wordt ingesteld (d.w.z. kleine k), is het verschil in Kp aanzienlijk.

Het komt er op neer, dat we lineariseren door de curve in fig. I-2 voor een klein deel als recht te beschouwen. Op zichzelf is dit toelaatbaar, maar de berekende helling van de raaklijn hangt, ook bij praktisch constante H sterk van Q af.

De verschillen in $1H_1$, $1H_2$ en $1H_3$ in fig. I-3 zullen met behulp van de blokschema's verklaarbaar zijn uit variaties in overbrengingsverhouding. Een ander voorbeeld kan gemakkelijk uitgewerkt worden aan de hand van fig. I-8.

4. Storingen

Het aantal soorten storingen, die in een regelcircuit kunnen optreden, kan zeer groot zijn. Ze worden daarom dikwijls behandeld in volgorde van de grootte van hun invloed op de geregelde grootheid of van de waarschijnlijkheid waarmede zij voorkomen.

Vele mogelijke storingen hebben een te verwaarlozen invloed: andere verlopen zo langzaam, dat het niet noodzakelijk is er rekening mee te houden. Dergelijke storingen worden natuurlijk niet in blokschema's aangegeven.

We zullen nu als voorbeeld enkele storingen in het proces van fig. II-6 aangeven.

De meest voor de hand liggende storing is hier de belastingsverandering. Als de toegevoerde vloeistofhoeveelheid Q_t toeneemt, treedt een daling van de temperatuur T op.

Bij een vergroting Q stellen we bij open kringloop, de invloed hiervan op de uitlaattemperatuur $-N_Q \cdot Q$, hetgeen op de in fig. II-10a getekende wijze kan worden aangegeven. In fig. II-10b is een andere methode getekend. Hier is natuurlijk

$$N'_Q = \frac{N_Q}{G_P}.$$

Op dezelfde wijze kunnen we de invloed verwerken van een verandering van de vloeistof-inlaattemperatuur, hetgeen overigens ook

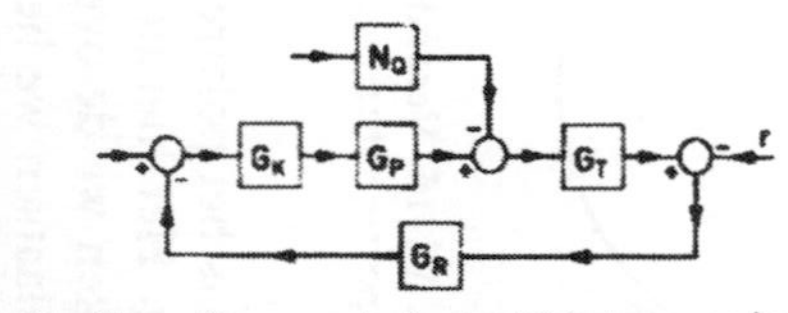

Fig. II-10a. Storingsinvloed in het blokschema van fig. II-9.

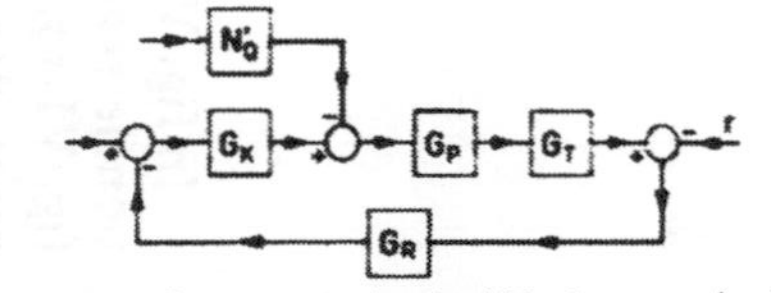

Fig. II-10b. Storingsinvloed in het blokschema van fig. II-9.

23

een belastingsverandering betekent. Als deze temperatuur met T_i C stijgt zal de uitlaattemperatuur bij open kringloop bijv. $N_T T_i$ C stijgen.

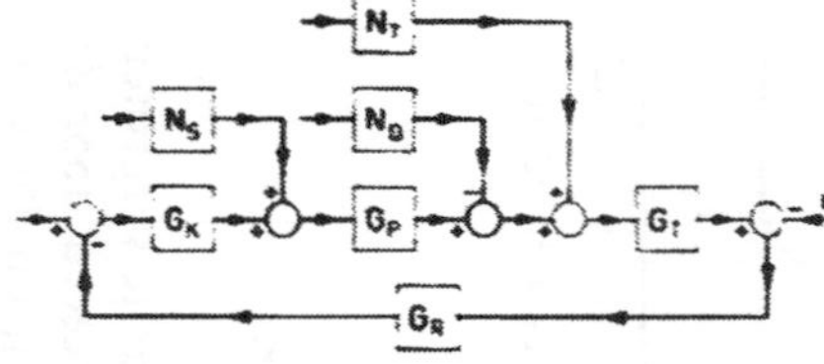

Fig. II-11. Blokschema bij fig. II-9 met storingen.

In fig. II-11 is dit bijgetekend, volgens de methode van fig. II-10a. Tevens is een andere belangrijke storing in het blokschema toegevoegd, nl. die door drukverandering in de voedingsstoom. Bij p kN m^{-2} drukstijging en dus toename van de warmtetoevoer bij constante klepheffing, stellen we de toename van het aantal kg stoom per seconde toegevoerd $N_S \cdot p$.

5. Terugkoppeling

Als het uitgaande signaal van een onderdeel van een circuit, al of niet omgevormd, weer in de baan van het ingaande signaal wordt gebracht, spreekt men van terugkoppeling.

Bij meekoppeling (fig. II-12a) wordt het teruggevoerde signaal bij het ingaande signaal opgeteld; bij tegenkoppeling (fig. II-12b) wordt het ervan afgetrokken.

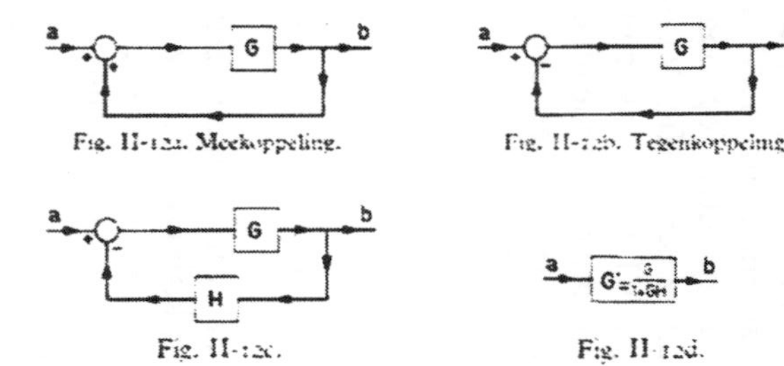

Fig. II-12a. Meekoppeling.

Fig. II-12b. Tegenkoppeling.

Fig. II-12c.

Fig. II-12d.

Het getekende schema levert in beide gevallen de eenvoudigste vorm van terugkoppeling. Men kan immers in de terugkoppelbaan een onderdeel met een bepaalde overbrengingsverhouding H aantreffen, zoals voorgesteld in fig. II-12c. In principe is dit zelfs noodzakelijk als de dimensie van G anders dan één is.

24

Is de hoekfrequentie van de storing ω_A of ω_B dan levert de regelaar een verbetering. Voor ω_A ω ω_B treedt verslechtering op. Een duidelijk beeld hiervan verkrijgt men door de afwijkingsverhouding A.V. als functie van de frequentie (op logaritmische schaal) uit te zetten (fig. VI-15).

De fig. VI-14 en 15 zijn getekend voor een kring bestaande uit een cascade van 4 tijdconstanten en een proportionele regelaar.

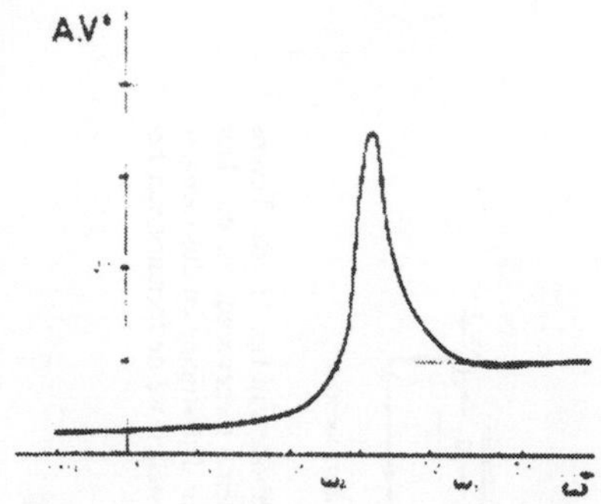

Fig. VI-15. Afwijkingsverhouding.

6. Verbetering van de stabiliteit

a. Door de constructie en de instelling van de regelaar.

Door de overbrengingsverhouding van een proportionele regelaar te vergroten bleek een kleinere offset te ontstaan. Deze wordt tot nul gereduceerd door de regelaar (eventueel bovendien) van integrerende actie te voorzien, omdat voor $\omega = 0$ de A.V. dan tot nul nadert.

Bezien we de dynamische eigenschappen van de regelkring dan blijkt, dat beide methoden een nadelige invloed kunnen hebben.

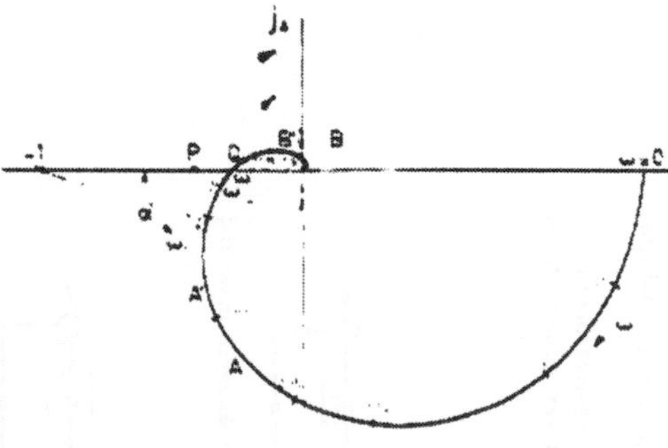

Fig. VI-16. Verandering proportionaliteitsgebied.

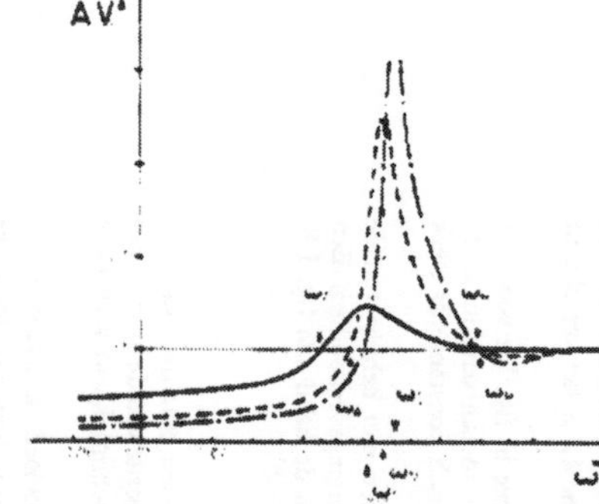

Fig. VI-17. Afwijkingsverhouding bij verandering van het proportionaliteitsgebied.

In fig. VI-16 is het polaire diagram getekend voor een regelkring met proportionele regeling. Het hieruit berekende verloop van de A.V. is weergegeven in fig. VI-17. Maken we de overbrengingsverhouding van de regelaar 2.5 maal zo groot, dan verplaatsen we het punt (—1, j 0) naar P. d.w.z. (—0,4, j 0). Vanuit dit punt wordt nu dus 1 G_r gemeten, waarbij de afstand OP 0,4 als nieuwe schaaleenheid te gebruiken is. In fig. VI-17 is het resultaat met gestippelde lijn weergegeven.

70

Regeltechniek en Automatisering in de Procesindustrie
(Control and Automation in Process Industry)

P.M.E.M. van der Grinten,
Prisma-Boeken, Utrecht/Antwerpen, 1968. 257pp.

This is the first of two books that Paul van der Grinten wrote on control. The second one, "Statistical Process Control" with J.M.H. Lenoir, written in 1973, was used as lecturing material by the author in his courses for students from Technical Highschool Eindhoven, until the beginning of the 80's.

Paul van der Grinten, finished his PhD thesis "Stochastic Processes in Measurement and Control" in 1962, under the supervision of Professor Onno Rademaker. Later, when working at DSM (Dutch State Mines), he was appointed as part-time professor in Eindhoven, where he was teaching courses on stabilizing control systems and statistical methods in control. He was also involved in the editorial board of Automatica in the very early days.

In the book, properties of dynamical processes are treated in terms of differential equations and Laplace domain representations. Basic control operations as PID control are treated and analyzed in their time response effects. Dynamic stabilizing control systems are considered under the influence of stochastic type of disturbances. Examples include typical processes from process industry: mixing and heating processes as well as (petro-)chemical processes.

DR. IR. P. M. E. M. VAN DER GRINTEN

REGELTECHNIEK EN AUTOMATISERING IN DE PROCESINDUSTRIE

PRISMA-BOEKEN
UTRECHT/ANTWERPEN

INHOUD

5

100 K^{-1}%, waarbij PB meestal kan worden ingesteld tussen 10 en 1000%, soms zelfs tussen 0 en ∞. De stapresponsie kan technisch gesproken nooit onmiddellijk de volledige eindwaarde bereiken, hetgeen symbolisch in de figuur is aangegeven door de stippellijn. In de frequentieresponsie, die theoretisch ook een horizontale rechte zou moeten opleveren, correspondeert dit met de gestippelde afval van de responsie voor zeer hoge frequenties (fig. 4.1*b*).

Fig. 4.1*c* geeft de stapresponsie van een PI-regelaar, waarbij behalve de met een factor K versterkte stap ook de integraal van de stap aan de uitgang verschijnt. Deze laatste term is een evenredig met de tijd stijgende lijn. De tijd benodigd om de startwaarde te verdubbelen wordt de *integratietijd* τ_i (Engels: reset time) genoemd en is meestal instelbaar tussen 0.1 en 50 minuten, soms zelfs tussen 0.1 en ∞. Dus:

$$R(t) = K(1 + t/\tau_i) \qquad (4.2)$$

De hierbij behorende overbrengingsverhouding is:

$$R(s) = K\left(1 + \frac{1}{s\tau_i}\right) \qquad (4.3)$$

Het Bode-diagram (fig. 4.1*d*) toont voor frequenties beneden de knikfrequentie $\omega_i = 1/\tau_i$ een toenemende versterking, tot theoretisch bij de frequentie nul een versterking oneindig zou zijn bereikt. In de praktijk is deze versterking natuurlijk beperkt zoals door stippellijnen in de figuur is aangegeven. In de stapresponsie komt deze beperking overeen met de afbuiging van de rechte bij grote t-waarden.

Fig. 4.1*e* geeft de stapresponsie van de PID-regelaar waarbij, behalve de eerder genoemde termen, ook de afgeleide naar de tijd van het ingangssignaal wordt opgeteld. Bij een stap zou dit theoretisch een oneindig hoge impuls $\delta(t)$ moeten geven*; in de praktijk is de impuls enigszins uitgedempt, bijvoorbeeld volgens de gestippelde kromme. In het frequentiediagram (fig. 4.1*f*) wil dit zeggen dat voor frequenties ver boven het tweede knikpunt $\omega_d = 1/\tau_d$ de versterking niet meer toeneemt, doch integendeel afneemt. De grootheid τ_d wordt *differentiatietijd* (Engels: rate time) genoemd en is meestal tussen 0.01 en 10 minuten instelbaar. De stapresponsie van de (theoretische) PID-regelaar is:

$$R(t) = K(1 + t/\tau_i + \tau_d\delta(t)) \qquad (4.4)$$

Deze zogenaamde *drietermregelaar* heeft als overbrengingsverhouding

$$R(s) = K\left(1 + \frac{1}{s\tau_i} + s\tau_d\right) \qquad (4.5)$$

Een variant hierop is de regelaar volgens de *serieschakeling*:

$$R(s) = K\left(1 + \frac{1}{s\tau_i}\right)(1 + s\tau_d) \qquad (4.6)$$

Beide overdrachten kunnen in de praktijk niet geheel verwezenlijkt worden. Overigens zou de 'oneindige' versterking voor de hoogste frequenties ook niet gunstig zijn voor een rustig systeem, zodat met opzet dikwijls een 'tamme' D-actie wordt aangebracht. De overbrengingsverhouding is dan bijvoorbeeld:

$$R(s) = K\left(1 + \frac{1}{s\tau_i}\right)\frac{1 + (1+\beta)s\tau_d}{1 + s\beta\tau_d} \qquad (4.7)$$

Hierin is β de *tamheidsfactor*, meestal vast ingesteld op de waarde 0.1.

*De zogenaamde Dirac-impuls $\delta(t)$ is per definitie oneindig hoog en smal en heeft als oppervlakte 1.

118

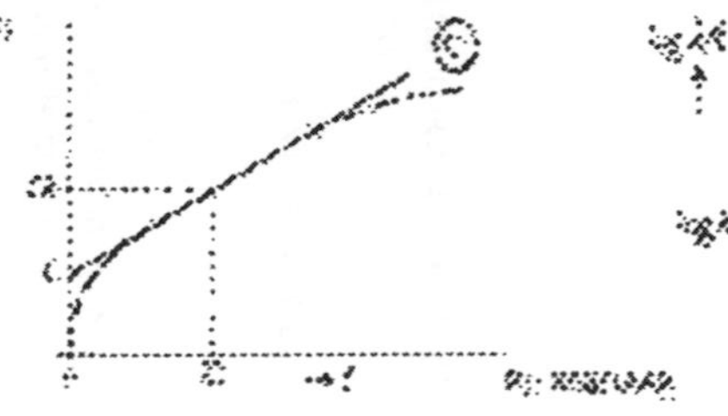

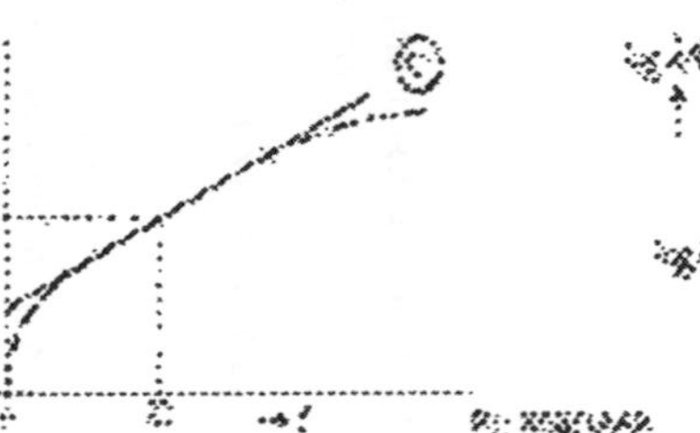

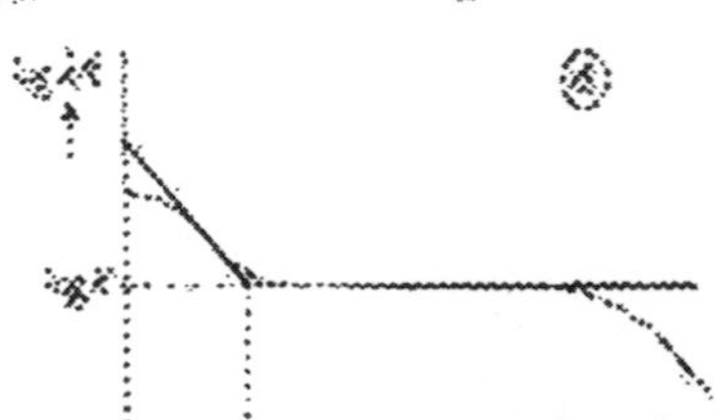

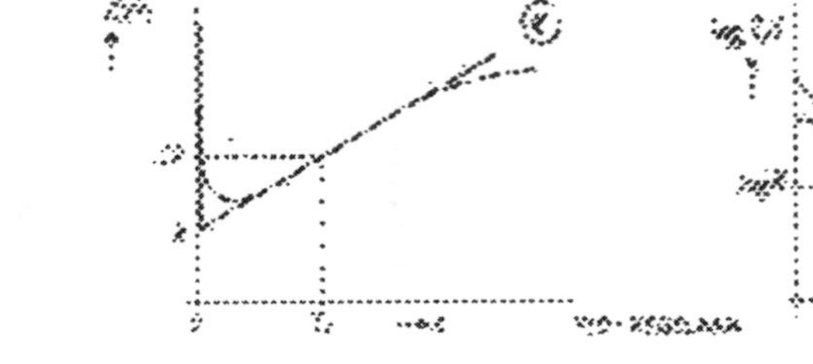

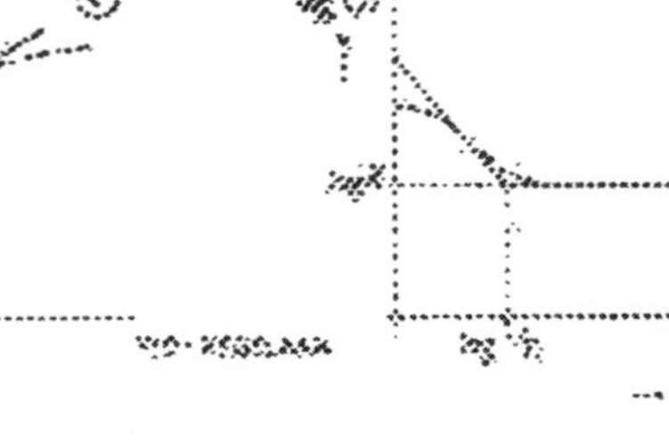

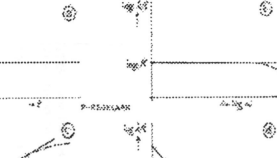

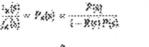

Fig. 4.1 Stap- en frequentieresponsies van P-, PI- en PID-regelaars

4.1.2. Kwaliteit van de regeling

We beschouwen de regeling van een eenvoudig eerste-orde proces, bijvoorbeeld de temperatuurregeling van een reactor volgens het schema van fig. 2.16. Het proces alleen heeft de overbrengingsverhouding volgens vergl. (2.23):

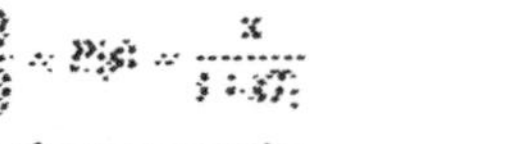

$$\frac{x_0(s)}{x_i(s)} = P(s) = \frac{\alpha}{1 + sT_1} \qquad (4.8)$$

met als stapresponsie:

$$H(t) = \alpha(1 - e^{-t/T_1}) \qquad (4.9)$$

Passen we een teruggekoppelde regeling toe met een regelaar $R(s)$ dan geldt volgens het blokschema van fig. 4.2:

$$\frac{x_0(s)}{x_i(s)} = P_R(s) = \frac{P(s)}{1 - R(s)P(s)} = \frac{\alpha}{1 + sT_1 - \alpha R(s)} \qquad (4.10)$$

119

gewezen. Daarbij is het nuttig om aan de hand van het signaalstroomdiagram reeds van te voren enig idee te krijgen van het soort optimum dat verwacht wordt. Voor de beschreven reactor vindt men als regel een optimale instelling voor F_r en F_g zoals in fig. 5.10 is geschetst. Voor een bepaalde temperatuur T_r en bij bepaalde waarden van de constanten geldt slechts één bepaalde combinatie van F_r en F_g die maximale winst geeft. Toename van F_r doet het spuiverlies te hoog oplopen. Afname van F_r doet F_g toenemen, waardoor wegens de te korte verblijftijd in de reactor C_p daalt en dus een verhoging volgt van de hoeveelheid grondstof die in de volgende scheidingstrappen moet worden gerecirculeerd. Toename in F_g heeft hetzelfde gevolg terwijl bovendien C_p nog ongunstig beïnvloed kan worden door de drukverhoging die hiermee gepaard gaat; afname geeft te weinig belasting. Uiteraard kunnen deze krommen geheel anders liggen bij andere inertgehalten in de grondstof, andere koeltemperaturen, enzovoorts. De gestippelde gedeelten in de grafieken geven het gedeelte aan dat niet bereikt kan worden omdat de druk P of de hoeveelheid F_g hun maxima bereikt hebben. Zou het optimum in dit gestippelde gedeelte liggen, dan is er kennelijk sprake van een niet-optimaal ontwerp. De vrijheidsgraad T_r tenslotte kan ook een optimale instelling tonen vanwege de sterke koppeling aan de druk. Deze zal echter veel minder kritisch zijn dan de twee andere vrijheidsgraden. Concluderend kan dus gezegd worden dat bij de optimalisering steeds moet getracht worden F_g en F_r in een zodanige verhouding te houden dat we op het 'rif', zeer smalle 'eilandjes' in fig. 5.10b, blijven. Juist dergelijke riffen kunnen zeer langzame convergenties veroorzaken indien ze niet van te

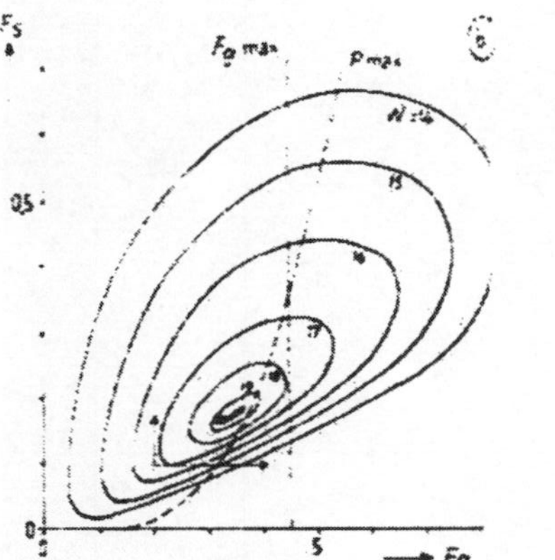

Fig. 5.10a en b. Winstcurves als functie van belasting en spui bij vaste overige omstandigheden.

voren worden opgemerkt.

Of de berekeningen van de optimale instelling eenmalig (off-line) kan geschieden of telkens herhaald moet worden na grote uitwendige veranderingen ('operators guide'), of

170

wellicht een voortdurend beschikbare rekenmachine (on-line) vereist wordt, moet in dergelijke gevallen zorgvuldig worden nagegaan. Dit is een van de hoofdonderwerpen van het volgende hoofdstuk.

5.4. DYNAMISCH OPTIMALISEREN

5.4.1. *Ladingsgewijze processen*

In par. 2.1 over procesdynamica is reeds gewezen op de moeilijkheden bij de regeling van ladingsgewijze en semi-continue processen. Deze processen kennen geen stationaire toestand, doch zijn voortdurend in beweging tussen het start- en stop-ogenblik. Het is natuurlijk mogelijk om met behulp van stabiliserende regelingen de condities gedurende die tijd constant te houden. De reden die ons dwingt om het proces van tijd tot tijd te onderbreken (nieuwe lading, katalysatorverwisseling, schoonmaken, enz.) geeft meestal ook aanleiding tot een achteruitgang van de winst die door het proces in de loop van een periode gemaakt wordt. Het doel is nu niet om deze op elk ogenblik maximaal te maken, doch om de totale winst over een gehele periode te optimaliseren. Vandaar dat meestal ook geen vaste doch juist variabele instellingen nodig zijn.

In het geval van *ladingsgewijze* of batchprocessen is het doel van de optimalisering meestal om zo snel mogelijk een lading van de begintoestand naar de gewenste eindtoestand te brengen. Dit leidt dan bijvoorbeeld tot de reeds besproken boem-boem regelingen, nu echter rekening houdend met niet-lineaire effecten.

Een tamelijk eenvoudig voorbeeld hiervan is een speciale PID-regelaar waarin de P- en D-regelacties vóór de I-regelactie zijn geschakeld. Als het batch-proces start, ontstaat aanvankelijk een dusdanig grote afwijking dat de I-actie geheel uitgestuurd raakt en de ingrijpelementen in hun uiterste stand gaan staan teneinde zo snel mogelijk de gewenste situatie te bereiken. Omdat de D-actie niet uitgestuurd is, zal deze reeds vóórdat de gewenste situatie ontstaat een negatief signaal geven en de I-actie uit zijn verzadiging halen. Bij goede instelling van zo'n regelaar kan het karakter van een volledige regeling (zie fig. 4.16b) bereikt worden. Een verdere uitbreiding van deze gedachtengang leidt zelfs tot het gebruik van regelaars met dubbel differentiërende – D^2 – actie. Bij het veelvuldig starten en stoppen van pieklastcentrales zijn deze met succes toegepast.

Bij meer ingewikkelde systemen moet echter ook na het bereiken van de gewenste waarde nog een dynamische weg afgelegd worden om zo snel mogelijk met de lading klaar te zijn. Dikwijls bestaat deze weg uit een verloop van de temperatuur T_R van hoge waarden, resulterend in een snel verloop van de reacties, naar lage waarden, die soms een betere ligging van het evenwicht opleveren. Als het criterium inderdaad is: de minimalisering van de tijd t_0 benodigd om van een beginconcentratie C_b naar een eindconcentratie C_e te komen, dan kunnen we dit als volgt formuleren: minimaliseer de functie

$$t_0 = \int_{C_b}^{C_e} f_1(T_R)\,dC,$$

waarbij:

$$\frac{dC}{dt} = f_2(T_R) \tag{5.28}$$

De functies $f_1(T_R)$ en $f_2(T_R)$ moeten hierbij uit de kinetische vergelijkingen gevonden worden. De oplossing van een dergelijk probleem vereist de toepassing van variatierekening[20].

171

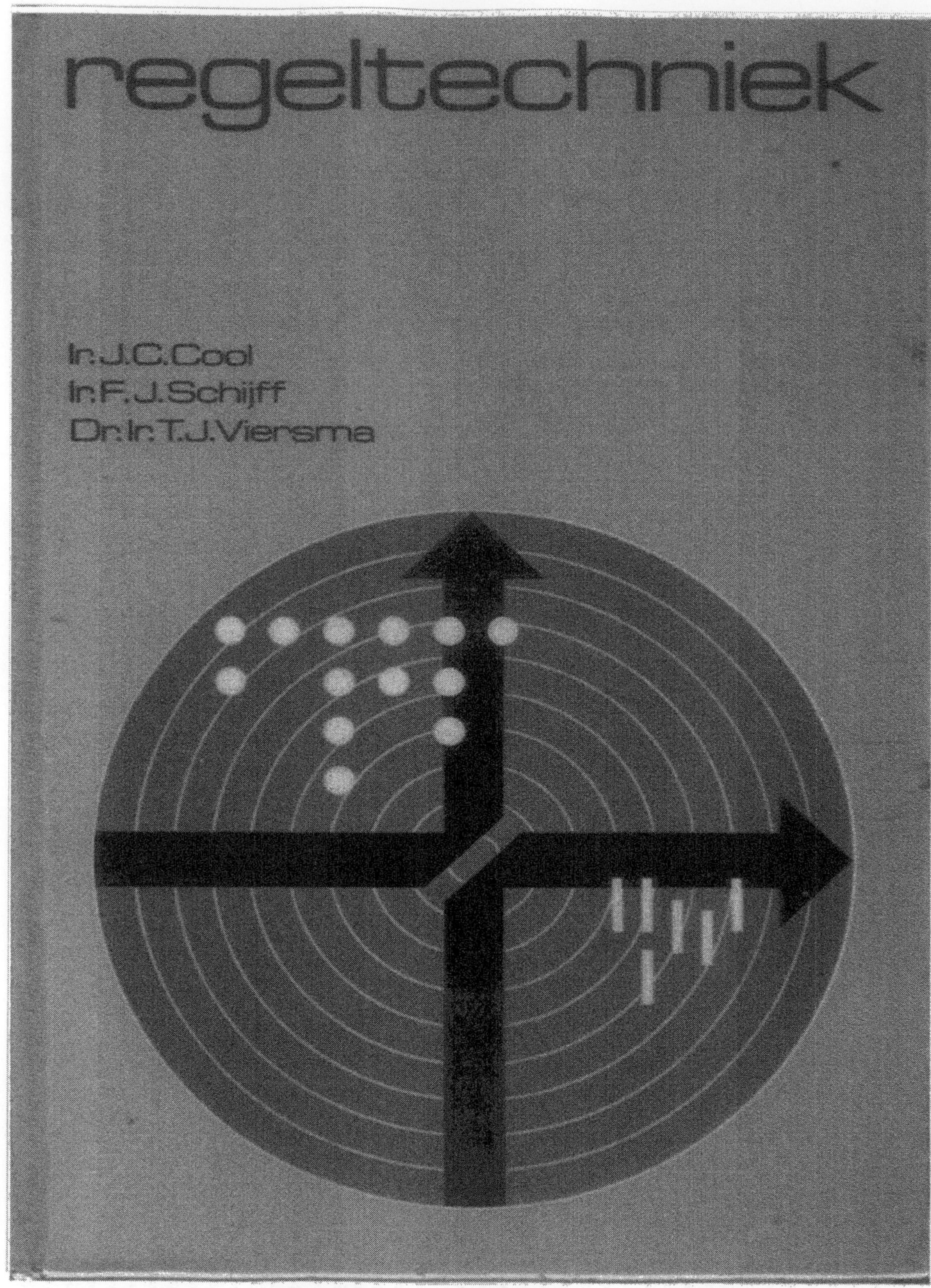
regeltechniek
Ir. J.C.Cool
Ir. F. J.Schijff
Dr. Ir. T.J.Viersma

Regeltechniek

J.C. Cool, F.J. Schijff, T.J. Viersma
Agon Elsevier, Amsterdam/Brussel, 1969.

This is a full-size textbook, 486pp, written by one of the early professors in control/hydraulics of Delft University (Viersma), and by one of his lecturers (Jan Cool) who later became a professor at the same university. It was used (in several new editions) at Dutch universities until the eighties, and in Dutch polytechnic schools all through the nineties.

It extensively discusses transfer function representation and properties of first order and second order systems with time delays, Bode and Nyquist diagrams and PID control, Nyquist stability criterion as well as the tuning of (PID-)controllers. Examples are from process control and from servo systems, among which hydraulic systems. Later chapters include stochastic processes and non-linear phenomena.

In woorden: De overbrengingsverhouding van een tegengekoppeld systeem is de overbrengingsverhouding van de rechtdoorgaande keten gedeeld door één plus de rondgaande overbrengingsverhouding.
Bij regelsystemen wordt steeds tegenkoppeling toegepast.

2.3.4 TEGENKOPPELING EN VOORWAARTSKOPPELING

In hoofdstuk 1 is reeds opgemerkt, dat indien een storing kan worden gemeten, het effekt van deze storing in het uitgangssignaal ook kan worden tegengegaan door voorwaartskoppeling (ook wel voorwaartsregeling of kompensatie genoemd). Beide mogelijkheden – tegenkoppeling en voorwaartskoppeling – zijn in fig. 2.10 aangegeven.

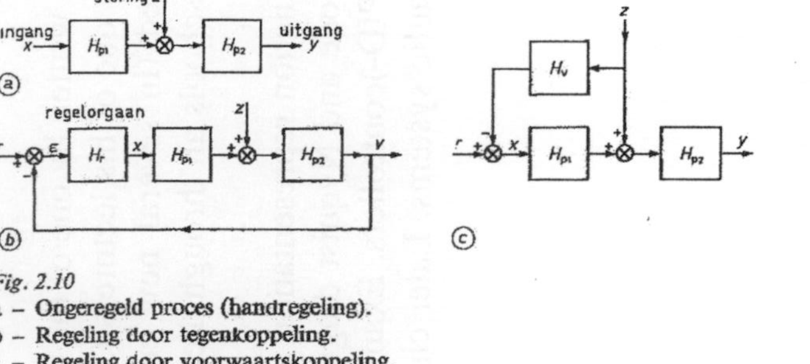

Fig. 2.10
a – Ongeregeld proces (handregeling).
b – Regeling door tegenkoppeling.
c – Regeling door voorwaartskoppeling.

Bij *tegenkoppeling* geldt in dit geval

$$H_{\text{regel}} = \frac{y}{r} = \frac{\breve{H}}{1 + \breve{H}} \tag{2.11}$$

met $\breve{H} = H_r H_{p1} H_{p2}$.
De overbrengingsverhouding uit vgl. (2.11) wordt het *regelgedrag* genoemd. Opdat y zo goed mogelijk r volgt, moet het regelgedrag zo goed mogelijk één zijn, hetgeen bereikt wordt als $\breve{H} \to \infty$.
Verder is het zgn. *stoorgedrag*

$$H_{\text{stoor}} = \frac{y}{z} = \frac{H_{p2}}{1 + \breve{H}} \tag{2.12}$$

38

Het streven is het stoorgedrag nul te maken, hetgeen voor eindige H_{p2} eveneens met $\breve{H} \to \infty$ bereikt wordt.

Bij *voorwaartskoppeling* (zie fig. 2.10c) geldt

$$H_{\text{regel}} = \frac{y}{r} = H_{p1} H_{p2}$$

Dit verschilt niet met het ongeregelde proces. Voor het stoorgedrag geldt nu

$$H_{\text{stoor}} = \frac{y}{z} = (1 - H_v H_{p1})\, H_{p2} \tag{2.13}$$

hetgeen exakt nul wordt indien voor $H_v = 1/H_{p1}$ gekozen wordt.
In de praktijk zal het vaak voorkomen dat H_{p1} zich wat wijzigt. Bij een afwijking in statisch of dynamisch gedrag van H_{p1} en H_v zal het stoorgedrag dan niet meer nul zijn.

Opmerking
In fig. 2.10 is uitgegaan van een één-op-één tegenkoppeling. Dit hoeft niet steeds het geval te zijn. In fig. 2.11 is aangegeven hoe een willekeurige kring

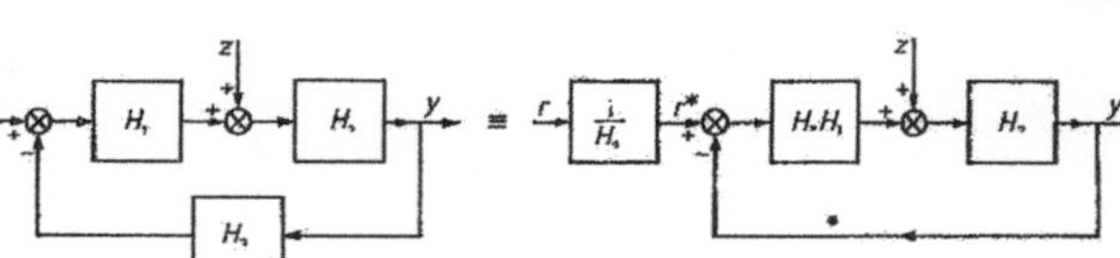

Fig. 2.11 Twee ekwivalente blokschema's. Regelgedrag (y/r) respektievelijk stoorgedrag (y/z) zijn voor beide schema's hetzelfde.

in een één-op-één tegenkoppeling getransformeerd kan worden. Voor het regelgedrag moet nu bij het rechter schema wel y/r beschouwd worden en niet y/r^*! Dit is onder meer van belang bij servosystemen. Regelsystemen in de procesindustrie hebben meestal van nature een één-op-één tegenkoppeling.

2.4 SAMENVATTING

Bij een vloeistof die in een vat verwarmd wordt, blijkt het mogelijk een diffe-

39

P-regeling maar wel nog een extra fase-achterstand. In het bodediagram (fig. 11.8) is hetzelfde af te leiden uit het feit dat voor $\omega > 1/\tau_i$ de amplitudekarakteristiek van de regelaar betrekkelijk snel naar de waarde K_r gaat terwijl de fasekarakteristiek betrekkelijk langzaam naar nul gaat. Bij $\omega = 5/\tau_i$ is α_r nog maar $1{,}02K_r$ terwijl de φ_r nog altijd $-11{,}3°$ is.

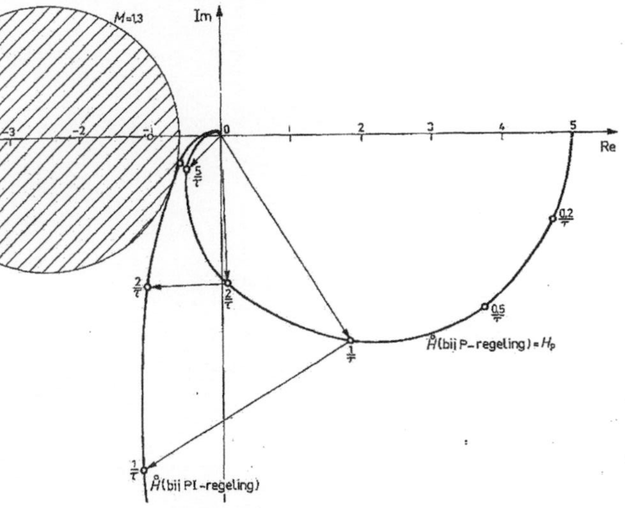

Fig. 12.5 Polair diagram van een proces met PI-regelaar.

$$H_p = \frac{5}{(\tau s + 1)(0{,}2\tau s + 1)(0{,}04\tau s + 1)}, \quad H_r = 1 + \frac{1}{\tau s}$$

De vektorkonstruktie laat duidelijk zien dat I-werking fase-achterstand geeft. Zie ook fig. 12.1.

12.2.2 I-WERKING IN HET TIJDDOMEIN

In fig. 12.6 is aangegeven welke invloed PI-regeling ten opzichte van P-regeling heeft. Voor het proces en de regelaar zijn dezelfde waarden gekozen als in fig. 12.5 met $\tau = 1$.

In de figuur is duidelijk te zien dat de wat geringere fasemarge bij PI-regeling

tot uiting komt in een wat groter doorschot. De statische afwijking bij PI-regeling is echter $\varepsilon_{st} = 0$. Bij de willekeurige storing, die een vrij hoogfrekwent karakter heeft, is het verschil tussen P- en PI-regeling nauwelijks zichtbaar.

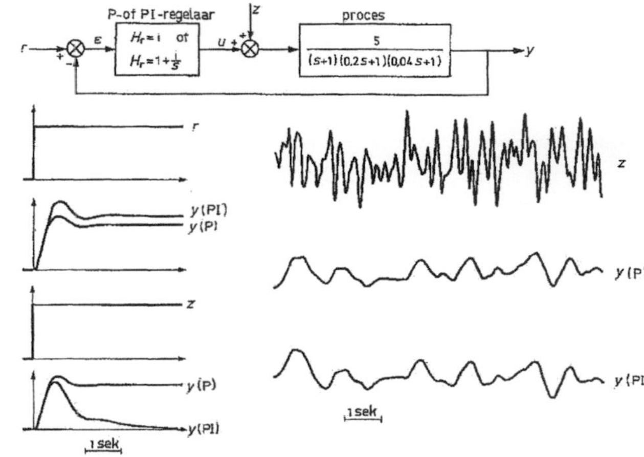

Fig. 12.6 Het effekt van P- en PI-regeling.

*12.2.3 I-WERKING BIJ PROCES MET INTEGRATOR

In het geval $\tau_1 \gg \tau_2$ kan $\tau_i < \tau_1$ gekozen worden zonder dat de $M = 1{,}3$-cirkel wordt doorsneden. De versterking voor lage frekwenties wordt ten opzichte van het geval $\tau_i = \tau_1$ verhoogd. De vraag is nu hoe groot τ_i in verhouding tot τ_2 gekozen moet worden. Een extreme situatie doet zich voor als het proces een integrator bevat. Wordt verder aangenomen dat $\tau_3 \ll \tau_2$ (het geval $\tau_2 = \tau_3$ zal hierna worden behandeld) dan is

$$\mathring{H} = K_r \frac{\tau_i s + 1}{\tau_i s} \frac{K_p}{s(\tau_2 s + 1)} = K_r K_p \frac{\tau_i s + 1}{\tau_i s^2 (\tau_2 s + 1)}$$

Bodediagram en polair diagram zijn in fig. 12.7 getekend. Opgemerkt moet worden dat K_p ***niet*** de statische versterking van het proces is.

In het polair diagram moet de kromme voor $\mathring{H}$ juist buiten de $M = 1{,}3$-cirkel

EARLY CONTROL TEXTBOOKS
in
NORWAY

Control Engineering (Reguleringsteknikk) Vol. 1

Jens G. Balchen
Tapir Book Company, Trondheim, 1963

Control Engineering (Reguleringsteknikk) Vol. 2

Jens G. Balchen
Tapir Book Company, Trondheim, 1967

Control Engineering (Reguleringsteknikk) Vol. 3

Jens G. Balchen, Magne Fjeld, Ole A. Solheim
Tapir Book Company, Trondheim, 1970

Contributed by

Jens G. Balchen

Norwegian University of Science and Technology
Trondheim, Norway

Control Engineering (Reguleringsteknikk) Vol. 1

Jens G. Balchen
Tapir Book Company, Trondheim, 1963

At the Norwegian Institute of Technology, Trondheim, Norway (now Norwegian University of Science and Technology, NTNU), an Associate Professorship of Control Engineering was established in 1954 and a Control Engineering Laboratory was instituted at the same time with a staff of laboratory and teaching assistants and technicians. Regular courses in Control Engineering Fundamentals were offered to students from various departments in 1955 based on Norwegian notes prepared as the course progressed. These notes were printed in two volumes at the Laboratory from 1957/58 and were used until 1963 when the first issue of Jens G. Balchen: Control Engineering (Reguleringsteknikk) Vol. 1, 1963, was published by Tapir Book Company, Trondheim.

This little book (256 pages) has been printed in a large number of editions with many modifications and additions and used as the text for the introductory course in Control Engineering.

From 1999 the book has been transformed into:
Jens G. Balchen, Trond Andersen, Bjarne A. Foss: Control Engineering, Vol. 1, Tapir Book Company, 1999, (568 pages).

The Control Engineering Laboratory was in 1962 converted into a Division (Institute) of Control Engineering within the Faculty of Electrical Engineering and in 1972 the Department of Engineering Cybernetics was established with 5 professorships.

Control Engineering, Vol. 1, 1963 (CE 1963) was based on classical theory of control systems with Laplace transforms and frequency response methods applied to analysis and synthesis. It contained some original contributions which became popular two decades later.

One such contribution was the development of the frequency response H ~ (H-infinity). This development is shown in CE 1963, pages 187-193.

Another unique contribution of CE 1963 was the frequency response based analysis of effects of limitations (saturation) in feedback control systems (see CE 1963, Ch. 11, pages 227-236).

To fill the need for a textbook on equipment for practical implementation of industrial control systems, Associate Professor Ole A. Solheim wrote the book:

Ole A. Solheim: Instrumentation Engineering, 1966, Tapir Book Company, which was reprinted in many updated editions.

REGULERINGSTEKNIKK

Grunnlag for dimensjonering av lineære reguleringssystemer

Jens G. Balchen

Professor i Reguleringsteknikk
Norges tekniske høgskole
Trondheim

Vol. 1

TAPIRS FORLAG
TRONDHEIM 1963

11.2. Metning. (Amplitudebegrensning).

I fig. 11.1 er det skissert noen statiske karakteristikker for typiske elementer fra reguleringsteknikken. De er alle karakterisert ved at de har en øvre og nedre begrensningsverdi, mer eller mindre skarpt definert. Slike elementer eller fysikalske relasjoner kan forekomme på de forskjelligste steder i reguleringssystemer.

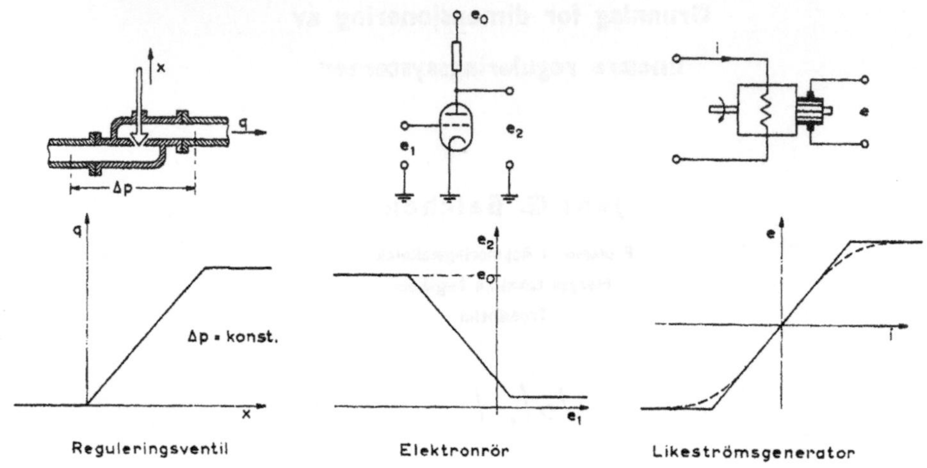

Fig. 11.1

Noen enkle eksempler på forekomst av ett slikt element i systemer, som forøvrig kan beskrives med lineære relasjoner (transferfunksjoner H), er vist i fig. 11.2. Samtidig er de mest betydningsfulle inngangs- og utgangsvariable antydet. Ved hjelp av analysemetoder for lineære systemer kan vi nå lettvint finne ut betingelsene for at metning <u>ikke</u> skal inntre. Vi kan derimot ikke uten videre finne ut hva som skjer dersom metning <u>inntrer</u>.

Kjenner vi formen på den inngangsvariable, kan vi, i alle fall for enkelte former, lettvint finne formen på signalet (x) foran det element der metningen inntrer. En vesentlig ting er at signalenes størrelse (amplitude) over alt i lineære systemer vil være proporsjonale. Har vi funnet formen av x for én størrelse av den inngangsvariable, kan vi derfor lettvint finne akkurat den amplitude av inngangssignalet som vil medføre metning på ett eller annet tidspunkt. I tilfeller der slik metning medfører dårligere ytelse av systemet har vi dermed fått en viktig opplysning.

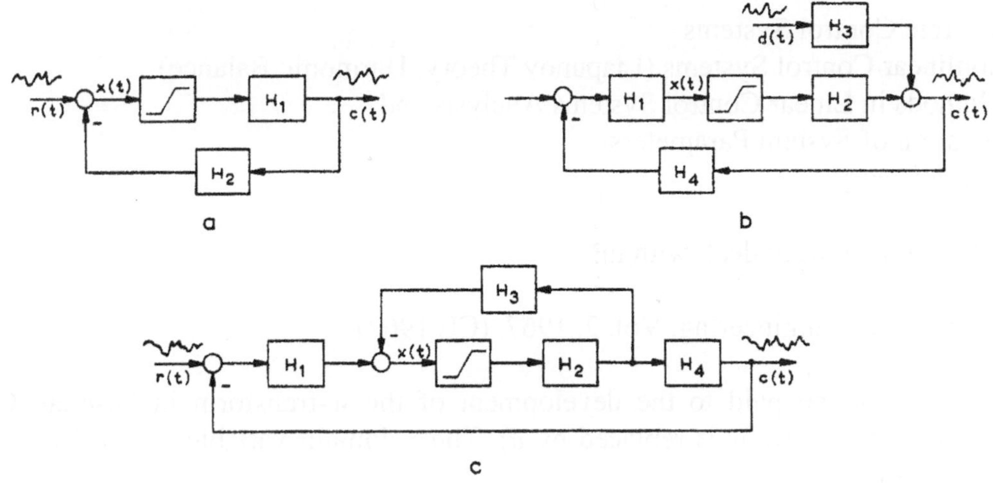

Fig. 11.2

11.3. Undersøkelse i tidsplanet.

I et lineært system kan vi, som nevnt, i prinsippet finne responsen hvor som helst, forutsatt at den (eller de) inngangsvariable er gitt. Selv om vi, for å bestemme forekomsten av metning, bare trenger å kjenne maksimalverdien av responsen, kan det imidlertid medføre svært tidskrevende beregninger, dersom disse foretas i tidsplanet.

Antar vi at det element som kan gå i metning er slik at det innenfor sitt lineære område har en forsterkning lik 1, finner vi for de tre systemene som er skissert i fig. 11.2:

System a: $\frac{x}{r}(s) = N(s) \quad \text{der} \quad N = \frac{1}{1 + H_1 H_2}$ (11.1)

System b: $\frac{x}{d}(s) = -H_3 \cdot H_4 \cdot H_1 \cdot N \quad \text{der} \quad N = \frac{1}{1 + H_1 H_2 H_4}$ (11.2)

Control Engineering (Reguleringsteknikk) Vol. 2

Jens G. Balchen
Tapir Book Company, Trondheim, 1967

During the 1960s there was a rapid growth in the enrolment of students who wrote their Masters- and Doctoral Theses at the Division of Control Engineering. Therefore a number of courses were developed covering advanced topics of control systems theory such as:

- Theory of Discrete Control Systems
- Theory of Nonlinear Control Systems (Liapunov Theory, Harmonic Balance)
- Stochastic Methods in Linear Control Systems Analysis and Identification of System Parameters
- Optimal Control Theory

The first 3 of these topics were dealt with in:

Jens G. Balchen: Control Engineering, Vol. 2, 1967. (CE 1967)

Special attention should be paid to the development of the *w*-transform in Discrete Control Systems Analysis (in CE 1967 *w* is replaced by *q*). The *z*-domain variable is transformed by a bilinear transformation

$$w = \frac{2}{T}\,\frac{z-1}{z+1} \quad \text{or} \quad z = \frac{1 + wT/2}{1 - wT/2}$$

Modeling the system in the *w*-domain rather than the *z*-domain leads to some very convenient results which make analysis and synthesis of discrete control systems nearly identical to that of continuous control systems (see CE 1967 pages 27-35).

REGULERINGSTEKNIKK

BIND 2

Videregående reguleringsteknisk teori

Jens G. Balchen

Professor i Reguleringsteknikk
Norges tekniske høgskole
Trondheim

TAPIRS FORLAG
Trondheim 1967

dier på enhetssirkelen, det vil si, $z = 1 \cdot e^{j\varphi}$ (hvilket er ensbetydende med å velge $s = j\omega$). Derved kan en tegne en kurve for vektoren $A(z)$, som selvsagt blir identisk med den som er vist i fig. 1.17 for $A^*(s)$. Stabilitet er da gitt av denne kurves beliggenhet i forhold til det kritiske punkt $-1 + j0$, som uttrykt i Nyquist's stabilitetskriterium. Innsetting av $z = 1 \cdot e^{j\varphi}$ i $A(z)$ (som for eksempel gitt i (1.48)) er imidlertid ikke særlig hensiktsmessig, idet det medfører endel tallregning. Denne siste metode til stabilitetsundersøkelse er derfor ikke særlig å foretrekke fremfor undersøkelse i s-planet.

Det finnes imidlertid et tredje alternativ som skal belyses nærmere.

1.9. q-transform.

Ved å innføre den bilineære transformasjonen

$$z = \frac{1 + \frac{T}{2} \cdot q}{1 - \frac{T}{2} \cdot q} \tag{1.65}$$

finner vi at utsiden av enhetssirkelen i z-planet blir transformert til h.h.p. i q-planet, og innsiden av enhetssirkelen til v.h.p. i q-planet. Ved denne transformasjon bibeholdes dessuten $A(q)$ som en rasjonal funksjon dersom $A(z)$ er en rasjonal funksjon.

Vi finner også ved sammenlikning av z-plan og q-plan at v.h.p. i s-planet blir transformert til v.h.p. i q-planet, og tilsvarende for h.h.p. Sammenhengen mellom s og q er vist grafisk i fig. 1.19, der koter for konstant absolutt dempning, frekvens og relativ dempning i s-planet er inntegnet i q-planet.

Ganske spesielt finner vi

$$q = u + jv = \frac{2}{T} \cdot \frac{z - 1}{z + 1} = \frac{2}{T} \cdot \frac{e^{Ts} - 1}{e^{Ts} + 1} = \frac{2}{T} \cdot \mathrm{Tg}\, \frac{T}{2} s \tag{1.66}$$

som ved innsetting av $s = j\omega$ gir

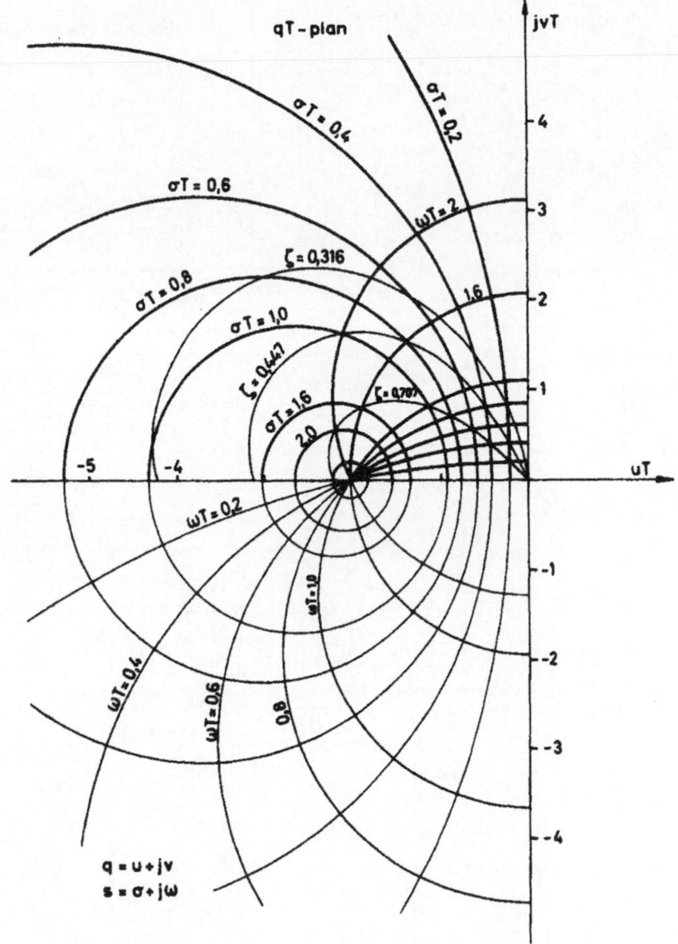

Fig. 1.19.

$$q = jv = \frac{2}{T} \cdot \frac{e^{j\omega T} - 1}{e^{j\omega T} + 1} = \frac{2}{T} \cdot \frac{e^{\frac{j\omega T}{2}} - e^{-\frac{j\omega T}{2}}}{e^{\frac{j\omega T}{2}} + e^{-\frac{j\omega T}{2}}} = j \cdot \frac{2}{T} \cdot \mathrm{tg}\, \frac{\omega T}{2} \qquad (1.67)$$

(1.67) gir

$$\omega \cdot \frac{T}{2} = \mathrm{arctg}\, (v \cdot \frac{T}{2}) \qquad (1.68)$$

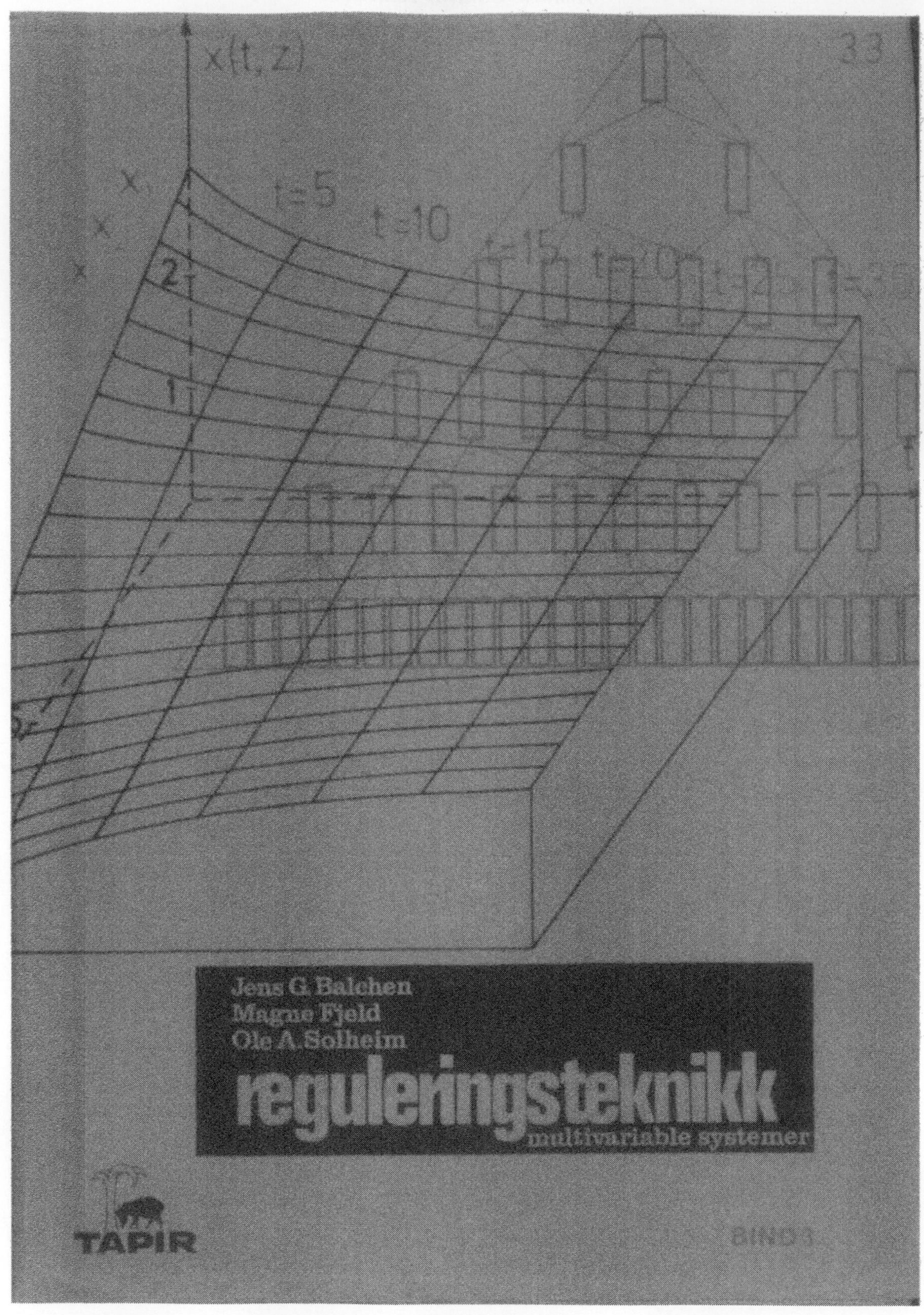
x(t, z)
t=5
t=10
Jens G. Balchen
Magne Fjeld
Ole A. Solheim
reguleringsteknikk
multivariable systemer
TAPIR

Control Engineering (Reguleringsteknikk) Vol. 3

Jens G. Balchen, Magne Fjeld, Ole A. Solheim
Tapir Book Company, Trondheim, 1970

After the (CE 1967) had been published in 1967, it was evident that there was a need for a new textbook covering "modern control theory" with particular emphasis on optimal control theory and aspects of practical implementations in computer control of multivariable processes. CE 1970 is completely based on state space descriptions of multivariable dynamic systems and has individual chapters on Modal Control, Optimal Control based on the Pontryagin Maximum Principle, Dynamic Programming, Optimal State Estimation (Kalman Filtering), Modal and Optimal Control of Distributed Parameter Systems.

CE 1970 was used for many years in graduate courses at the Department of Engineering Cybernetics together with a variety of books from international publishers. It was regarded among the students to have high pedagogical quality. An example of this is the derivation of the optimal discrete state estimator (Kalman Filter) in CE 1970 pages 269-273.

It has been a declared policy of the Department of Engineering Cybernetics to have available textbooks like CE 1963, CE 1967 and CE 1970 in the Norwegian language together with international books in order that the mother tongue professional language shall stay competitive.

6.3. Optimal diskret tilstandsestimering. [8][9][10]

I kapittel 5.9 har vi utviklet den optimale tilstandsestimator (Kalmanfilter) for et kontinuerlig system. Samme type betraktninger som der er gjort kan legges til grunn for utvikling av en optimal tilstandsestimator for et diskret system beskrevet med likningene

$$\underline{x}(k+1) = \Phi(k)\underline{x}(k) + \Delta(k)\underline{u}(k) + \Omega(k)\underline{v}(k) \tag{6.3-1}$$

$$\underline{y}(k) = D(k)\underline{x}(k) + \underline{w}(k)$$

For et slikt system kan vi, tilsvarende det som er gjort i kapittel 5.9, definere kovariansmatriser som følger

$$E((\underline{x}(k)-\overline{\underline{x}})(\underline{x}(k)-\overline{\underline{x}})^T) = X(k) \qquad \text{der } E(\underline{x}(k)) = \overline{\underline{x}}$$

$$E((\underline{v}(k)-\overline{\underline{v}})(\underline{v}(j)-\overline{\underline{v}})^T) = V(k)\delta_{kj} \qquad \text{der } E(\underline{v}(k)) = \overline{\underline{v}} \tag{6.3-2}$$

$$E((\underline{w}(k)-\overline{\underline{w}})(\underline{w}(j)-\overline{\underline{w}})^T) = W(k)\delta_{kj} \qquad \text{der } E(\underline{w}(k)) = \overline{\underline{w}}$$

I (6.3-2) har vi antatt at $\underline{v}(k)$ og $\underline{w}(k)$ er diskrete hvitstøyprosesser. Dette kommer til uttrykk ved funksjonen

$$\delta_{kj} = \begin{cases} 1 & , k = j \\ 0 & , k \neq j \end{cases}$$

Vi skal også anta at $\underline{v}(k)$ og $\underline{w}(k)$ er ukorrelerte. Dette går fram av at

$$E[\underline{w}(k)\underline{v}(j)^T] = 0$$

I (3.2-77) er det utledet et uttrykk for hvordan kovariansmatrisen av responsen av et kontinerlig dynamisk system vil bli når det utsettes for en stokastisk påvirkning. Vi får da

$$X(t) = \Phi(t-t_0)X(t_0)\,^T(t-t_0) + \int_{t_0}^{t} \Phi(t-\tau)CVC^T\Phi^T(t-\tau)d\tau \quad (3.2-77) \quad (6.3-3)$$

Dersom prosessen er diskret, kan vi ved å innføre $t_0 = kT$ og $t = (k+1)T$ der T = tasteperiode, få

$$X(k+1) = \Phi(k)X(k)\Phi(k)^T + \Omega(k)V(k)\Omega(k)^T \quad (6.3-4)$$

I et diskret system vil en gjerne definere to forskjellige tilstandsestimater:

$\bar{\underline{x}}$ = aprioriestimatet av $\underline{x}(k)$ før målingen $\underline{y}(k)$ utnyttes

$\hat{\underline{x}}$ = aposterioriestimatet av $\underline{x}(k)$ etter at målingen $\underline{y}(k)$ er utnyttet

Med dette som utgangspunkt kan en tilsvarende det som er gjort i kapittel 5.9 finne et optimalt estimat $\hat{\underline{x}}$ ved å minimalisere objektfunksjonalen

$$J = E[(\underline{x}(k)-\bar{\underline{x}})^T(\bar{X})^{-1}(\underline{x}(k)-\bar{\underline{x}}) + (\underline{y}(k)-D\bar{\underline{x}})^T W^{-1}(\underline{y}(k)-D\bar{\underline{x}})] \quad (6.3-5)$$

Det estimat som derved fremkommer refereres ofte til som et "minste kvadraters-minimum varians estimat". Derved kommer en frem til et resultat som kan presenteres i følgende sekvens:

1. Et apriori-estimat $\bar{\underline{x}}(k)$ og den tilsvarende kovariansmatrise $\bar{X}(k)$ antas kjent fra det foregående tidspunkt $(k-1)$.

2. Når målingen $(\underline{y}(k)$ foreligger ved tidspunktet k, beregnes aposteriori-estimatet $\hat{\underline{x}}(k)$ og den tilsvarende kovariansmatrise $\hat{X}(k)$ ved hjelp av formlene

$$\hat{\underline{x}}(k) = \bar{\underline{x}}(k) + K(k)(\underline{y}(k)-D(k)\bar{\underline{x}}(k)) \quad (6.3-6)$$

der $\quad K(k) = \bar{X}(k)D^T(k)[D(k)\bar{X}(k)D^T(k)+W(k)]^{-1}$ (6.3-7)

$$\hat{X}(k) = [I-K(k)D(k)]\bar{X}(k) \qquad (6.3\text{-}8)$$

3. Apriori-estimatet $\bar{\underline{x}}(k+1)$ og tilsvarende kovariansmatrise $\bar{X}(k+1)$ ved neste tidspunkt $(k+1)$ beregnes av

$$\bar{\underline{x}}(k+1) = \Phi(k)\hat{\underline{x}}(k) + \Delta(k)\underline{u}(k) + \Omega(k)\bar{\underline{v}}(k) \qquad (6.3\text{-}9)$$

som er en anvendelse av (6.3-1), og

$$\bar{X}(k+1) = \Phi(k)\hat{X}(k)\Phi^T(k) + \Omega(k)V(k)\Omega^T(k) \qquad (6.3\text{-}10)$$

som er en anvendelse av (6.3-4).

Det diskrete Kalmanfilter kan med fordel illustreres med et blokkdiagram som vist i figur 6.3-1. Figuren viser tydelig at filteret består av en diskret modell av prosessen basert på (6.3-1) som løper i paralell med den virkelige prosessen eksitert av pådragsvektoren. Modellen genererer et estimat $(\bar{\underline{y}}(k))$ av målevektoren og det er differansen mellom dette estimatet og den virkelige målevektor $(\underline{y}(k))$ ved samme tidspunkt som gjennom den tidsvarierende matrisen $K(k)$ sørger for oppdateringen av tilstandsestimatet ved tilbakekopling. Dersom $K(k)$ skulle vært null, ville vi fått det tilfelle som ofte refereres til med betegnelsen "ballistisk" som er karakterisert ved at modellens beregninger er fullstendig uavhengige av den virkelige prosessens respons. Det er lett å vise at dette tilfelle vil oppstå etter en tid dersom prosessforstyrrelsene $\underline{v}(k) = 0$ og når modellen er en perfekt representasjon av den virkelige prosess. Det eneste som da er forskjellen mellom prosessens og modellens oppførsel er at begynnelsesbetingelsene for prosessen er ukjente, men virkningen av disse vil etter hvert dø ut.

EARLY CONTROL TEXTBOOKS
in
POLAND

Fundamentals of Control Systems - Volume I: Linear Systems

Paweł Nowacki , Ludger Szklarski and Henryk Górecki
Polish Scientific Publishers, Warszawa 1958

Fundamentals of Automatic Control

Stefan Węgrzyn
Polish Scientific Publishers, Łódź – Warszawa – Kraków, 1960

Contributed by

Wladislaw Findeisen

Warsaw University of Technology
Warsaw, Poland

Ryszard Gessing

Silesian Technical University
Gliwice, Poland

Fundamentals of Control Systems - Volume I: Linear Systems

Paweł Nowacki , Ludger Szklarski and Henryk Górecki
Polish Scientific Publishers, Warszawa 1958

The idea of this textbook, entitled in Polish „Podstawy teorii układów regulacji automatycznej", came from Professor Paweł Jan Nowacki, who was later to become president of IFAC (1969 – 1972). The book had to be a comprehensive text for students of control engineering, from the mathematical background up to practical examples.
Indeed, the first two chapters of the book introduce the reader to Fourier and Laplace transforms along with some elements of the theory of complex variables, permitting to understand the use of transfer functions and block diagrams for the description of control systems. The relations between frequency response and time response are elaborated here in practical detail.

The next chapter covers stability assessment of feedback systems; it is in this respect quite comprehensive, presenting not only the Routh-Hurwitz and the Nyquist criteria, but also the Russian approaches of Mikhailov and Neimark.

Chapter four is devoted to design methods of control loops on the basis of frequency response, using predominantly logarithmic characteristics. The authors pay attention to practical tools, such as the Nichols chart, as well as to the synthesis of series and parallel correctors for improvement of the control loop behaviour.

The last two chapters of the Nowacki & al. textbook are hardware oriented. The student gets introduced to components of electrical control systems, such as sensors, amplifiers and servomotors, and then receives a description of two examples: a Ward-Leonard drive controlled by an amplidyne and the control system of an electric mine-winder with asynchronous motor.

The book in its whole was written in the years 1956 to 1958. Professor Nowacki suggested that chapters 1 and 2 be written in collaboration with Dr. Henryk Górecki from the Mining Academy in Cracow, while he himself was a professor at the Warsaw University of Technology, and remained to be the main author. Dr. Górecki used to travel by train from Cracow to Warsaw and stayed with Professor Nowacki as his guest. The co-authors discussed the content of the book and worked till small hours. Then Dr. Górecki wrote chapters 3 and 4 at home and Professor Nowacki read and approved the manuscript. Chapters 5 and 6 were composed by Dr. Górecki and Professor Ludger Szklarski of the Mining Academy in Cracow.

Later on the three authors worked jointly on Volume II of the book, which was to appear in 1962 with the same publisher. The volume covers the issues of performance criteria, theory of multidimensional and non-stationary systems, as well as discrete systems, nonlinear systems, and some elements of optimal control.

Back in 1958 the Nowacki – Szklarski – Górecki textbook on control systems was practically the first to be written by Polish authors; it was, however, not alone on the market. A parallel and considerable effort was put into bringing to the Polish reader several important foreign books on control. The following translations into Polish were published in roughly the same time:

G.S. Brown and D.P.Campbell, Principles of Servomechanisms (Polish 1957)

A.A. Feldbaum, Electrical Systems of Automatic Control (original in Russian, Polish 1958)

W.Oppelt, Kleines Handbuch technischer Regelvorgänge (Polish 1958)

J.C. Gille, M. Pélegrin, P. Decaulne, Théorie et technique des asservissements (Polish 1961).

As a result, the student, as well as the engineer in Poland had a fair insight into different schools and approaches to control systems and a good choice. This fact makes it also difficult to assess the impact of the Nowacki-Szklarski-Gorecki textbook alone; it is worth mentioning, however, that it was printed in 3000 copies.

Table of Contents

4.8. Introduction to the synthesis of correctors
4.9. Log amplitude and phase characteristics for an optimal transient process
4.10. Determination of desired logarithmic amplitude characteristics
4.11. Typical logarithmic amplitude characteristics and their classification
4.12. Nomograms for the synthesis of correctors
4.13. Determination of the relations needed for the synthesis of correctors
4.14. Application of nomograms to synthesis of non-typical logarithmic amplitude and phase characteristics
4.15. Synthesis of series correctors
4.16. Properties of the logarithmic amplitude characteristic of a series loop element shunted by a corrector
4.17. Synthesis of parallel correctors
4.18. Synthesis of mixed (series and parallel) correctors

5. Components of control systems
5.1. Electro-mechanical analogies
5.2. Typical components of control systems
5.2.1. Introduction
5.3. Complex components of control systems
5.3.1. Sensors
5.3.2. Summing devices
5.3.3. Amplifiers
5.3.4. Servomotors
5.3.5. Stabilizers, dampers and accelerators
5.3.6. Feedback elements
5.3.7. Controllers

6. Practical examples of control systems
6.1. Ward-Leonard drive controlled by an amplidyne
6.2 Amplidyne controller applied to electric mine-winder with an asynchronous motor

Bibliography

Contributed by

Wladislaw Findeisen
Warsaw University of Technology
Warsaw, Poland

PAWEŁ NOWACKI, LUDGER SZKLARSKI, HENRYK GÓRECKI

PODSTAWY TEORII UKŁADÓW REGULACJI AUTOMATYCZNEJ

TOM I

UKŁADY LINIOWE

PAŃSTWOWE WYDAWNICTWO NAUKOWE

Obliczamy równanie (3.5.27):

$$T(s) = -\frac{M(s)}{N(s)} = -G(s)$$

i podstawiamy w nim $s = j\omega$.

Punkt $T_0 = 1 + j0$ powinien leżeć na zewnątrz krzywej $-G(j\omega)$, czyli wewnątrz obszaru stabilności, jeśli układ zamknięty jest stabilny. Warunek ten oczywiście jest ważny wówczas, gdy układ otwarty jest stabilny.

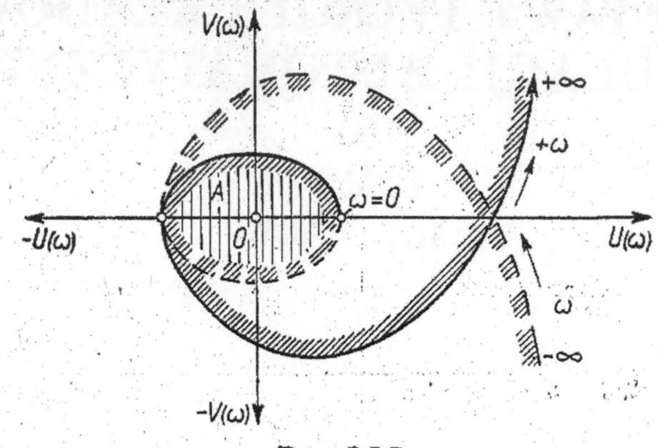

Rys. 3.5.7

Jak widać z rys. 3.5.6 i ze wzoru (3.5.26), rozważaliśmy tutaj równanie

$$-G(s) = +1 \tag{3.5.30}$$

zamiast, jak to się czyni normalnie przy kryterium Nyquista—Michajłowa, równania

$$G(s) = -1. \tag{3.5.31}$$

W naszym przypadku otrzymaliśmy krzywą będącą zwierciadlanym odbiciem krzywej wyrażonej równaniem (3.5.31) względem osi jV i naszym punktem krytycznym jest punkt $(+1+j0)$ zamiast $(-1+j0)$.

Przejdziemy teraz do interpretacji kryterium Michajłowa za pomocą analizy obszaru stabilności.

Załóżmy, że dane jest równanie charakterystyczne układu zamkniętego

$$M(s) + N(s) = 0. \tag{3.5.32}$$

Wprowadzimy teraz parametr $T(s)$ zamiast zera:

$$M(s) + N(s) = T(s). \tag{3.5.33}$$

Równanie (3.5.32) jest oczywiście wówczas szczególnym przypadkiem równania (3.5.33). Następnie zakładamy w równaniu (3.5.33) $s = j\omega$ i wykreślamy krzywą

$$T(j\omega) = U(\omega) + jV(\omega) = M(j\omega) + N(j\omega). \tag{3.5.34}$$

Fundamentals of Automatic Control

Stefan Węgrzyn
Polish Scientific Publishers, Łódź – Warszawa – Kraków, 1960

Carrying the Polish title “Podstawy Automatyki” it was one of the first books written by Polish authors, devoted to Automatic Control. The first edition appeared in 1960 and was authored by Stefan Węgrzyn in cooperation with his graduate students and collaborators Z. Bubnicki, A. Bukowy, Z. Cichowska, R. Gessing, Z. Pogoda, A. Macura, A. Skrzywan and T. Szweda; it was issued as a paperback textbook. The content was based on lectures read by Professor Stefan Węgrzyn at the Electrical Engineering Faculty of the Silesian University of Technology in the period 1954 – 1959. The textbook covered basic notions of control such as dynamics of linear systems, systems with feedback, stability, quality of control, continuous–time and sampled data systems, multivariable systems, nonlinear control, systems with dead zone relay, on–off control, peak-holding control, an introduction to digital control, as well as stochastic models and their application to automatic control.

A corrected second edition was printed in a similar paperback form in 1961; its cover is shown in the attached picture. A copy of the first edition was hard to come by.

On the basis of the 1960 and 1961 textbook editions a wide-circulation handbook “Control Fundamentals” was elaborated by Stefan Węgrzyn, the first edition of which appeared in 1963. The content of the book was in principle the same as that of the textbook previously mentioned; however, the individual chapters were elaborated more carefully, corrected and modified. In particular, the chapter devoted to digital control was enlarged so as to become a separate part of the book. As a whole, the 1963 book is divided into four parts: the first is devoted to linear systems, the second – to nonlinear systems, the third – to digital control and the fourth – to stochastic control.

A second edition of the 1963 book, amended and enlarged, appeared in 1972. It contains a new part covering complex control systems, where problems related to identification, digital control algorithms and structures of complex control systems are described. A chapter devoted to optimal control was also added. The new problems were elaborated by Professor Węgrzyn in cooperation with M. Bargielski, K. Nałęcki and O. Palusiński.

Following 1972, the book had several new editions or just new printings, the last in 1980. In 1978 entirely new material was added, namely a chapter devoted to real-time computer control. It was elaborated in cooperation with A. Wolisz and T. Czachórski.

In general, it is beyond any doubt that the work of Stefan Węgrzyn and his collaborators has influenced the control education in Poland more than any other single achievement. There is no formal evidence, but it certainly was used by students in places other than the Silesian University of Technology, as well as by control engineers all over the country. Due to continuous changes and amendments the book was on the market for a period of over 20 years, up to the1980’s.

Contributed by

Ryszard Gessing
Silesian Technical University
Gliwice, Poland

PODSTAWY AUTOMATYKI

Praca zbiorowa

pod redakcją

STEFANA WĘGRZYNA

Wydanie drugie

PODSTAWY AUTOMATYKI

1961

PAŃSTWOWE WYDAWNICTWO NAUKOWE

ŁÓDŹ — WARSZAWA — KRAKÓW

$$K(p) = k \tag{2.50}$$

Element taki nazywamy elementem bezinercyjnym. W układach rzeczywistych każdy element posiada pewną inercję i w zasadzie nie ma idealnego elementu bezinercyjnego o funkcji przejścia będącej liczbą stałą rzeczywistą k. Pojęciem tym operujemy jednak bardzo często. Na przykład dane są dwa elementy pracujące w układzie łańcuchowym, przy czym stała czasowa jednego z nich jest dużo większa, aniżeli drugiego. Możemy wówczas przyjąć, że jeden z elementów jest bezinercyjny i przy obliczeniu wypadkowej funkcji przejścia uwzględnić tylko stałą czasową drugiego.

Zgodnie z wzorem (2.39) charakterystyka częstości elementu inercyjnego pierwszego rzędu jest równa

$$\hat{K}(j\omega) = \frac{k}{1 + j\omega T} = \frac{k}{\sqrt{1 + \omega^2 T^2}}\, e^{-j\,\mathrm{arc\,tg}\,\omega T} \tag{2.51}$$

Wykres tej charakterystyki na płaszczyźnie K(p) przedstawiony jest na rys. 2.23.

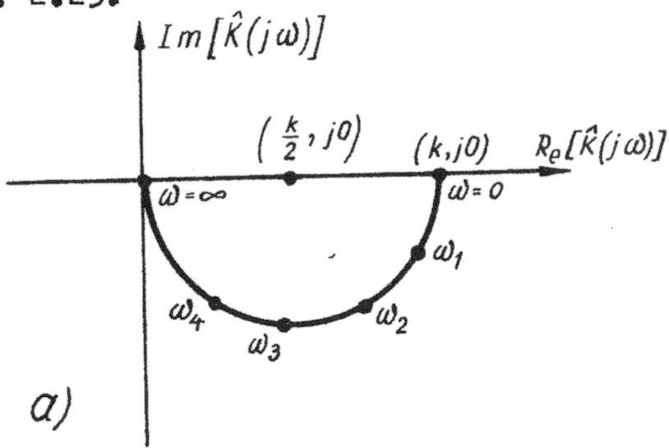

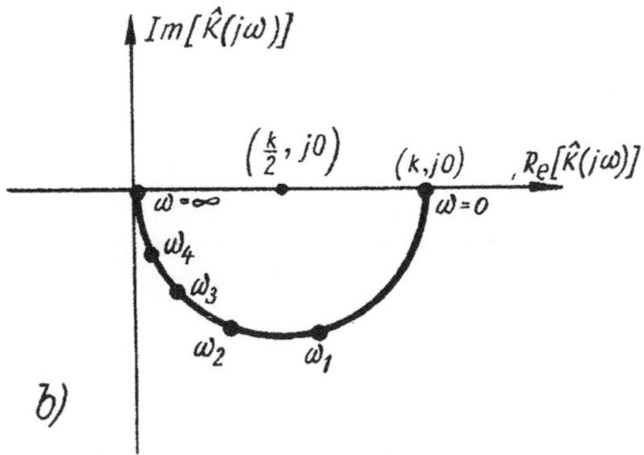

Rys. 2.23. Charakterystyki częstotliwości elementów inercyjnych pierwszego rzędu, $T_1 < T_2$: a) stała czasowa T_1, b) stała czasowa T_2

30

EARLY CONTROL TEXTBOOKS
in
RUSSIA
(Soviet Union)

Foundations of automatic regulation: theory

V. V. Solodovnikov (editor),
Moscow, Mashgiz, 1954, 1117 p.

Lectures on the theory of automatic control

M. A. Aizerman
Moscow, Fizmatgiz, 1958, 520 p.

Elements of the theory of automatic regulation

A. A. Voronov,
Moscow, Voenizdat, 1954, 471 p.

Some problems in the theory of stability of motion

N. N. Krasovski,
Moscow, Fizmatgiz, 1959, 211 p.

Contributed by

Boris Polyak

Institute for Control Problems
Russian Academy of Sciences
Moscow, Russia

Introduction

Studies in the field of automatic control in Russia have a long history, starting from the seminal paper by I. Vyshnegradski (1876) and work of the famous mathematician P. Chebyshev (1871). The fundamental contribution to the theory is due to V. Lyapunov (1892), who developed the basic approach to the analysis of stability of nonlinear systems. Perhaps the first textbook on automatic regulators was the lecture course by the great Russian expert in mechanics N. Jukovski (1909). In the Soviet period automatic regulation attracted a lot of attention due to its role in industrialization of the country. In 1934, the Commission on Remote Control and Automation was founded, later it was transformed into the Institute of Automation and Remote Control of the Soviet Academy of Sciences (1939); the journal "Automation and Remote Control" was established in 1936. Some textbooks on automation in specific areas were published, e.g. for power machines (E. Nikolai, 1930), thermal processes (S. Gerasimov, 1937), electrical devices (B. Domanski, 1938), heating equipment (Yu. Kornilov, 1938), construction of automatic regulators (V. Lossievski, 1942). After the war, numerous Western books on control (such as MacColl, 1947; Lauer, Lesnik and Matson, 1948; Oldenburg and Sartorius, 1949; Gardner and Burns, 1949; James, Nichols and Phillips, 1951) were translated into Russian (we refer to the year of the translation); at the same time, original Russian textbooks appeared. Among the first were the following:

1. Yu. Kornilov, V. Piven', Foundations of the theory of automatic regulation, Moscow, Mashgiz, 1947.
2. M. Aizerman, Introduction to dynamics of automatic regulation of motors, Moscow, Mashgiz, 1950.
3. A. Voronov, Elements of the theory of automatic regulation, Moscow, Voenizdat, 1950.
4. Z. Blokh, Regulation of engines, Moscow, Gostechteorizdat, 1950.
5. B. Domanski, Introduction to automation and remote control, Moscow, Gosenergoizdat, 1950.
6. K. Krug, Transition processes in linear electrical circuits, Moscow, Gosenergoizdat, 1948.
7. M. Meerov, Foundations of automatic regulation of electrical machines, Moscow, Gosenergoizdat, 1952.
8. E. Popov, Theory of automatic regulation, Leningrad, LKVVIA, 1952.

Below, we describe three textbooks of the early 1950's which played a fundamental role in teaching and research, being the stimulating examples for imitation for many years; the fourth sample is a textbook of 1986, which represents the high standards of that period.

We survey some Russian textbooks which played the main role in the development of research and education in this country. There were many others; some of which were even more widespread. The textbooks authored by E. Popov, M. Meerov, A. Netushil, V. Besekerski, Ya. Tsypkin, Ya. Roitenberg are worth mentioning. There also existed numerous textbooks covering some selected topics of automatic control. Of course, the attempt to survey all of them requires much more effort...

Foundations of automatic regulation: theory

V. V. Solodovnikov (editor),
Moscow, Mashgiz, 1954, 1117 p. (Fig. 1.)

The contributors to the book include M. Aizerman, V. Voronov, L. Goldfarb, A. Krasovski, A .Lerner, A. Letov, F. Mikhailov, B. Petrov, V. Petrov, V. Solodovnikov and Ya. Tsypkin (for the full list of authors see Fig.2). Actually this is not an ordinary textbook – this huge volume can be called Encyclopedia of Automatic Regulation, and for a long time it played the role of THE BOOK for Soviet researchers and students. Most of the distinguished Soviet scientists in the field of control (with few exceptions – e.g. A. Feldbaum, A. Lur'e) were invited to join the team, which had a common understanding of the subject, techniques and tools of this science.

The monograph has some real-life examples and applications, but mainly it is focused on theory. Two other volumes are intended to cover practical recommendations and applications to various areas of industry.

The book consists of two parts – linear and nonlinear systems, respectively. Each part is divided into four sections, while a section includes several chapters. The structure looks very reasonable; all principal topics of the theory are addressed, and even a contemporary reader can encounter a lot of stimulating ideas.

Section 1 contains basic notions of control; the language of differential equations and transfer functions is used in parallel, and Laplace transform is exploited as the main technique.

Section 2 is devoted to stability of linear systems. Graphical tools play a significant role, and Mikhailov, Nyquist and Bode plots (known is Russian as the method of logarithmic frequency characteristics) are described in detail. The topic which is not well known for Western researchers is the so called D-decomposition (Chapter 11) – the separation of the stability domain in the parameter space. The approach goes back to Vyshnegradski and was developed by Yu.Nejmark (1948); the boundary of the stability domain in the 2D parameter plane can be expressed explicitly.

> One of the results by A. Aizerman and F Gantmacher (Fig.3) reads as follows. Consider the characteristic polynomial $P(s)=D(s)+M(s)$, where $\deg D=n$, $\deg M= m$; q,t,f are numbers of zero, real positive and unstable roots of D; r is the integer part of f/2; M is Hurwitz polynomial. Under these assumptions there exist D, M such that P is stable if and only if the following conditions hold: $m\geq q+t-1$, $n+m>4r-1$ (provided $m>0$ is even and f is even; for other situations the last condition vary). This result implies the existence of block diagrams which do not possess stability under arbitrary values of parameters involved (structural instability).

Very interesting is Section 3 – "Performance analysis and design". Of course, in 1954, no theory of optimal control existed, but the role of the performance index and specifications is expressed very clearly. Mainly the suggested tools rely on the Bode plot or on pole-zero location. Nevertheless Chapters 20 and 25 deal with integral quadratic performance index. For instance, the plot in Fig.4 exhibits the level sets of such an index as functions of the parameters k and T, describing a model of the electrical device (potentiometer); this allows to choose optimal values of the parameters k=7.5, T=0.15. Chapter 21 addresses the theory of pulse systems and provides stability criteria for discrete-time models.

The theory of regulation under random disturbances is studied in Section 4. The basic notions of stationary random processes and their applications to control systems are given first (Chapters 22, 23). Less traditional is Chapter 24, – the disturbances are not assumed to be random, but rather arbitrary bounded; the problem is to find the worst possible disturbances that lead to the largest deviation of the output. This so called problem of accumulation of disturbances was first studied by Bulgakov (1946). Of course, it is a typical problem of constrained optimal control.

Nearly half of the volume (Part 2) is devoted to nonlinear systems. This line of research was traditional for the Russian school; classical contributions of Lyapunov, Andronov and Krylov-Bogolubov can be mentioned. We focus just on a few challenging results addressed in this part of the book.

Chapter 31 "Improving dynamic properties of automatic regulation by use of nonlinear controls" is written by A. Lerner; note that it is published a few years earlier than the first papers on the Pontryagin maximum principle. The author starts with the formulation of the problem of time-optimal control for linear or nonlinear systems subjected to constraints on some of the derivatives (which may be considered as controls). Without rigorous analysis, he claims that the control variable should take boundary values only (the bang-bang principle); the instants of switching are found from a system of nonlinear equations (it is assumed that there are just n switching instants, where n is the order of the system). No general theory is provided; however several examples are discussed and optimal processes in the phase plane are obtained (see, e.g. Fig. 5). Note that simultaneously the papers on optimal control of linear systems were published by A. Feldbaum.

Various approximate methods for analysis of periodic regimes are considered in Section 6 (the method of small parameters, method of harmonic balance). Relay systems and self-oscillations in them are the subject of Section 7. The final Section 8 addresses graphical and numerical methods for calculation of time-response in time-varying or nonlinear systems.

ОСНОВЫ АВТОМАТИЧЕСКОГО РЕГУЛИРОВАНИЯ

ТЕОРИЯ

Под редакцией
доктора технических наук
профессора
В. В. СОЛОДОВНИКОВА

*

ГОСУДАРСТВЕННОЕ НАУЧНО-ТЕХНИЧЕСКОЕ ИЗДАТЕЛЬСТВО
МАШИНОСТРОИТЕЛЬНОЙ ЛИТЕРАТУРЫ
Москва 1954

Figure 1

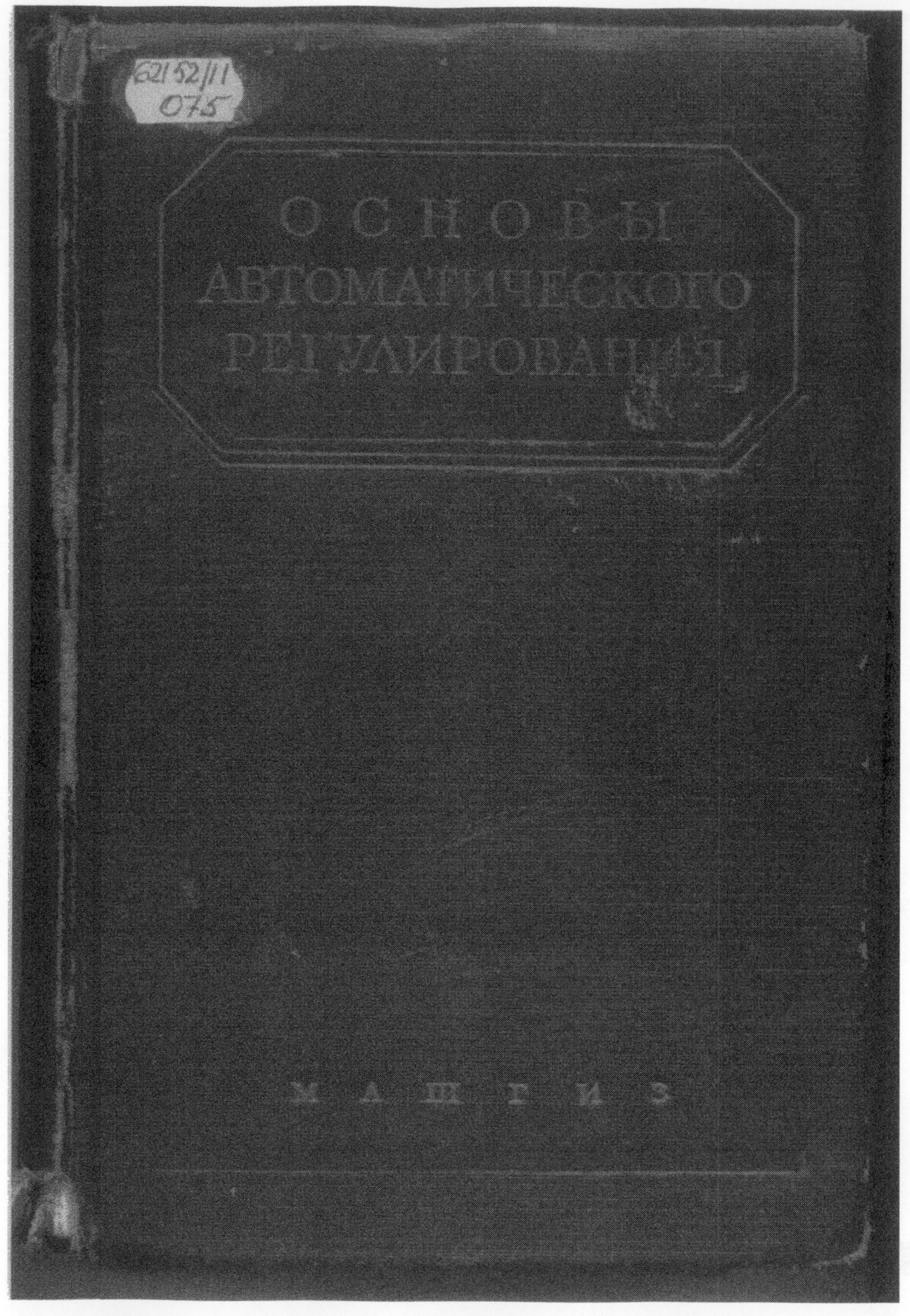
ОСНОВЫ
АВТОМАТИЧЕСКОГО
РЕГУЛИРОВАНИЯ
МАШГИЗ

АВТОРЫ КНИГИ

М. А. АЙЗЕРМАН, д-р техн. наук (гл. XI, XXXIV), Д. А. БАШКИРОВ, канд. техн. наук (гл. XXXVIII), П. В. БРОМБЕРГ, канд. техн. наук (гл. XXXVI), А. А. ВОРОНОВ, канд. техн. наук, доц. (гл. III), Л. С. ГОЛЬДФАРБ, д-р техн. наук, проф. (гл. XXXIII), В. В. КАЗАКЕВИЧ, д-р техн. наук (гл. XXVII, XXVIII), А. А. КРАСОВСКИЙ, канд. техн. наук, доц. (гл. XX), А. Я. ЛЕРНЕР, канд. техн. наук (гл. VII, XXXI), А. М. ЛЕТОВ, д-р физико-матем. наук, проф. (гл. IX, XXXII, XXXVI), П. С. МАТВЕЕВ, инж. (гл. XXXIX), Ф. А. МИХАЙЛОВ, канд. техн. наук (гл. XXV), Б. Н. ПЕТРОВ, чл.-корр. АН СССР (гл. IV, XIX), В. В. ПЕТРОВ, канд. техн. наук (гл. XXVI, XXIX, XXX), Г. С. ПОСПЕЛОВ, канд. техн. наук, доц. (гл. XXXV), В. В. СОЛОДОВНИКОВ, д-р техн. наук, проф. (гл. II, V, VI, XII, XV, XVI, XVII, XXII, XXIII, Введение и вводные замечания к разделам), Ю. И. ТОПЧЕЕВ, инж. (гл. VIII, XIV, XVIII), Г. М. УЛАНОВ, канд. техн. наук (гл. XIII, XXIV, XXX), А. В. ХРАМОЙ, канд. техн. наук (гл. I), Я. З. ЦЫПКИН, д-р техн. наук, проф. (гл. X, XXI, XXXVII).

1*

Figure 2

В этом случае в первом квадранте $\left(T_r'^2 > 0;\ T_\kappa > 0\right)$ заведомо нет точек, принадлежащих области устойчивости. Если же

$$KT_u(1 + K_cT_u) > T_u(1 + \overline{K}_cT_g) + T_g,$$

то в соответствии с фиг. 38 и 39 Д-разбиение плоскости $T_r'^2$, T_κ имеет вид, графически представленный на фиг. 42.

В этом случае в первом квадранте плоскости $T_r'^2$, T_κ есть точки, принадлежащие области устойчивости. Выбором достаточного трения в чувствительном элементе можно скомпенсировать вредное действие масс чувствительного элемента. Надо лишь, чтобы T_κ было больше некоторого порога, увеличивающегося монотонно с ростом $T_r'^2$.

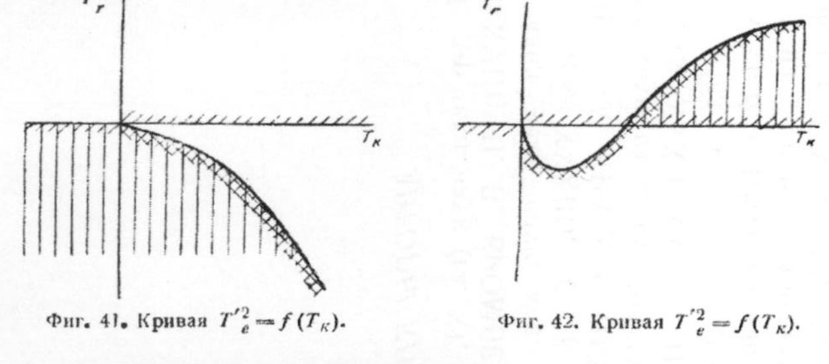

Фиг. 41. Кривая $T'^2_e = f(T_\kappa)$.

Фиг. 42. Кривая $T'^2_e = f(T_\kappa)$.

5. УСЛОВИЯ СУЩЕСТВОВАНИЯ ОБЛАСТИ УСТОЙЧИВОСТИ

В ряде случаев, применяя построения областей устойчивости, описанные выше, приходят к выводу, что пространство параметров рассматриваемой системы вообще области устойчивости не содержит и что сами построения были проделаны поэтому напрасно.

В некоторых случаях можно по схеме установки или по виду уравнений движения установить, содержит ли пространство параметров системы область устойчивости, и сразу же отбраковать системы, не содержащие областей устойчивости, избегая в таких случаях напрасных построений [1].

Рассмотрим системы, характеристическое уравнение которых сводится к виду

$$D_p(\lambda) + M_p(\lambda) = 0,$$

где $D_p(\lambda)$ — произведение любого числа множителей вида

$$T\lambda + 1,\ \lambda,\ T\lambda - 1,\ T_\kappa^2\lambda^2 + 2\zeta_\kappa T_\kappa\lambda + 1,\ T_\kappa^2\lambda^2 + 1;$$

$M_p(\lambda)$ — произведение любого числа множителей вида

$$K,\ \tau\lambda + 1,\ \tau_d^2\lambda^2 + 2\zeta_d\tau_d\lambda + 1.$$

Здесь T, ζ_κ, T_κ, K, τ, τ_d и ζ_d — независимые друг от други положительные числа (параметры системы).

280

Введем теперь следующие обозначения: q — число нулевых, а t — правых действительных корней у полинома $D_p(\lambda)$; f — число корней полинома $D_p(\lambda)$, расположенных на мнимой оси и справа от нее; r — целая часть дроби $\frac{1}{2}f$; n и m — соответственно степени полиномов $D_p(\lambda)$ и $M_p(\lambda)$, а $V = n + m$. Тогда условия существования области устойчивости устанавливаются следующей теоремой.

Для того чтобы в пространстве параметров системы, имеющей характеристическое уравнение [1] существовала область устойчивости, необходимо и достаточно, чтобы удовлетворялись два неравенства:

1) неравенство $m \geqslant q + t - 1$;

2) одно из неравенств, приведенных в табл. 1 и выбираемое в зависимости от m и f.

Таблица 1

	$m = 0$	$m > 0$ четно	m нечетно
f четно	$v > 4r$	$v > 4r - 1$	$v > 4r - 2$
f нечетно	$v > 4r$	$v > 4r$	$v > 4r + 1$

Для частного случая одноконтурной системы без воздействий по производным $M_p(\lambda) = K = \mathrm{const}$, следовательно, $m = 0$ и в этом случае условия теоремы сводятся к неравенствам

$$q + t \leqslant 1;$$
$$n > 4 \cdot r.$$

Доказательство сформулированной выше теоремы и перечень литературы по этому вопросу содержатся в работе [7].

ЛИТЕРАТУРА

1. Fraser R. A., Duncan W. I., On the criteria for the stability of small motions, Proc Royal Soc., v. 124, London, 1929, p. 642.
2. Соколов А. А., Критерий устойчивости линейных систем регулирования с распределенными параметрами и его применение, Инженерный сборник, т. II, вып. 2, Изд. АН СССР, 1946.
3. Кабаков И. П., О процессе регулирования давления пара, Инженерный сборник т. II, вып. 2, Изд. АН СССР, 1946.
4. Воробьев Ю. В., Кац А. М., Ремизов В. М., Соколов А. А., Фатеева И. А., Исследования в области регулирования паровых турбин, сборник статей под ред. М. З. Хейфица, Госэнергоиздат, 1950.
5. Неймарк Ю. И., Устойчивость линеаризованных систем, ЛКВВИА, 1949.
6. Неймарк Ю. И., Об определении значений параметров, при которых система автоматического регулирования устойчива, „Автоматика и телемеханика", т. IX, № 3, 1948.
7. Айзерман М. А. и Гантмахер Ф. Р., Условия существования областей устойчивости у одноконтурных систем автоматического регулирования, содержащих воздействия по производным, „Прикладная математика и механика" № 1, 1954.

Figure 3

где

$$a_0 = k; \quad a_1 = 1 + kT_2; \quad a_2 = 0{,}25; \quad a_3 = 0{,}025; \quad a_4 = 0{,}001;$$

$$b_0 = k; \quad b_1 = kT_2.$$

Формула (419) дает

$$J_0 = \int_0^\infty [x(t) - x(\infty)]^2 dt =$$

$$= \frac{1}{2a_0^2}\left(\frac{b_0^2 \begin{vmatrix} a_1 & -a_2 & a_4 & 0 \\ a_0 & a_1 & -a_3 & 0 \\ 0 & -a_0 & a_2 & -a_4 \\ 0 & 0 & -a_1 & a_3 \end{vmatrix} + b_1^2 \begin{vmatrix} a_0 & a_1 & a_1 & 0 \\ 0 & a_0 & -a_3 & 0 \\ 0 & 0 & a_2 & -a_4 \\ 0 & 0 & -a_1 & a_3 \end{vmatrix}}{\begin{vmatrix} a_0 & -a_2 & a_4 & 0 \\ 0 & a_1 & -a_3 & 0 \\ 0 & -a_0 & a_2 & -a_4 \\ 0 & 0 & -a_1 & a_3 \end{vmatrix}} - 2b_0 b_1 \right) =$$

$$= \frac{1}{2k} + \frac{1}{2k^2} \cdot \frac{1{,}31 - 0{,}25kT_2 - 0{,}025k + 5{,}25kT_2^2 - k^2T_2^3}{5{,}25 + 4{,}25kT_2 - k^2T_2^2 - 0{,}625k} - \frac{T_2}{2}.$$

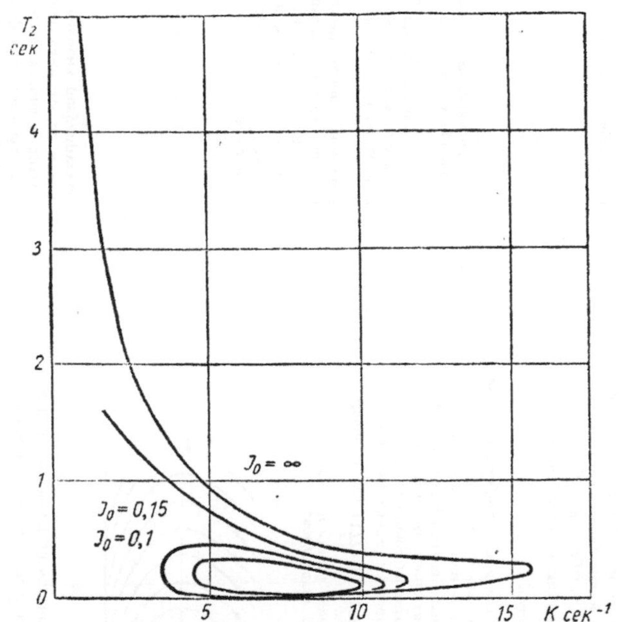

Фиг. 114. Зависимость интегральной квадратической оценки J_0 от передаточного коэфициента разомкнутой системы k и постоянной времени диференцирующего контура T_2.

На фиг. 114 изображена зависимость интегральной квадратической оценки J_0 от передаточного коэфициента разомкнутой системы k и постоянной времени диференцирующего контура T_2. Указанная зависимость изображается семейством кривых, вдоль которых $J_0 = \text{const}$.

551

Figure 4

в) Изображение совокупности оптимальных процессов на фазовой плоскости

Наглядное представление о всей совокупности оптимальных процессов при различных начальных состояниях системы может дать их изображение на фазовой плоскости или в фазовом пространстве.

Поскольку ограничения, наложенные на значения производных или координат системы, однозначно определяют форму оптимального процесса перехода системы из одного (начального) состояния в другое (заданное) состояние, то изложенными выше методами всегда может быть определен оптимальный закон изменения во времени регулируемой величины, а следовательно, и ее производной:

$$x = f_{опт}(t);$$

$$\dot{x} = y = f'_{опт}(t).$$

Исключая из этих зависимостей время, находим зависимость

$$y = \varphi_{опт}(x),$$

представляющую уравнение траектории оптимального процесса на плоскости x, y. Иногда оказывается более удобным представлять движение системы при помощи других величин: первой и второй производных регулируемой величины, значений двух координат системы и т. п. В таких случаях по осям фазовой плоскости откладываются значения этих величин и время исключается из законов оптимального изменения этих величин.

В тех случаях, когда состояние системы определяется более чем двумя переменными, фазовые траектории являются кривыми в пространстве с числом измерений, равным числу величин, определяющих состояние системы.

Рассмотрим в качестве примера совокупность оптимальных процессов для системы, в которой значение второй производной ограничено пределами

$$\ddot{x}_{k1} < \ddot{x} < \ddot{x}_{k2},$$

а значение первой производной — пределами

$$\dot{x}_{k1} < \dot{x} < \dot{x}_{k2}.$$

Разница в абсолютных значениях пределов изменения ограниченных величин может возникнуть, например, из-за влияния нагрузки.

Согласно изложенному в п. а ограничение второй производной приводит к тому, что участки фазовых траекторий, соответствующих каждому этапу движения системы, имеют вид парабол второй степени. Если нагрузка системы создается сухим трением, то предельное значение ускорения будет зависеть от того, увеличивается или уменьшается по абсолютной величине скорость изменения выходной величины системы. Полагая в уравнениях (171) и (172) $i = 2$; $j = 1$; $y = \dot{x}$; $x_{y=0} = \overset{\circ}{x}$ и исключая из этих уравнений время, найдем, что увеличению абсолютного значения скорости будут соответствовать участки фазовых траекторий вида

$$y = -\operatorname{sgn}\overset{\circ}{x}\sqrt{2\ddot{x}_{k2}(\overset{\circ}{x} - x)\operatorname{sgn}\overset{\circ}{x}}, \qquad (177)$$

а уменьшение абсолютного значения скорости

$$y = +\operatorname{sgn}\overset{\circ}{x}\sqrt{2\ddot{x}_{k1}\operatorname{sgn}x(\overset{\circ}{x} - x)\operatorname{sgn}\overset{\circ}{x}}. \qquad (178)$$

Здесь множители $\operatorname{sgn}\overset{\circ}{x}$ введены для учета изменения знака сил трения при изменении знака скорости $\dot{x} = y$.

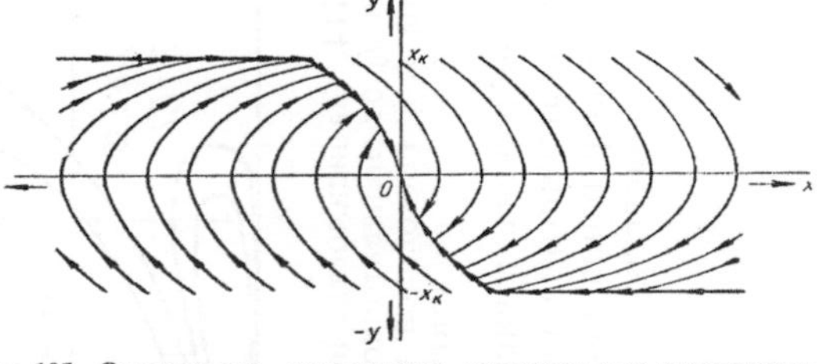

Фиг. 105. Совокупность оптимальных процессов при ограничении первой и второй производных от регулируемой величины.

Зависимости (177) и (178) представляют семейство парабол, отличающихся значением $\overset{\circ}{x}$, зависящим от начального состояния системы. По этим траекториям движется изображающая точка в течение первого этапа оптимального процесса.

Форма второго этапа оптимального процесса — приближения к заданному состоянию системы — не зависит от начальных условий. Уравнение фазовой траектории этого этапа движения системы получается аналогично предыдущему и имеет в данном случае вид

$$y = -\operatorname{sgn}x\sqrt{2\ddot{x}_{k2}|x|}. \qquad (179)$$

Ограничение скорости движения системы выделяет на плоскости x, y полосу $-\dot{x}_{k1} < y < \dot{x}_{k2}$.

Диаграмма совокупности оптимальных процессов для системы, ограниченной по скорости и по ускорению, показана на фиг. 105.

2. ОСУЩЕСТВЛЕНИЕ ПРОЦЕССОВ ОПТИМАЛЬНОЙ ФОРМЫ ПРИ ПОМОЩИ НЕЛИНЕЙНЫХ СВЯЗЕЙ

Из изложенного выше следует, что, во-первых, каждому начальному и конечному состоянию системы соответствует определенная форма процесса регулирования, которая при наложенных на систему ограничениях обеспечивает минимальную длительность процесса регулирования,

Figure 5

Lectures on the theory of automatic control

M. A. Aizerman
Moscow, Fizmatgiz, 1958, 520 p. (second edition), (Fig.6).

The first edition of the book was published in 1956, its material was originally selected and used in lecture courses for engineers and was oriented towards practical applications. Later, the content of the course was revised to serve as a textbook for students. It became a standard for numerous other textbooks on automatic regulation.

The book consists of five chapters and two appendices. Four chapters are devoted to linear systems, the last one deals with nonlinear systems. The input-output description of single-input single-output systems and the frequency-domain approach are prevailing; however the state space language arises in some cases. Consider, for instance, Fig.7; this is the analysis of a linear system plus one scalar nonlinearity (the so called absolute stability problem). The linear part is given in state space form.

> In 1949, M. Aizerman formulated the following conjecture: if nonlinearity belongs to a sector (Fig. 8) and the linearized system is stable for all linear characteristics in this sector, then the nonlinear system is stable as well. This Aizerman conjecture is true for $n \leq 2$, however in 1958 (the year of publication of the textbook) V .Pliss constructed a counterexample for $n=3$.

The contents of the chapters are as follows. Chapter 1 deals with automatic regulators (examples, terminology, feedback, types of regulators are provided). Basic notions are presented in Chapter 2 (block diagrams, transfer functions, frequency response, the construction of mathematical model from empirical data). Stability of linear systems is analyzed in Chapter 3; the main techniques are based on D-decomposition (see above); Michailov and Nyquist plots are considered as particular cases of the approach. Performance is the subject of Chapter 4. There are many practical recommendations on how to deduce properties of the time response from the transfer function characteristics; a lot of attention is paid to numerical and graphical construction of time response. The last Chapter 5 deals with self-oscillations and forced oscillations in nonlinear systems. Relay systems are considered as examples.

The presentation is clear and rigorous, however, at the engineering level – not all statements are validated, while all of them are explained and illustrated by examples.

М. А. АЙЗЕРМАН

ЛЕКЦИИ ПО ТЕОРИИ АВТОМАТИЧЕСКОГО РЕГУЛИРОВАНИЯ

ИЗДАНИЕ ВТОРОЕ, ДОПОЛНЕННОЕ И ПЕРЕРАБОТАННОЕ

ГОСУДАРСТВЕННОЕ ИЗДАТЕЛЬСТВО ФИЗИКО-МАТЕМАТИЧЕСКОЙ ЛИТЕРАТУРЫ
МОСКВА 1958

Figure 6

действия по второй производной величина $K_{кр}$ сначала увеличивается до некоторого максимального значения, а затем уменьшается до нуля. Отрицательное воздействие по второй производной лишь уменьшает $K_{кр}$.

§ 4. Суждение об устойчивости исходной системы по устойчивости ее линейной модели

В предыдущих параграфах был рассмотрен вопрос об устойчивости линейной модели системы автоматического регулирования.

В начале главы уже указывалось, что устойчивость линейной модели свидетельствует в лучшем случае об устойчивости исследуемой нелинейной системы по отношению к достаточно малым возмущениям. В настоящем параграфе это утверждение будет уточнено. Кроме того, в некоторых случаях можно сделать более сильные утверждения об устойчивости реальной системы, если установлено, что ее линейная модель устойчива.

Чтобы пояснить это, уточним понятие устойчивости, введенное в начале главы III.

Систему, содержащую нелинейные элементы, условимся называть *устойчивой «в малом»*, если можно указать столь малую область начальных отклонений, что после любого отклонения из этой области регулируемый режим восстанавливается (за конечное время или в пределе при $t \to \infty$).

Таким образом, ***говоря, что регулируемый режим устойчив «в малом», мы лишь констатируем наличие области начальных отклонений, по отношению к которым система устойчива*** (*т. е. наличие области устойчивости*), ***но не определяем как-либо ее границ.***

Разумеется, устойчивость системы «в малом» не препятствует тому, что при реальных начальных отклонениях система может себя вести как неустойчивая, так как понятие «устойчивости в малом» не учитывает того, что область устойчивости системы может быть ограничена. Чтобы говорить об устойчивости реальной системы, надо сопоставить область устойчивости и область начальных отклонений, реально возможных во время эксплуатации установки.

Условимся говорить, что система *устойчива «в большом»* в том случае, когда определены границы области начальных

отклонений, после которых регулируемый режим восстанавливается, и выяснено, что реальные начальные отклонения принадлежат этой области.

Наконец, условимся говорить, что система *декрементна* или ***неограниченно устойчива*** (или *устойчива «в целом»*) в том случае, когда область начальных отклонений, после которых восстанавливается положение равновесия, вообще не ограничена. В этом случае исходная система обладает теми же свойствами, что и ее линейная модель: из факта ее устойчивости «в малом» следует ее устойчивость по отношению к любым начальным отклонениям.

Ограничимся пока случаем системы, которая отличается от линейной наличием одной нелинейности. В общем случае процесс в такой системе описывается уравнениями

$$\left.\begin{aligned} \dot{x}_1 &= \sum_{j=1}^{n} a_{1j}x_j + f(x_k), \\ \dot{x}_i &= \sum_{j=1}^{n} a_{ij}x_j, \quad i = 2, 3, \ldots, n, \end{aligned}\right\} \qquad (3.42)$$

отличающимися от линейной системы

$$\left.\begin{aligned} \dot{x}_1 &= \sum_{j=1}^{n} a_{1j}x_j + ax_k, \\ \dot{x}_i &= \sum_{j=1}^{n} a_{ij}x_j, \quad i = 2, 3, \ldots, n, \end{aligned}\right\} \qquad (3.43)$$

лишь наличием одной нелинейной функции $f(x_k)$, стоящей в первом уравнении вместо ax_k.

Найдем теперь область значений a, при которых система (3.43) устойчива. Пусть установлено, например, что система (3.43) устойчива при

$$a^* < a < a^{**}$$

и неустойчива, если

$$a < a^* - \varepsilon \quad \text{или} \quad a > a^{**} + \varepsilon,$$

Figure 7

по крайней мере при достаточно малом положительном числе ε.

Иначе говоря, значения $a = a^*$ и $a = a^{**}$ служат границей области устойчивости по a.

Пусть, далее, при составлении линейной модели исходной системы (3.42) нелинейная функция $f(x_k)$ заменена линейной функцией $a_0 x_k$, причем a_0 выбрано так, что

$$a^* < a_0 < a^{**}. \qquad (3.44)$$

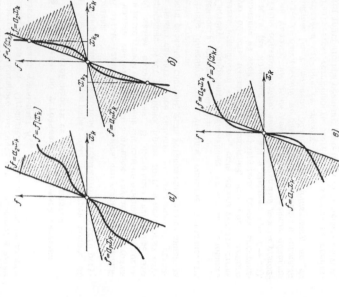

При этом несущественно, каким образом определялось число a_0, переходом ли к малым колебаниям, т. е. заменой кривой $f = f(x_k)$ касательной в точке $x_k = 0$, или же экспериментальным усреднением, т. е. заменой этой кривой прямой, хотя и проходящей через точку $x_k = 0$, но не совпадающей с касательной. Если выполняется неравенство (3.44), то построенная линейная модель устойчива. Какое заключение может быть сделано тогда об устойчивости исходной системы (3.42)?

Рассмотрим два числа a_1 и a_2, удовлетворяющих неравенству

$$a^* < a_1 < a_0 < a_2 < a^{**}. \qquad (3.45)$$

Построим теперь в плоскости f, x_k два луча $f = a_1 x_k$ и $f = a_2 x_k$ и сопоставим их с кривой $f = f(x_k)$.

Возможны три случая, представленных на рис. 144.

1. Кривая $f = f(x_k)$ целиком, т. е. при любых достижимых во время эксплуатации системы значениях x_k, лежит между лучом $f = a_1 x_k$ и лучом $f = a_2 x_k$ (рис. 144, *а*).

2. Кривая $f = f(x_k)$ лежит между лучами $f = a_1 x_k$ и $f = a_2 x_k$ лишь при достаточно малых x_k и пересекает при каком-либо значении x_k, например при $x_k = x_{k2}$, один из лучей (рис. 144, *б*).

3. Кривая при достаточно малом x_k не лежит между лучами (рис. 144, *в*).

С помощью метода Ляпунова установлено, что всегда, как бы ни было выбрано число a_0 при замене системы (3.42) ее линейной моделью (3.43), можно найти такие два числа a_1 и a_2, чтобы были верны следующие утверждения:

1. В первом случае (рис. 144, *а*) устойчивость линейной модели свидетельствует о том, что исходная система неограниченно устойчива.

2. Во втором случае (рис. 144, *б*) из факта устойчивости линейной модели следует лишь, что исходная система устойчива «в малом».

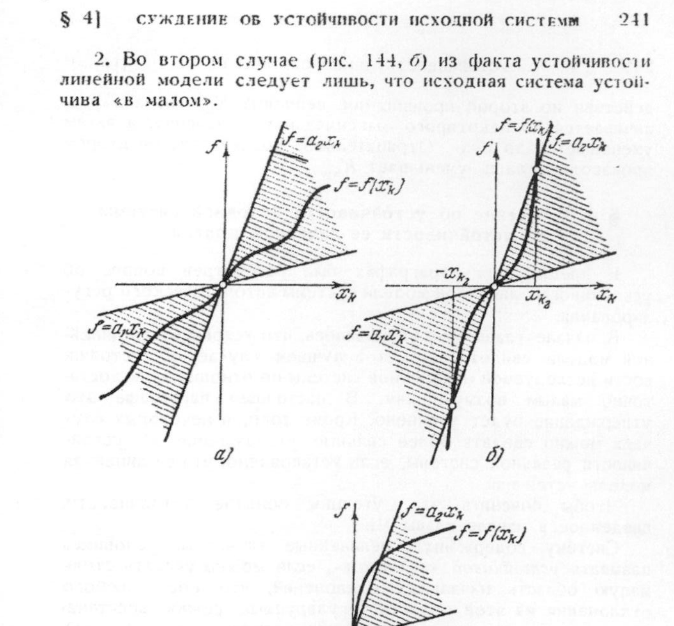

Рис. 144.

Более того, зная наименьшее значение x_k, при котором кривая пересекает один из лучей, можно определить область,

Figure 8

Elements of the theory of automatic regulation

A. A. Voronov,
Moscow, Voenizdat, 1954, 471 p. (second edition) (Fig. 9).

The first edition was published in 1950; note that "Voenizdat" means "Military Publishing House." The role of automation was well understood in military applications, this is why research in the field of automatic regulation had strong support from defense institutions. The author had a rich experience in giving lectures at Moscow Higher Technical University (Bauman MVTU) and Moscow Power Institute. These two Moscow universities were leading institutions in teaching automatic regulation. Later, the author published several textbooks on special topics in control, and the book under consideration was revised and divided into two volumes. To summarize, A. A. Voronov was one of the most popular and acknowledged authors in the field of automatic control. His books are well written and combine simplicity and rigor.

The book has peculiar features of the time when it was created (the last years of the Stalin era). The first phrase in Introduction reads: "I. V. Stalin in his work of genius 'Economic problems of socialism in the USSR' points out..." and a long citation (having no direct relation to automatic regulation) follows. On the same page we find reference to Marx and Engels, the next two contain three citations from Stalin again. Later, a long survey of Russian contribution to the field is presented; Western inventors and researchers are mentioned only in the negative aspect (for instance, Watt is said to have reinvented his regulator 18 years after Polzunov). At that time, this was the mandatory requirement for any scientific book...

The content of the textbook under consideration is close enough to Aizerman's (see above), but the nonlinear part is lacking. There are also many simplified real-life examples, and a lot of pages contain figures of specific devices and regulators (Fig. 10).

А. А. ВОРОНОВ

ЭЛЕМЕНТЫ ТЕОРИИ АВТОМАТИЧЕСКОГО РЕГУЛИРОВАНИЯ

ИЗДАНИЕ ВТОРОЕ,
ПЕРЕРАБОТАННОЕ И ДОПОЛНЕННОЕ

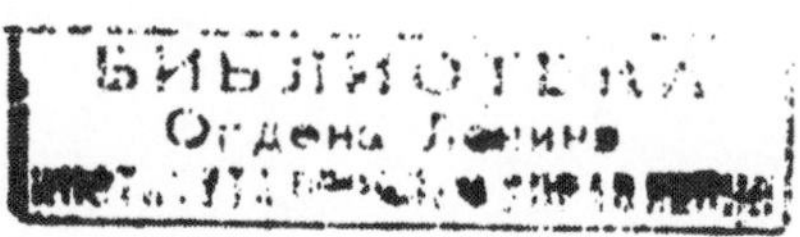

ВОЕННОЕ ИЗДАТЕЛЬСТВО
МИНИСТЕРСТВА ОБОРОНЫ СОЮЗА ССР
МОСКВА — 1954

Figure 9

Такой регулятор при вращении его шпинделя с постоянной скоростью оказывается неустойчивым. Регуляторы с отрицательным статизмом применяются в системах релейного типа.

§ 18. УГОЛЬНЫЙ РЕГУЛЯТОР НАПРЯЖЕНИЯ

На рис. 68, *а* представлена схема авиационного регулятора напряжения с угольным столбом.

Столб $R_у$, представляющий собой набор тонких угольных шайб, сжимается пружиной F через ряд промежуточных звеньев (гибкая лента $Л$, профильный рычаг, якорь A, нажимная шайба).

Обмотка электромагнита $W_э$ регулятора включается на регулируемое напряжение. При прохождении тока через электромагнит его якорь A стремится повернуться в направлении против хода часовой стрелки, т. е. в направлении, противоположном тому, в котором действует пружина.

С якорем жестко связан воздушный успокоитель K. Угольный столб включается последовательно с обмоткой возбуждения регулируемого генератора.

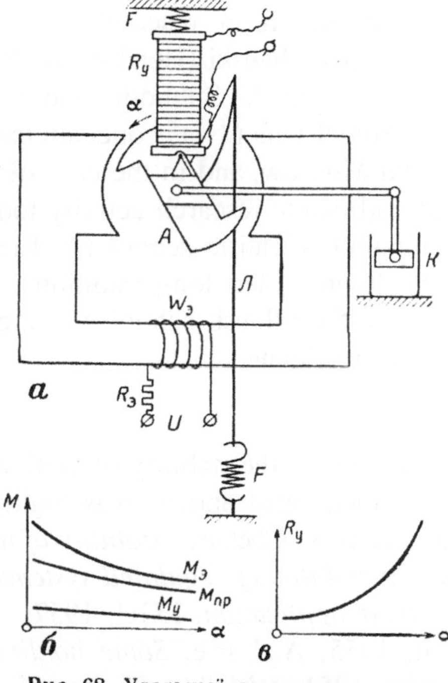

Рис. 68. Угольный регулятор напряжения

Уравнение движения якоря электромагнита имеет вид:

$$J\frac{d^2\alpha}{dt^2} + D\frac{d\alpha}{dt} = M_э - M_{пр} - M_у,$$

где α — угол поворота якоря;
J — момент инерции подвижных частей, приведенный к оси якоря;
D — коэффициент демпфирования;
$M_э$ — момент, развиваемый электромагнитом;
$M_{пр}$ — момент, создаваемый пружиной;
$M_у$ — момент, обусловленный реакцией угольного столба.

109

Figure 10

Some problems in the theory of stability of motion

N. N. Krasovski,

Moscow, Fizmatgiz, 1959 (211 p.)

There are several reasons to distinguish this book among many others, which are in between a research monograph and a textbook (such compromise was highly typical for the Soviet literature of that period).

The author is the founder of the famous Sverdlovsk (now Ekaterinburg) school on automatic control. It includes such prominent researchers as Yu. Osipov (President of the Russian Academy of Sciences), A .Kurzhanski (Head of the Russian National Committee on Automatic Control), the late Academician A. Subbotin and many others. It is worth mentioning that research activity in the Soviet Union was concentrated mainly in Moscow. However there were scientific centers beyond Moscow, and in the control field two of them were of the top level – Sverdlovsk and Leningrad (some research activity took place also in Kiev, Gor'kii, Kazan and other cities). The Leningrad scientific school (A. Lur'e, V. Yakubovich, A. Pervozvanski, V. Fomin, A .Fradkov, G. Leonov) has long traditions, going back to the brilliant St.-Petersburg mathematicians, while the Sverdlovsk school was created from the beginning, and the role of N.Krasovski is hard to overestimate.

The book (Fig. 11) is devoted to the stability of motion. This direction of research (originated by Lyapunov, as has been mentioned above) was highly popular in the Soviet literature of that period. Such publications as N. Chetaev, *Stability of motion*, Gostekhizdat, 1956; M. Aizerman, F. Gantmacher, *Absolute stability of regulated systems*, AN SSSR, 1961; V. Zubov, *Methods of A. M. Lyapunov and their application*, LGU, 1957; A. Letov, *Stability of nonlinear regulated systems*, Gostekhizdat, 1955; A. Lur'e, *Some nonlinear problems in the theory of automatic regulation*, Gostekhizdat, 1951; I. Malkin, *Theory of stability of motion*, Gostekhizdat, 1952; V. Pliss, *Some problems in the theory of stability in whole*, LGU, 1958; E. Barbashin, *Introduction to stability theory*, Nauka, 1967, played a fundamental role in the development of stability theory, and even among them the book under consideration takes a top-level position.

Indeed, the book contains a lot of facts and techniques available in stability theory by that time. They are presented in simple and explicit form. All the statements are accompanied with proofs; however, the style is very far from formal. Many new results are presented, among them the important Barbashin-Krasovski theorem on the existence of Lyapunov function, stability under disturbances bounded in L_1 norm, stability of time-delay systems (Fig. 12) and so on.

Later N. Krasovski contributed to the theory of optimal control (see his book *Theory of control of motion,* Nauka, 1968), to game-theoretical approach to control and to many other areas.

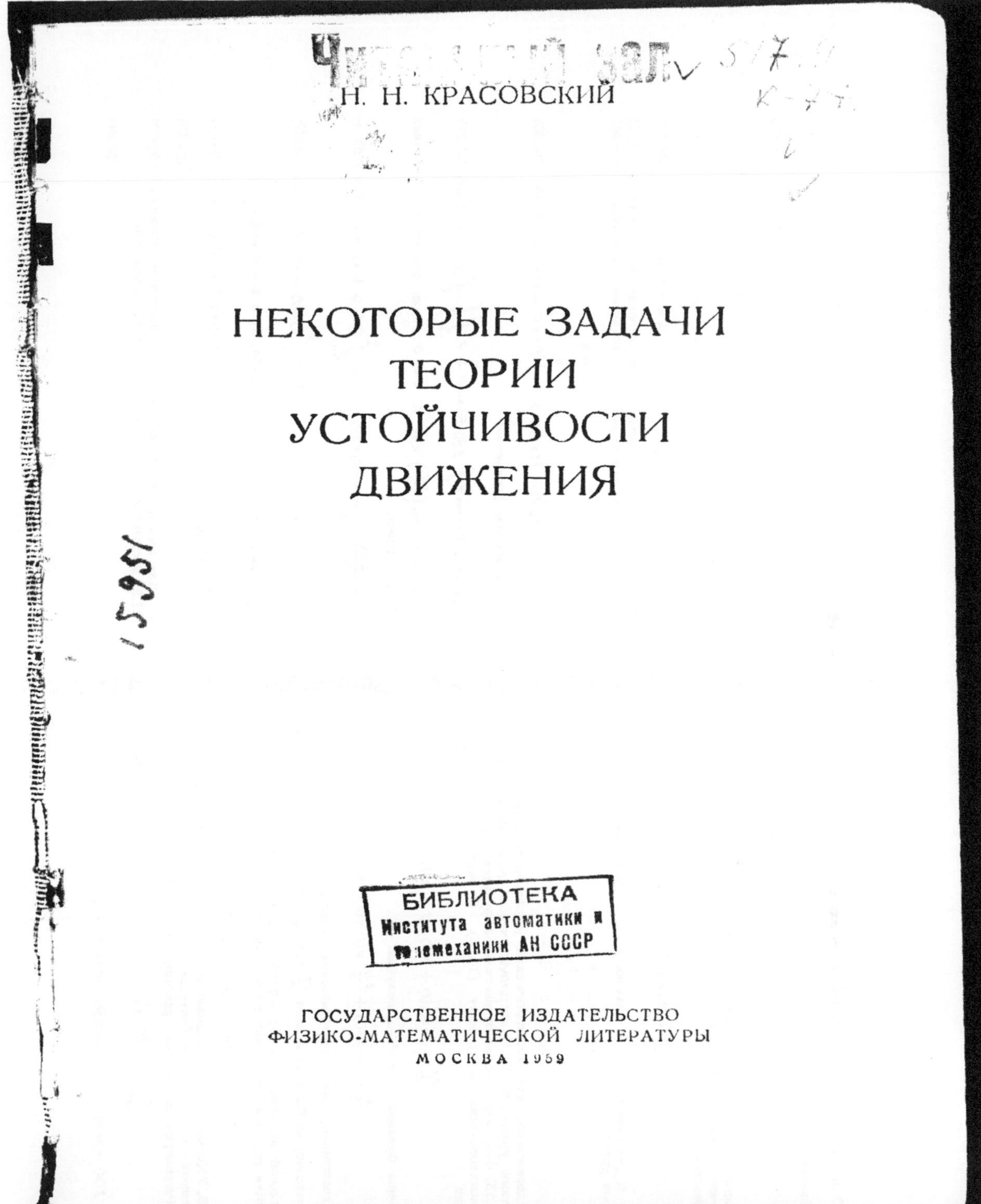
Н. Н. КРАСОВСКИЙ

НЕКОТОРЫЕ ЗАДАЧИ ТЕОРИИ УСТОЙЧИВОСТИ ДВИЖЕНИЯ

ГОСУДАРСТВЕННОЕ ИЗДАТЕЛЬСТВО
ФИЗИКО-МАТЕМАТИЧЕСКОЙ ЛИТЕРАТУРЫ
МОСКВА 1959

Figure 11

удовлетворяющих условию

$$v(x_1(0), \ldots, x_n(0)) > qv(x_1(\xi), \ldots, x_n(\xi)) \quad \text{при} \quad \xi < 0, \; 0 < q < 1, \; q = \text{const}. \tag{34.14}$$

Кривые $x_i(\xi)$, удовлетворяющие неравенству (34.14), во всяком случае содержатся в семействе кривых $x_i(\xi)$, удовлетворяющих неравенству

$$\sum_{i=1}^{n} x_i^2(0) \geqslant q\frac{\rho_{\min}}{\rho_{\max}} \sum_{i=1}^{n} x_i^2(\xi), \tag{34.15}$$

где $\rho_{\min}$ — наименьший по модулю, а $\rho_{\max}$ — наибольший по модулю корни уравнения

$$\begin{vmatrix} b_{11}-\rho & \ldots & b_{1n} \\ b_{21} & \ldots & b_{2n} \\ \ldots & \ldots & \ldots \\ b_{n1} & \ldots & b_{nn}-\rho \end{vmatrix} = 0.$$

Следовательно, для асимптотической устойчивости решения $x = 0$ системы уравнений (34.1) достаточно, чтобы правая часть равенства (34.13) была определенно-отрицательной функцией на кривых $x_i(\xi)$, удовлетворяющих неравенству (34.15). В частности, для уравнения

$$\frac{dx}{dt} = -ax(t) + b(t)\,x(t - h(t)),$$

выбирая функцию $v = x^2/2a$, получим:

$$\frac{dv}{dt} = -x^2(t) + \frac{b(t)}{a}x(t)\,x(t - h(t)).$$

Следовательно, для асимптотической устойчивости решения $x = 0$ достаточно, чтобы производная $\frac{dv}{dt}$ была функцией определенно-отрицательной на кривых $x^2(0) \geqslant qx^2(\xi)$, и для асимптотической устойчивости решения достаточно выполнения неравенства

$$|b(t)| < qa \quad (0 < q < 1.$$

3. Рассмотрим еще несколько примеров исследования простых нелинейных систем методом Ляпунова.

Рассмотрим нелинейное уравнение второго порядка

$$\frac{d^2x}{dt^2} + \varphi\left(\frac{dx}{dt}, t\right) + f(x(t - h(t))) = 0, \tag{34.16}$$

где $f(x)$ — непрерывно дифференцируемая функция, удовлетворяю-

щая условиям

$$\frac{f(x)}{x} > a > 0, \quad |f'(x)| < L \quad \text{при} \quad x \neq 0, \tag{34.17}$$

$\varphi(y, t)$, $h(t)$ — непрерывные, периодические функции времени t, удовлетворяющие неравенствам:

$$\frac{\varphi(y, t)}{y} > b > 0 \quad \text{при} \quad y \neq 0, \tag{34.18}$$

$$h(t) \geqslant 0, \quad h(t) \leqslant h \quad (h = \text{const}). \tag{34.19}$$

При $t > h$ (считая, что начальное возмущение имело место при $t \leqslant 0$) уравнение (34.16) можно записать в следующем эквивалентном виде:

$$\frac{dx}{dt} = y,$$

$$\frac{dy}{dt} = -\varphi(y(t), t) - f(x(t)) + \int_{-h(t)}^{0} f'(x(t+\vartheta))\,y(t+\vartheta)\,d\vartheta. \tag{34.20}$$

Рассмотрим функционал

$$V(x(\vartheta), y(\vartheta)) = 2\int_0^x f(\xi)\,d\xi + y^2 + \nu^2 \int_{-h}^{0}\left(\int_{\xi}^{0} y^2(\vartheta)\,d\vartheta\right)d\vartheta, \tag{34.21}$$

и вычислим $\lim\left(\frac{\Delta V}{\Delta t}\right)$ при $\Delta t \to +0$ вдоль траекторий уравнения (34.16). Имеем:

$$\lim_{\Delta t \to +0}\left(\frac{\Delta V}{\Delta t}\right)_{(34.16)} = -2y\varphi(y, t) + 2\int_{-h(t)}^{0} f'(x(t+\vartheta))\,y(t+\vartheta)\,y(t)\,d\vartheta + \nu^2\int_{-h}^{0}(y^2(t) - y^2(t+\vartheta))\,d\vartheta. \tag{34.22}$$

При $\nu = a/h$ из условий (34.17), (34.18) получим теперь оценку

$$\lim_{\Delta t \to +0}\left(\frac{\Delta V}{\Delta t}\right)_{(34.16)} < -\int_{-h}^{0}\left(\frac{a}{h}y^2 - 2L|y(t)\,y(t+\vartheta)| + \frac{a}{h}y^2(t+\vartheta)\right)d\vartheta. \tag{34.23}$$

Для того чтобы функционал V удовлетворял условиям теоремы 31.3, достаточно потребовать выполнения неравенства

$$h < \frac{a}{L}. \tag{34.24}$$

Figure 12

EARLY CONTROL TEXTBOOKS in SPAIN

Teoría de los Servomecanismos (Servomechanism Theory)

Antonio Colino
Published in 1950

Contributed by

Javier Aracil

Escuela Superior de Ingenieros
Universidad de Sevilla, Spain

Teoría de los Servomecanismos (Servomechanism Theory)

Antonio Colino
Published in 1950

The first book published in Spanish (at least as far as I am aware) on what is known today as Control Engineering is *Teoría de los Servomecanismos (Servomechanism Theory)* by Antonio Colino. It is well known that the field of Engineering that has come to be known as Automatic Control came about as a result of the study of a type of electromechanical feedback system, servomechanisms. It is worth noting that this book was published in 1950, just a few years after the first books on this subject came out in English (e.g. the classic *Theory of Servomechanisms*, by Hubert M. James, Nathaniel B. Nichols and Ralph E. Phillips, published in 1947).

Colino's book was an introduction to this new field of engineering for the Spanish-speaker scholar. He had to deal with the translation of new terms, especially *feedback*, for which he aptly proposed the Spanish word *realimentación*. It should be mentioned that, since 1972, Antonio Colino has been a member of the Spanish Royal Academy of the Language, the most important Spanish linguistic institution. The prologue to the book was written by one of the most illustrious Spanish engineers of the 20th century, Esteban Terradas, who perceptively pointed out how important automatic control would be for engineering in the 20th century. Colino's proposal was accepted and since then engineers in Spain –automatic control engineers and electrical engineers alike– have used the term *realimentación*.

This acceptance was based as much on the use of the term (which, ultimately, justifies a term's adoption) as its appropriateness given that the prefix *re-* in *realimentación* means "once more". Some authors use the word *retroalimentación*. I think, however, that *realimentación* gives a better sense of what happens in this process. As *realimentación* connotes, the system is fed again with the results of the actions previously taken. This is exactly what we want the term to represent; thus, I think Colino's proposal was excellent and that it demonstrates great awareness of the Spanish language.

Even today, the index of the book comprises the programme for an introductory course on the theory of automatic control of linear systems – specifically, on the classic methods based on the study in the frequency domain – and it includes the typical subjects for such a course: transfer functions and their graphic representations, stability criteria (specifically, Nyquist's), and other such topics. The book was used as a textbook and from it the Spanish engineers of the 1950's learned the fundamentals of automatic control.

CONSEJO SUPERIOR DE INVESTIGACIONES CIENTÍFICAS
PATRONATO JUAN DE LA CIERVA
INSTITUTO NACIONAL DE ELECTRONICA

TEORIA
DE LOS
SERVOMECANISMOS

POR

A. COLINO
Ingeniero Industria

Con prólogo de E. TERRADAS

MADRID
1950

INDICE

Tendremos en un servo sencillo

$$\theta_s = G\,\varepsilon \quad ; \quad \varepsilon = \theta_e - \theta_s$$

luego en función de la entrada se obtendrá

$$\varepsilon = \frac{\theta_e}{1 + G} \qquad \theta_s = \frac{G}{1 + G}\,\theta_e$$

La función $G_0(p) = \dfrac{G(p)}{1 + G(p)}$ es la transmisión conjunta del servo, que liga la salida con la entrada, teniendo en cuenta los efectos de la realimentación.

c) Servo con circuito en la realimentación (figura 3.3):

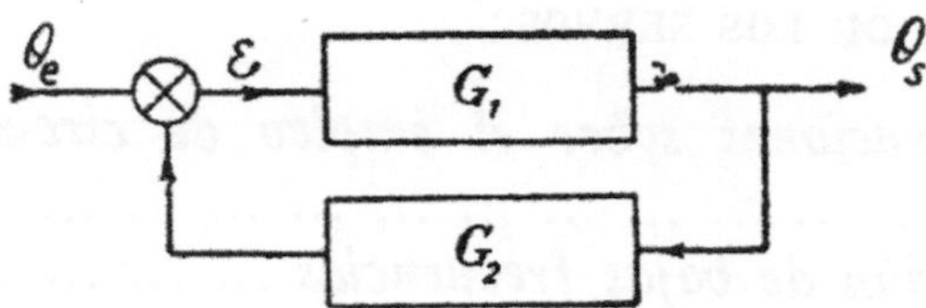

Fig. 3.3.—*Servo con circuito en la realimentación.*

De la figura se deduce

$$\theta_s = G_1\,\varepsilon \quad ; \quad \varepsilon = \theta_e - G_2\,\theta_s$$

$$\theta_s = \frac{G_1}{1 + G_1 G_2}\,\theta_e$$

Este tipo de servo no es muy empleado, porque puede producir error permanente (véase la teoría de errores en los servos).

d) Servo con doble realimentación.—Este tipo de servo es mucho más empleado (recuérdese la estabilización con tacómetro). (Figura. 3.4.)

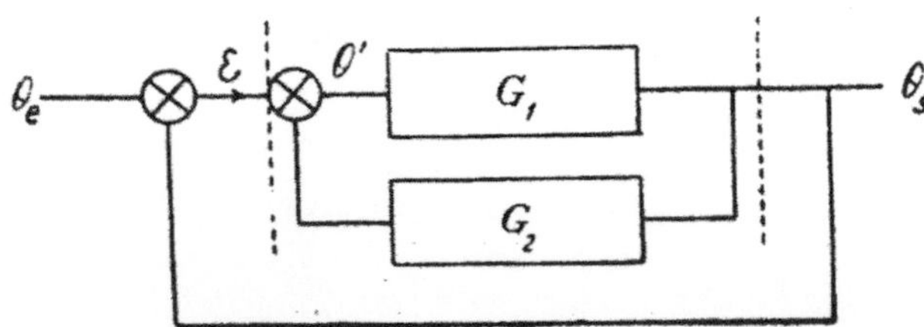

Fig. 3.4.—*Servo con doble realimentación.*

34

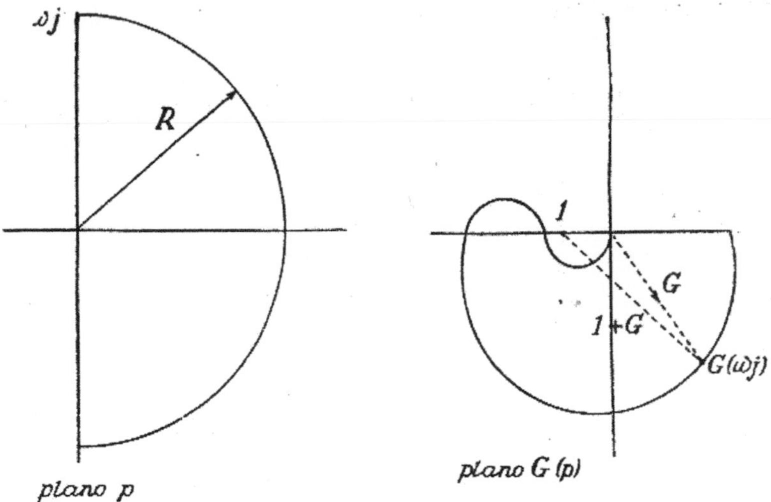

Fig. 3.7.—*Diagrama de Nyquist.*

En resumen: el criterio de Nyquist permite simplemente, trazando el lugar geométrico de G (ωj), conocer si un servo es estable o inestable, según encierre o no el punto —1. Por ser G (p) una función de p con coeficientes reales G (ωj) y G $(-\omega j)$, son sencillamente conjugadas, y basta trazar G (ωj) de $\omega = 0$ a $\omega = \infty$ ya que la figura de $\omega = 0$ a $\omega = -\infty$ es la mera imagen de la curva anterior en el eje real.

Hay que hacer observar que G (ωj) no es más que la respuesta a una función exponencial $e^{j\omega t}$ y, por lo tanto, medible experimentalmente en sus dos componentes de amplitud y fase, lo que permite el trazado experimental del diagrama de Nyquist de un servo ya construído.

6. *Condiciones complementarias a la de estabilidad en un servo.*—Para enjuiciar la actuación de un servo no es suficiente conocer si es o no estable; es además preciso conocer otras características, como son el tiempo de formación de la salida, forma de ésta, etc.

En realidad, para llegar a un conocimiento exacto sería preciso hallar la inversión de

$$\theta_s(p) = \frac{G(p)}{1 + G(p)} \theta_e(p)$$

39

EARLY CONTROL TEXTBOOKS

in

SWEDEN

Reglerteori (Control Theory)

Laszlo von Hamos
Svenska bokförlaget Bonniers. Stockholm 1968.

Reglerteori (Control Theory)

Karl Johan Åström
Almqvist \& Wiksell, Stockholm 1968.

Contributed by

Karl Johan Åström

Lund Institute of Technology
Lund, Sweden

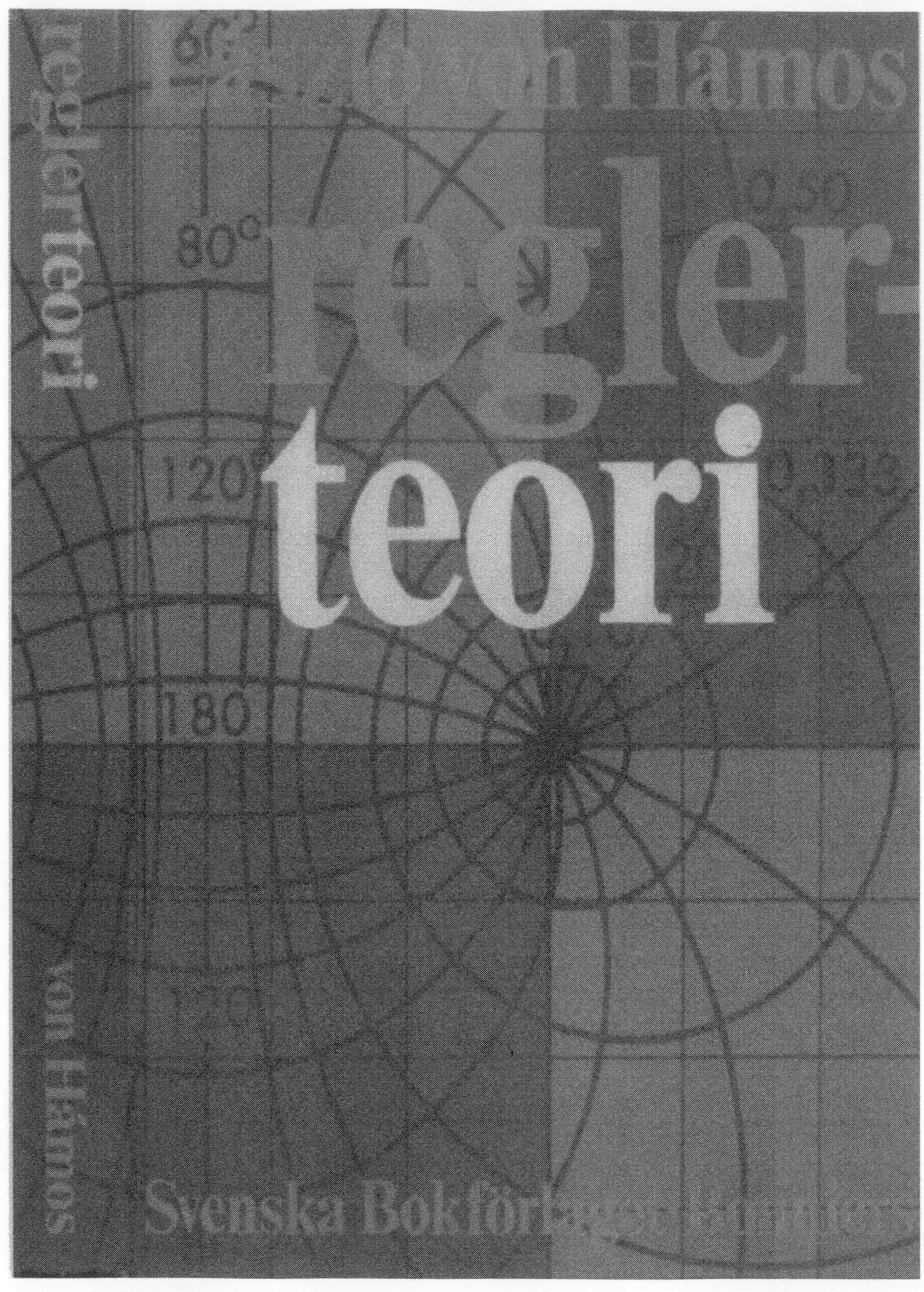
Hámos
regler-
teori
reglerteori
von Hámos
80°
120
180

Reglerteori (Control Theory)

By Laszlo von Hamos
Svenska bokförlaget Bonniers. Stockholm 1968.

Swedish education in automatic control began in 1949 at the Royal Institute of Technology (KTH) in Stockholm. The first courses were given by Laszlo von Hamos, who worked for the Robot Weapons Bureau (Robotvapenbyrån) at the Swedish Air Force. Laszlo von Hamos became the first professor of automatic control in Sweden in 1959. He also the Swedish representative who signed the Heidelberg resolution that led to the creation of IFAC. The control group at KTH had two experienced teaching assistants, Gunnar Attebo from Källe, a Swedish process control company and Bengt Sjöberg from Saab Aircraft. The broad industrial base of the teachers gave a sound industrial perspective inspired from both aerospace and process control applications. The course was also complemented by laboratory exercises.

The book Reglerteori developed from the course material that von Hamos taught at KTH since the early 1949s. The book has 18 chapters: Introduction. Classification of control systems. Analysis of control systems by solving differential equations. Alternative ways to describe dynamics of feedback loops. Differential equations for control components. Combination of linear blocks. Graphical representations of transfer functions (Bode, Nyquist and Nichols plots). Overview of dynamical properties of blocks with simple differential equations. Methods for computing transient responses. Algebraic stability analysis (Routh-Hurwitz criteria). Root-locus. Stability analysis based on transfer functions (the Nyquist theorem). Static accuracy. Dynamic accuracy. Improvement of system properties (compensation). Overview of nonlinear systems. Simulation using analog computing. Overview of advanced control methods (multivariable, adaptation, optimization, learning, computer control). There are appendices on Laplace transforms and exercises. In recommendations for further reading there are references to the textbooks by D'Azzo and Houpis, Dorf, Naslin, Oppelt and Takai. The book gives a good coverage of classical control theory emphasizing modeling in terms of block diagrams and Laplace transforms, design based on root locus, Bode and Nichols charts and simulation based on analog computing.

It is interesting that in the early 1950s there was a controversy among the teachers at KTH about the best way to teach dynamics of linear systems. One professor in electromagnetics was a strong advocate of the Heaviside operation calculus, other professors in mechanical engineering advocated the Laplace transform. A hybrid, called the s-multiplied Laplace transform was also used by one professor. A reflection of the controversy was that von Hamos used the variable p instead of s as the argument of the transforms.

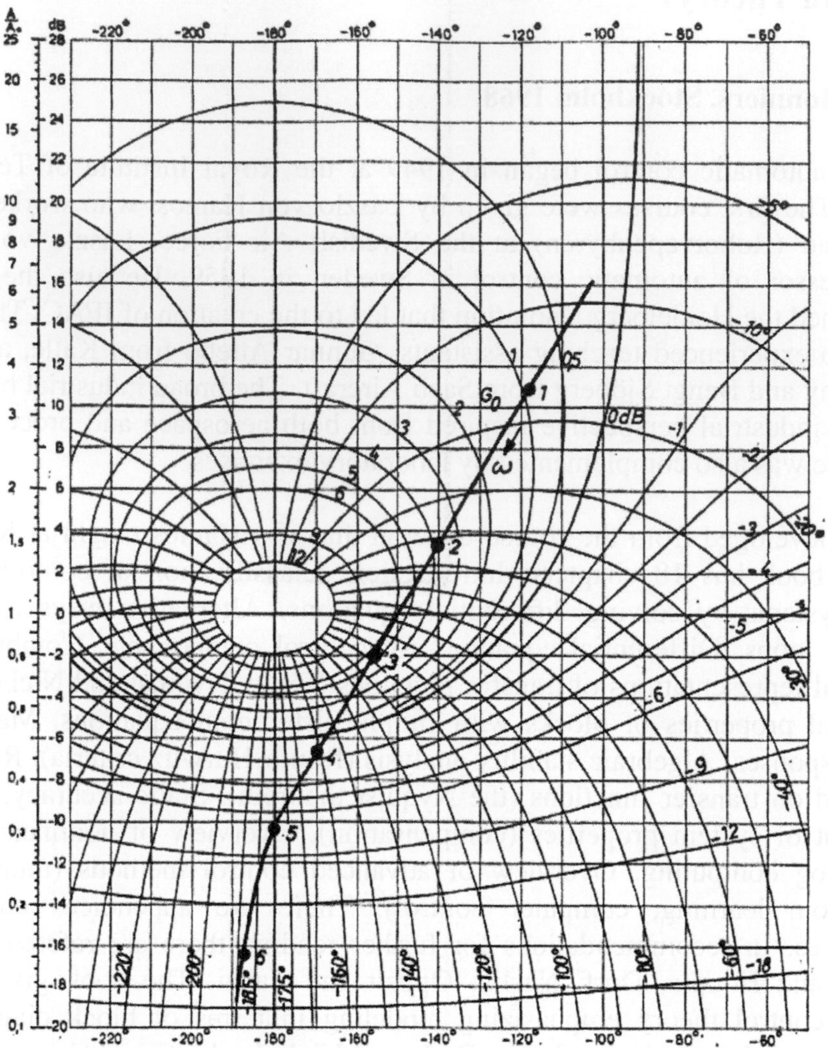

Fig. 7 : 11. Nichols-diagram med kurvskaror för konstant M i dB och konstant N i grader. Kurvan G_0 härrör från typexempel 3 : 2 så att figuren är en motsvarighet till fig. 7 : 5.

således endast rita $G_0(j\omega)$ på ett genomskinligt papper och translatera bladet parallellt med ln $|G|$-axeln tills G_0-kurvan kommer att beröra M-kurvan med det önskade värdet. Anmärkas bör, att kurvan G_0 lämpligen konstrueras i ett Bode-diagram på ovan angivet sätt, varefter den överförs till Nicholsdiagrammet.

102

Reglerteori

KJ Åström Reglerteori

Reglerteori (Control Theory)

By Karl Johan Åström
Almqvist \& Wiksell, Stockholm 1968.

Karl Johan Åström was appointed professor in Automatic Control at the newly established engineering school Lund Institute of Technology in 1965. Prior to that, he had worked for IBM Research under Jack Bertram, a class-mate of Kalman, Franklin and others from Columbia University. Returning to university Åström carried with him early experiences from computer control and a perspective on the American university scene. Returning to the university Åström developed a new course in control that integrated classical control with the emerging state space approach. The advantage with such an approach is that it introduces two key approaches to control in a unified framework. The course was supplemented with exercises, and laboratories that were integrated with lectures. A book based on the new course was published in 1968. The book has five chapters: Introduction, Dynamical Systems, Reachability and observability, Analysis of Feedback Systems, Stability and Design. The introduction paints a broad perspective on control, its ideas and practical application. The section of dynamical systems starts with state space models and continues with input-output descriptions. The treatment of reachability and observability leads up to Kalman's representation theorem which gives a nice way to present the classical difficulty of pole-zero cancellation. The chapter on analysis presents classical concepts of sensitivity functions, Bodes relations and error coefficients. Stability is treated both from the Lyapunov point of view and as input-output stability. The particular methods include Routh-Hurwitz criterion, root-locus and the Nyquist theorem, practical stability and margins are also discussed. The chapter on Design treats specifications and control system design both from the viewpoints of classical compensation as well as state feedback and observers.

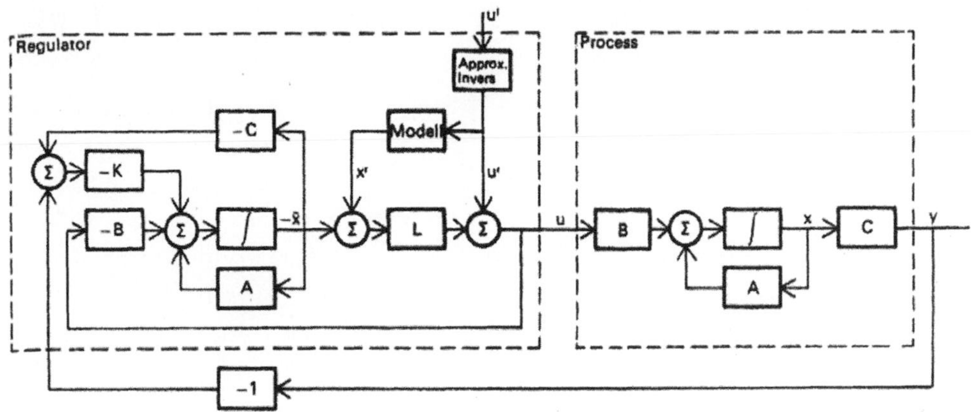

Fig. 5.1. Blockschema för en regulator baserad på tillståndsåterkoppling och rekonstruktion av icke mätbara tillståndsvariabler med ett Kalmanfilter.

$$\frac{d}{dt}\begin{pmatrix} x \\ \tilde{x} \end{pmatrix} = \begin{pmatrix} A-BL & BL \\ 0 & A-KC \end{pmatrix}\begin{pmatrix} x \\ \tilde{x} \end{pmatrix} + \begin{pmatrix} B \\ 0 \end{pmatrix} u^r + \begin{pmatrix} BL \\ 0 \end{pmatrix} x^r$$

$$y = Cx$$

$$u = -Lx + L\tilde{x} + u^r + Lx^r \tag{5.25}$$

Vi finner således att det slutna systemet är av ordningen $2n$. Systemets poler är egenvärdena till matriserna $A-BL$ och $A-KC$. Det följer vidare av (5.25) att tillståndet $\tilde{x}$ ej är styrbart vare sig från u^r eller x^r. Jämför med analysen i avsnitt 3:6. De moder som genereras vid rekonstruktionen av tillståndsvariablerna är således ej styrbara.

Enligt Kalmans uppdelningssats (sats 3:6.1) kan det återkopplade systemet uppdelas på det sätt som visas i fig. 5.2. Insignal-utsignal-relationen från u^r och x^r till y ges av

$$Y(s) = C[sI - A + BL]^{-1}B[U^r(s) + LX^r(s)]$$

Överföringsfunktionen är således entydigt bestämd av A, B, C och L. Den är helt oberoende av hur matrisen K väljs. Man kan visa att analoga förhållanden gäller oavsett hur tillståndsvariablerna rekonstrueras. De moder som införs i observeraren blir alltid icke styrbara.

Den regulatorstruktur som visas i fig. 5.1 är mycket flexibel. Den kan också användas vid reglering av flervariabla system. För att dimensionera regulatorn krävs kunskap om matriserna A, B och C, som kan erhållas ur en matematisk modell av processen. Vidare måste matriserna K och L bestämmas. För ett styrbart system med en insignal ger sats 5.1 ett L sådant att det slutna systemet får givna poler. Det finns också utvidgningar av sats 5.1 till flervariabla system. Val av matrisen K påverkar ej det slutna systemets över-

EARLY CONTROL TEXTBOOKS
in
SWITZERLAND

Die Regelung von Dampfanlagen (Control of Thermal Power Plants)

Paul Profos
Springer-Verlag Berlin/Göttingen/Heidelberg 1962

Contributed by

Adolf H. Glattfelder and Walter Schaufelberger

ETH (Federal Technical University) Zürich
Switzerland

Die Regelung von Dampfanlagen (Control of Thermal Power Plants)

Paul Profos
Springer-Verlag Berlin/Göttingen/Heidelberg 1962

Paul Profos received his Diploma Degree at ETH in 1936 and his Ph. D. in 1943 with a dissertation on "Vektorielle Regeltheorie". In this early work, he used vectors in the complex plane and frequency response techniques. Paul Profos worked for about twenty years in industry, already during his Ph. D., with Sulzer Brothers Limited. He joined ETH in 1958 as Full Professor.

In his book on "Control of Thermal Power Plants" he summarises the practical and theoretical work done in industry with the goal to make it explicitly available to practicing engineers.

The motivations to this book were the rapid changes in the construction of thermal power plants from 1945 to around 1960. It marks the transition from traditional drum boilers with natural circulation to once-through steam generators which opened the way to a substantial increase in steam pressure, steam temperature and steam mass flow and thus in unit power output. However, such once-through steam generators could no longer be operated safely in manual control mode, but required a high performance control system. It could no longer be designed by the traditional rules of thumb, but required a systematic approach based on mathematical models.

A key case was the super-heater control. As the steam temperature came close to the limit of the materials, dynamic responses became much more critical. Aging or the threat of ruptures required better theoretical insight and consequently Profos developed his "kappa-D" theory to improve temperature control. Possibilities of oscillations and instability in other parts of the plant (evaporators, parallel tubes) were also investigated. Already at that time, control engineering was seen as an integrating technology.

Soon after the book appeared, systematic simulation studies became routine in order to guarantee performance of the controlled systems.

The book focuses on modelling. Its main elements are transparent physics, and well chosen assumptions leading to simple mathematical models. They are given as low order differential-difference equations and/or transfer functions. The control systems are described by structures and block-diagrams where Profos uses the then new IFAC graphic symbols. Typical controller settings and simulations of closed loop responses are not presented, as the needed technologies were not yet available at the time.

The book is addressed to practicing engineers and advanced students and quickly became the standard in the field. It adheres to the slogan of Profos "Es gibt nichts praktischeres als eine gute Theorie" („Nothing is more practical than a good theory.").

Die Regelung von Dampfanlagen

Von

Dr. Paul Profos
o. Professor
an der Eidgenössischen Technischen Hochschule
Zürich

Mit 320 Abbildungen und Tabellen

Springer-Verlag
Berlin / Göttingen / Heidelberg
1962

neben aber alle noch ein Glied enthalten, das die schon früher erwähnte komplizierte transzendente „Formfunktion" nun in der Bildung

$$F = \frac{1}{\sigma} e^{-\varkappa_D \frac{\sigma}{1+\sigma}} \tag{7.145}$$

enthält ($\zeta = 1$).

Die Anwendung der LAPLACE-Rücktransformation darauf liefert (über die Lösung von Faltungsintegralen)

$$\gamma[t] = L^{-1}\{F\} = e^{-\varkappa_D}\left\{e^{-\tau} J_0(2\sqrt{\varkappa_D \tau}) + \int_0^{\tau} e^{-\tau} J_0(2\sqrt{\varkappa_D \tau})\, d\tau\right\}. \tag{7.146}$$

Die darin enthaltene BESSEL-Funktion J_0 ist nur über einen Teil des praktisch benötigten Argumentbereiches tabelliert. Für den übrigen Bereich sind Reihenentwicklungen zu Hilfe zu nehmen.

Die der *Formfunktion* entsprechende Teilübergangsfunktion legt im wesentlichen den Verlauf der Übergangsfunktion bei Temperatur-

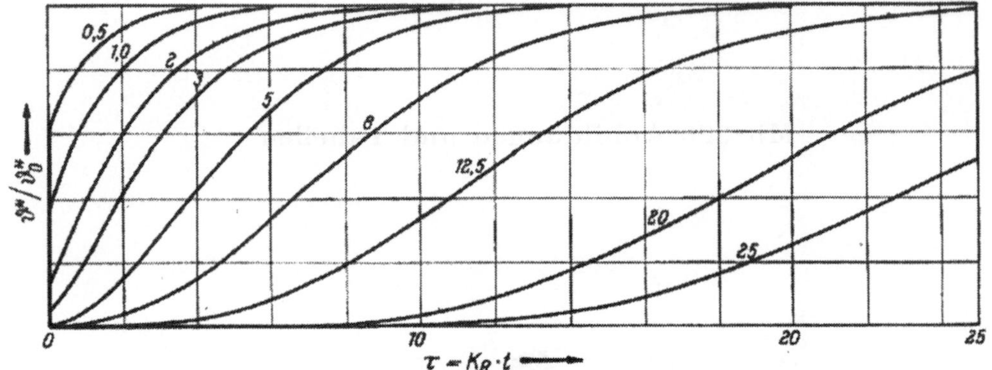

Abb. 7.40 Schar der Übergangsfunktionen entsprechend der Formfunktion $\gamma_{II}(t)$, mit $\varkappa_D$ als Parameter

störung fest. Sie ist, analog dem Frequenzgang der Formfunktion, nur von dem *einzigen Parameter* $\varkappa_D$ abhängig. Es läßt sich daher auch hier die Gesamtheit der auftretenden Teilübergangsfunktionen durch eine Schar von Kurven erfassen. Diese ist in Abb. 7.40 für bezogenes Zeitmaß dargestellt.

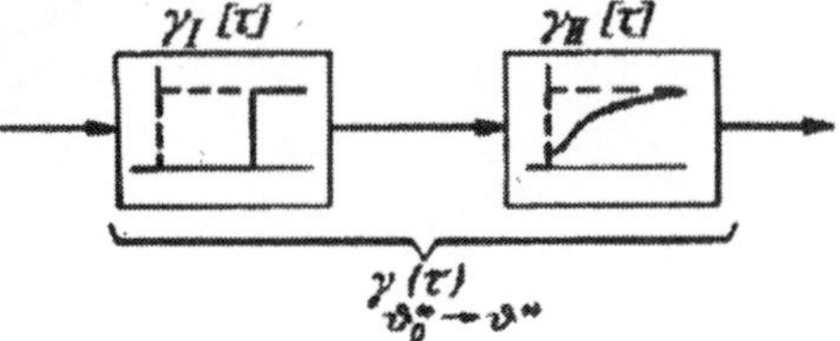

Abb. 7.41 Blockschema zur Bildung der Übergangsfunktion für Temperaturstörung aus den Teilübergangsfunktionen $\gamma_I(t)$ und $\gamma_{II}(t)$

Zur Berechnung der Übergangsfunktion für Temperaturstörung ist nun noch die dem Teilfrequenzgang G_I (s. Gl. (7.135)) entsprechende Teilübergangsfunktion $\gamma_I[t]$ zu berücksichtigen. Diese hat — wie bereits bei der Betrachtung des Frequenzganges konstatiert — reinen Totzeitcharakter (vgl. Abb. 7.41).

EARLY CONTROL TEXTBOOKS
in
TURKEY

Servomekanizmalarin Prensipleri

Translation of Principles of Servomechanisms
G. S. Brown and D. P. Campbell, John Wiley, 1948
Translated by Münir Ülgür
Istanbul Technical University, 1962

Otomatik Kontrol Ders Notlari

(Automatic Control Course Notes)
Mehmet Nimet Özdaş
Istanbul Technical University, 1978

Contributed by

A. Talha Dinibütün

Doğuş University
Istanbul, Turkey

History. Education in Automatic Control in Turkey started about 50 years ago at Istanbul Technical University (ITU). In 1953, Prof. Münir Ülgür started teaching the course "Servomekanizma" at the Faculty of Electrical Engineering at ITU. In 1955, Prof. Mehmet Nimet Özdaş started teaching the course "Automatic Control" at the Mechanical Engineering Faculty of ITU. At that time, there were no textbooks written in Turkish for these courses. Both professors, therefore, initially prepared and used their personal course notes, and those students who knew foreign languages made use of relevant textbooks written in foreign languages.

Servomekanizmalarin Prensipleri

Translation of Principles of Servomechanisms
G. S. Brown and D. P. Campbell, John Wiley, 1948
Translated by Münir Ülgür
Istanbul Technical University, 1962

Professor Ülgür recommended the textbook "Principles of Servomechanisms by G. S. Brown and D. P. Campbell, John Wiley, 1948" to his students who knew English. He later translated this book and had it printed by ITU in 1962. Starting in 1962, he used this book as a textbook for the course "Servomekanizma". The first textbook in the field of Automatic Control in Turkish to be used in Turkey was this translated book.

The Turkish title of this 400-page textbook "Principles of Servomechanism", translated by Prof. Ülgür is "Servomekanizmalarin Prensipleri" and it is a 427-page book.

The *contents* of the translated textbook are the same as the original one. The main titles of these contents are:

1. Outline of Subject
2. Dynamics of Elementary Control Systems
3. Transient Response Using the Laplace Transform
4. Sinusoidal Response of Closed-Loop Systems
5. Systems Diagrams, Equivalent Circuits and Block Diagrams
6. Introduction to Synthesis, Determining the Gain Constant K
7. Methods of G Function Synthesis – Linear Coordinates
8. Methods of G Function Synthesis – Logarithmic Coordinates
9. Systems Subjected to Multiple Disturbances
10. Experimental Studies in Servomechanisms
11. Method for Approximating the Transient Response from the Frequency Response

    Table of (sin x)/x

    Problems

    Bibliography

    Index

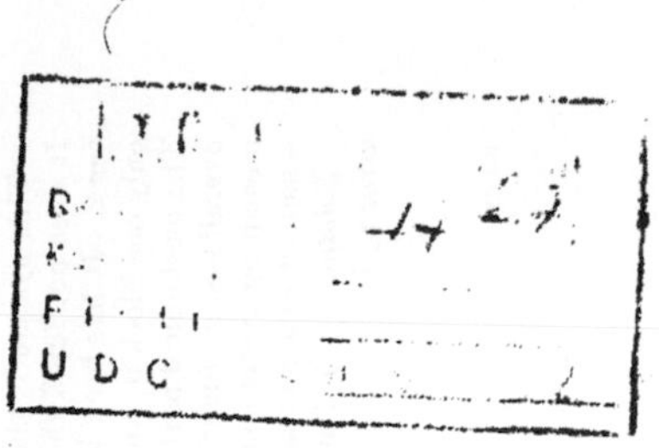

SERVOMEKANİZMALARIN PRENSİPLERİ

Kapalı-Çevrim Kontrol Sistemlerinin Dinamikleri ve Sentezleri

YAZANLAR:
GORDON S. BROWN, PROFESSOR OF ELECTRICAL ENGINEERING
DIRECTOR, SERVOMECHANISMS LABORATORY
MASSACHUSETTS INSTITUTE OF TECHNOLOGY
DONALD P. CAMPBELL, ASSISTANT PROFESSOR OF ELECTRICAL ENGINEERING
MASSACHUSETTS INSTITUTE OF TECHNOLOGY

ÇEVİREN :
M. MÜNİR ÜLGÜR, PROFESÖR, ELEKTRİK FAKÜLTESİ,
İSTANBUL TEKNİK ÜNİVERSİTESİ

BERKSOY MATBAASI
İSTANBUL — 1962

tachometer (Table 3), item 5. The components are arranged to form a closed loop as shown in Figure 7.

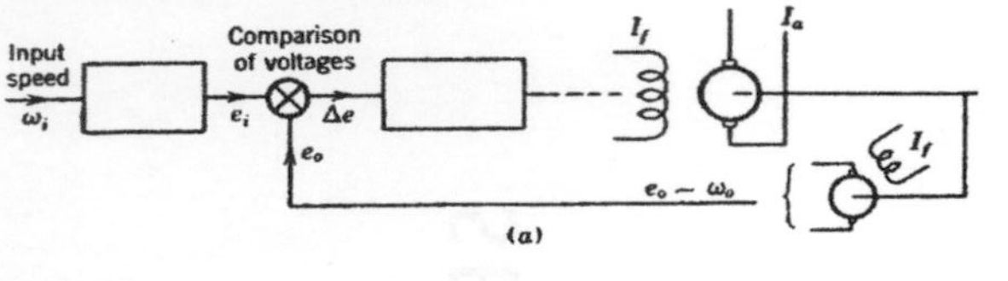

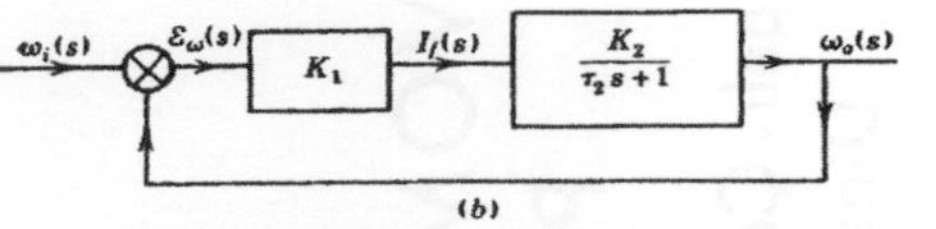

Figure 7. Closed-loop speed control system.

The object of this study is:

(1) To formulate the overall transfer function.
(2) To draw the locus for the transfer function.
(3) To investigate the possible steady-state performance of this closed-loop speed control system without resorting to elaborate mathematical treatment.
(4) To determine the frequency range over which the output-to-input magnitude and phase response fall within acceptable limits.

This system is the same as the one used in Chapter 2, Figure 2·6. It does not fulfill the criteria for a servomechanism as stated in Chapter 2. It has been idealized so that the diagram of Figure 7*b* contains only one block in which energy-storing and dissipating elements appear. The mechanical portion of the shunt motor contributes this effect. The overall system function ignores any possible energy storages in the error-measuring system and the tachometer attached to the output. Their sensitivities are lumped in the constant term K_1.

Solution. (1) The transfer function relating output speed to speed error $\mathcal{E}_\omega$ is

$$\frac{\omega_o}{\mathcal{E}_\omega}(j\omega) = \frac{K_1K_2}{j\omega\tau_2 + 1} \tag{35}$$

where

$$\mathcal{E}_\omega(j\omega) = \omega_i(j\omega) - \omega_o(j\omega) \tag{36}$$

Misal 1. Hızları karşılaştırarak farkı gerilim şeklinde ifade eden hata ölçen bir cihaz; bir amplifikatör (Tablo 3, 10. sıra); ve şönt ikaz akımı kontrol edilen bir doğru akım motoru (Tablo 3, 3. sıra); bir takometre (Tablo 3, 5. sıra)'den müteşekkil kapalı-çevrimli hız kontrol sistemini nazarı itibara alalım. Elemanlar, şekil 7 de gösterildiği gibi, bir kapalı çevrim teşkil edecek şekilde düzenlenmişlerdir.

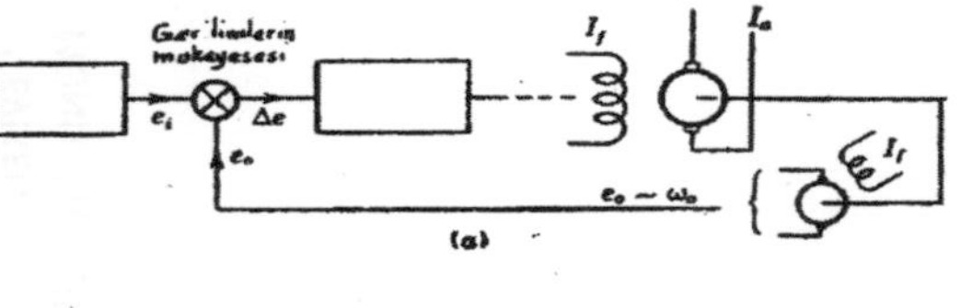

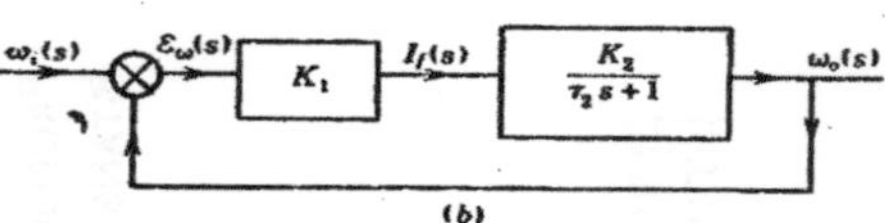

Şekil 7. Kapalı-çevrimli hız kontrol sistemi.

Bu etüdün konusu şunlardır:

(1) Tüm transfer fonksiyonunun formüle edilmesi,
(2) Transfer fonksiyonunun geometrik yerinin çizilmesi,
(3) Kapalı-çevrimli bu hız kontrol sisteminin mümkün kararlı hal özelliğinin, matematik metodun teferruatına baş vurmaksızın, araştırılması,
(4) Çıkışın-girişe oranının büyüklük ve faz cevabının kabul edilebilen sınırlar içine düştüğü frekans domeninin tayin edilmesidir.

Bu sistem 2. bölümde şekil 2.6 da kullanılmış olanın aynıdır. Bu sistem 2. bölümde ifade edildiği gibi servomekanizmaya ait kriteryumları tahkik etmez. Bu, şekil 7*b* deki diyagramda, yalnız enerji toplayıcı ve sarfedici elemanları bir araya getiren bir bloku ihtiva edecek şekilde idealize edilmiştir. Şönt motorun mekanik kısmı bu tesiri sağlar. Tüm sistem fonksiyonu, hata ölçen sistem ve çıkışa bağlı takometrede meydana gelmesi mümkün her hangi bir enerji toplanışını ihmal etmektedir. Bunların duyarlıkları K_1 sabit terminde toplanmıştır.

Otomatik Kontrol Ders Notlari (Automatic Control Course Notes)

Mehmet Nimet Özdaş

Istanbul Technical University, 1978

Professor Özdaş developed his personal course notes and then the Mechanical Engineering Faculty of the Istanbul Technical University printed those notes under the title "Otomatik Kontrol Ders Notlari" in 1978. This was the first printed material in the Turkish language for automatic control education.

Otomatik Kontrol Ders Notlari is 254 pages long and contains 262 figures and 5 tables. The main section titles are:

1. Introduction to automatic control
2. Modeling of dynamic systems
3. Transient response analysis
4. The Laplace transform
5. Analog computers
6. Measurement
7. Comparison
8. The controller
9. Performance of closed-loop systems
10. Frequency response methods
11. Closed-loop frequency response
12. The stability of linear feedback systems
13. The Root-Locus Method
14. The Nyquist Criterion
15. Performance criteria

İ.T.Ü. MAKİNA FAKÜLTESİ
OTOMATİK KONTROL KÜRSÜSÜ

OTOMATİK KONTROL DERS NOTLARI

Prof. Dr. M. Nimet ÖZDAŞ

İ.T.Ü. MAKİNA FAKÜLTESİ
OFSET ATÖLYESİ

1978

Şu halde amplitüdler oranı $|G|$ i elde etmek için, logaritmik diyagram da her bir terimin cebrik toplamı alınacak demektir.

$(1+j\omega\tau_1)$ veya $\left(\dfrac{1}{1+j\omega\tau_2}\right)$ gibi terimlerin yukarıda belirtilen esaslar dahilinde asimptotik diyagramları çizilebilir. 0 ve −1 eğimli asimptotların kesiştiği KESİM FREKANSI $\omega\tau=1$ veya $\dfrac{1}{\tau}=\omega$ bağıntısıyle bulunur. Yukarıdaki üç terimli sistem fonksiyonunda kesim frekansları $\omega_1, \omega_2, \omega_3$ olursa asmptotik ve hakiki diyagram Şek.10.5 deki gibi olur.

Amplitüdler oranı diyagramı altına faz açısı diyagramı çizilirken her bir terim için faz açısı diyagramı çizilerek, bunların cebrik toplamı alınır. Bu diyagramın çiziminde dikkat edilecek husus $\omega=\dfrac{1}{\tau}$ frekansında faz açısı $\pm\dfrac{\pi}{4}$ olup faz açısı eğrisinin bu noktaya göre simetrik olmasıdır, Şek.10.5.

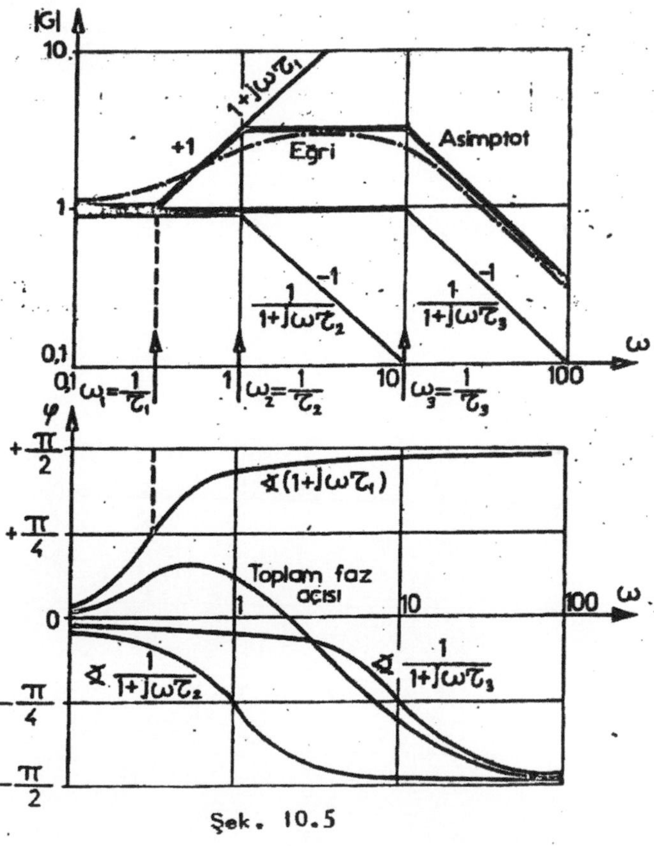

Şek. 10.5

Şu halde amplitüdler oranı $|G|$ yi elde etmek için, logaritmik diyagramda her bir terimin cebrik toplamı alınacak demektir.

$(1 + j\omega\tau_1)$ veya $\left(\frac{1}{1 + j\omega\tau_3}\right)$ gibi terimlerin yukarıda belirtilen esaslar dahilinde asimptotik diyagramları çizilebilir. 0 ve -1 eğimli asimptotların kesiştiği KESİM FREKANSI $\omega\tau = 1$ veya $\frac{1}{\tau} = \omega$ bağıntısıyla bulunur. Yukarıdaki üç terimli sistem fonksiyonunda kesim frekansları $\omega_1, \omega_2, \omega_3$ olursa asimptotik ve hakiki diyagram Şek.10.5 deki gibi olur.

Amplitüdler oranı diyagramı altına faz açısı diyagramı çizilirken her bir terim için faz açısı diyagramı çizilerek bunların cebrik toplamı alınır. Bu diyagramın çiziminde dikkat edilecek husus $\omega = \frac{1}{\tau}$ frekansında faz açısı $\pm\frac{\pi}{4}$ olup faz açısı eğrisinin bu noktaya göre simetrik olmasıdır. Şek.10.5.

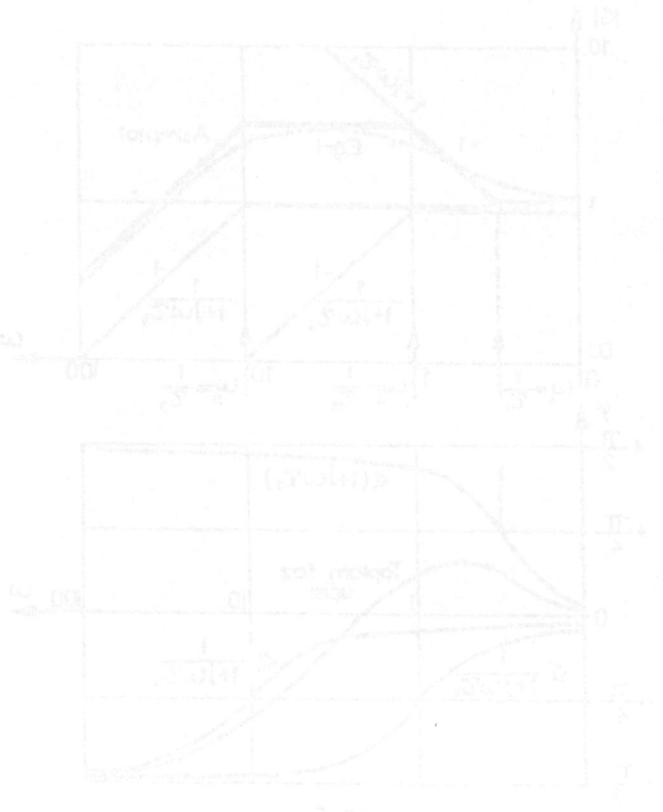

EARLY CONTROL TEXTBOOKS
in the
UNITED KINGDOM

An Introduction to Servomechanisms

Porter, A.,
London: Methuen, 1950, 1952, 1954, 1957

An Introduction to the Theory of Control in Mechanical Engineering

Macmillan, R.H.,
Cambridge, England: Cambridge University Press, 1951

Servomechanisms

West, J.C.,
London: The English Universities Press, 1953

Contributed by

Stuart Bennett

The University of Sheffield
Sheffield, United Kingdom

An Introduction to Servomechanisms

Porter, A.,
1950, London: Methuen.
(Second edition 1952, reprinted 1954 and 1957)

This was the first text book on control systems published in the UK and it was widely used on university and technical college courses. Arthur Porter, an influential figure in the development of automatic control in the UK during the 1940s, served as secretary to the Servomechanisms Panel, a wartime government organization responsible for coordinating work on servomechanisms [1]. As secretary to the panel he was aware of the major developments in the UK and met most of leading control people including Arnold Tustin, A L (John) Whiteley, L. Jofeh and Frederic C Williams. Immediately prior to the war Porter had been awarded a Commonwealth Fellowship which enabled him to spend two years working at MIT with Vannevar Bush's group. And in the early part of 1945 he spent some time in the USA gathering information for a government report on industrial applications of automatic control [2].

Porter, like Harold Locke Hazen and Gordon S Brown, had come to control systems through his involvement with the building and operation of differential analyzers. He recalls that in 1932, as an undergraduate reading physics at the University of Manchester, he read a paper by E C Bullard about solving second order differential equations using a moving coil galvanometer and decided that for his final year project he wanted to do something related to "calculating machines" [3]. Douglas R Hartree had recently returned from a visit to Bush at MIT and wanted to build a mechanical version of the Bush differential analyzer and Porter's interest was welcomed: he was given the task of building a capstan torque amplifier. His involvement with the differential analyzer continued during his MSc and PhD and his interest in control was strengthened through an enquiry from Albert Callender of Imperial Chemical Industries about using the differential analyzer to solve equations relating to automatic control of temperature in a chemical process.

On returning to the UK, in 1939, he had hoped to join P M S Blackett, then at Manchester, but was directed instead to join the Admiralty Research Establishment at Teddington, he subsequently moved to join Blackett's Operational Research Group and then in March 1942 was made head of the fire-control section at the Air Defence Research and Development Establishment at Malvern.

In 1946 he was appointed as Professor and Head of Instrument Technology at the Royal Military College of Science (RMCS), Shrivenham, and it is from this period that the text book emerged. In 1946, with L Jofeh, he gave a series of lectures on servomechanisms at Northampton Polytechnic Institute in London and at the RMCS he gave a course on servomechanisms to second year students studying Engineering Physics. These lectures formed the basis of the book. The first edition of the book had eight chapters: Closed-sequence control systems; Basic equations of linear servo systems; Transient and steady-state behaviour; Harmonic response diagrams; Lag correction and stabilization; Notes on some non-linear problems. In the second edition a chapter on Digital Servomechanisms was added. He avoided the use of the Laplace transform because he wrote "...it tends to surround the subject with a mathematical cloak which sometimes mystifies rather than clarifies", instead he used an "essentially classical treatment" with an emphasis on physical principles.

The structure of the book is one which was to be followed many times in subsequent years: system modelling, transient and steady state behaviour; frequency response; methods of compensation and stabilization; and some discussion of non-linear systems. In dealing with transient and steady state behaviour he covers classical and operational methods of solution and error classes and also, not surprisingly, suggests that the differential analyser and the newer forms of electronic servo-simulators are valuable means of determining transient behaviour.

The section on harmonic response covers the use of the "response vector locus" diagram (Nyquist) and the "log-gain" diagram (Bode), and he mentions the use of the inverse response vector diagram. He uses the diagrams to deal with stability and gain and phase margins but does not mention M and N circles or the Nichols approach. When it comes to modifying the system to improve performance he covers "lag compensation by pre-correction" i.e. feed-forward; "integrals of correction" i.e. use of a lag network; "derivatives of correction" i.e. lead network; and "subsidiary loops". The favoured method, in Britain, for stabilising and modifying servo-system performance was the use of subsidiary feedback loops which "provide a powerful method of stabilizing servo systems" Porter argued, and which can be used to "compensate dynamic lags", "stabilize the system" and "increase the speed of response" and such systems used feedback signals from generator field windings or a tachometer generator.

The addition of servo-control to previously manually operated systems during the early years of the war had quickly shown up problems of non-linear behaviour arising from static friction, saturation of amplifiers and servo-motors, and backlash in gearing throughout the loop. The chapter on non-linear problems derives from work on these problems: the major contributions made by Tustin and the several studies of the problems were done on the differential analyser at University of Manchester are mentioned.

Although Porter's book remained in print and was in use until the 1960s it dated quickly with respect to methods of analysis and design. As a text book its value was in some of the descriptions and modelling of components and in the physical insight into the behaviour. I also think on re-reading the book that what must have come through to students was the enthusiasm and authenticity of the author: this is a book written by someone who had struggled to make things work and knew intimately the problems and limitations.

Before publication of the book, Porter left RCMS and took up the post of Director of Science Research at Ferranti Ltd, Canada. He returned to England briefly in 1955 when he took up the post of Professor of Light Electrical Engineering at Imperial College, London returning to Canada in 1958 as Dean of Engineering at the University of Saskatchewan. In 1961 he took up the post as Head of Industrial Engineering at the University of Toronto which he held until 1975. From 1975 until 1981 he was chair of the Royal Commission on Electrical Power Planning, Ontario.

[1] A. Porter, "The servo-panel - a unique contribution to control systems engineering," *Electronics and Power*, pp. 330-333, 1965.

[2] A. Porter, "The application of control systems in industry," Ministry of Supply, London Interdepartmental Committee on Servomechanisms Publication 2, October 1945.

[3] A. Porter, "Building the Manchester differential analyzers: a personal reflection," *IEE Annals of the History of Computing*, pp. 86-92, 2003.

Introduction to SERVOMECHANISMS

by

A. PORTER
M.Sc., Ph.D.

Professor of Light Electrical Engineering at the Imperial College of Science

WITH 79 DIAGRAMS

LONDON · METHUEN & CO. LTD.
NEW YORK · JOHN WILEY & SONS, INC.

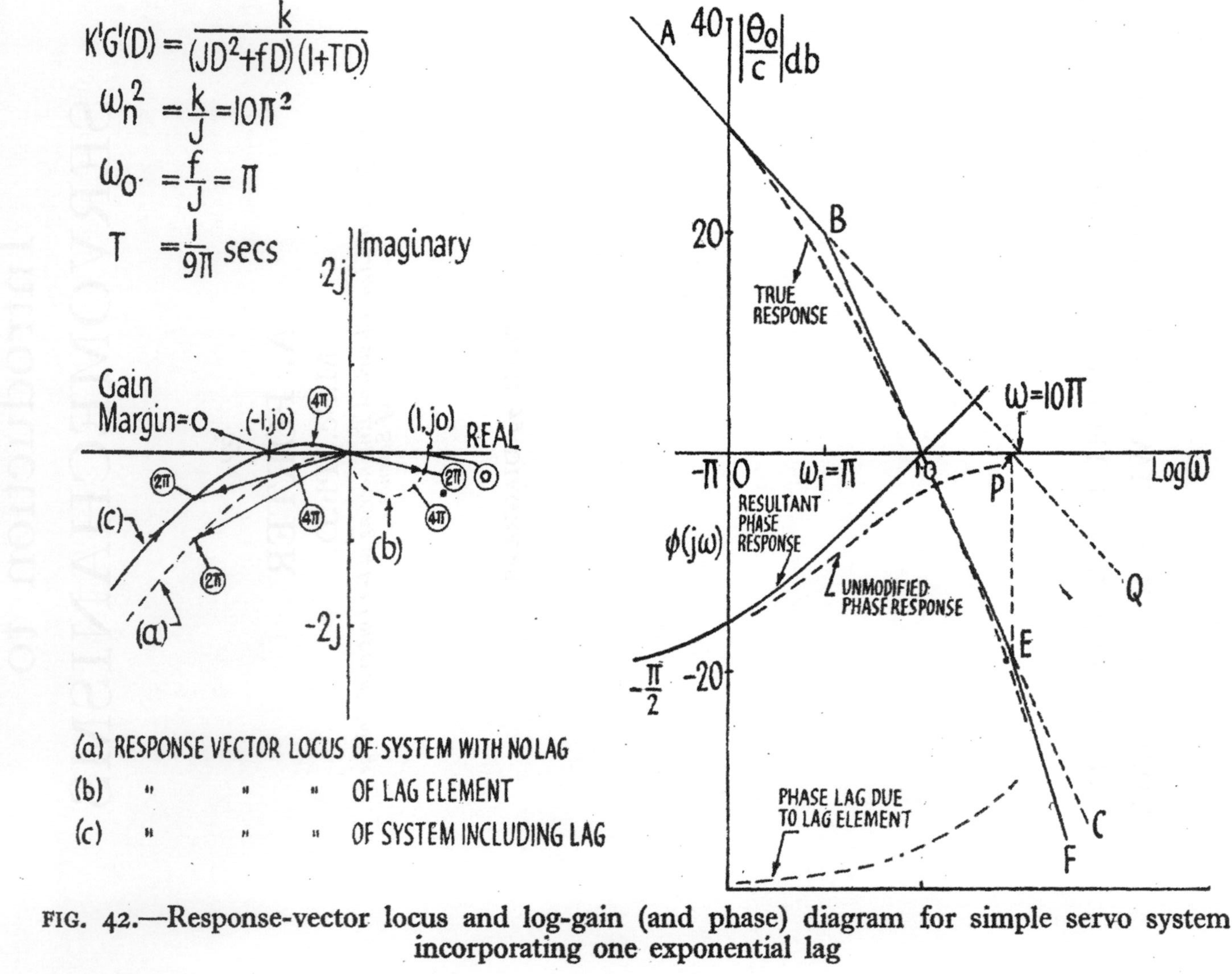

FIG. 42.—Response-vector locus and log-gain (and phase) diagram for simple servo system incorporating one exponential lag

An Introduction to the Theory of Control in Mechanical Engineering

Macmillan, R.H.,
1951, Cambridge, England: Cambridge University Press.

Robert Hugh Macmillan graduated in 1941 from the University of Cambridge in Mechanical Engineering Sciences. He then served in the Technical Branch of the RAF until 1947 when he joined the staff of Cambridge University where he established a small control research group which he directed until 1953 when he was succeeded by John F Coales. In 1946 Macmillan had attended the series of lectures on servomechanisms at Northampton Polytechnic Institute in London given by Arthur Porter and L. Jofeh. This book appears to have been written during 1948-9 in that, although not published until 1951, the preface is dated 1949. Porter read the first draft of the book and Macmillan acknowledges the help received. The style of the book and the author's voice is that of an academic: it is clear that Macmillan had studied carefully the literature on the subject, and attempted to draw from it, as he states in his introduction, "the methods, principles and philosophy of control" rather than being concerned with "specific applications." He says that he has not made any attempt "to include those facts and figures which would be necessary…to provide a handbook for the technician or the designer."

As he was teaching students who were studying Engineering Science it is no surprise he argued that all engineers should study automatic controls and that "the theory of control…by touching upon so many branches of applied engineering science…acts as a unifying influence, which is welcome in these days of excessive specialization." He comments that "once the basic principles have been grasped" then the methods of analysis can be widely applied even to such diverse subjects as "economic controls and the control of infectious diseases." By way of an explanation of the title Macmillan says that he has used examples largely drawn from mechanical systems because electrical systems "…tended to monopolize the present literature."

In arguing for a study of the theory of control he writes:

> "There seems…to be a regrettable tendency amongst some practising engineers to regard the mathematical analysis of control systems with distrust; they prefer the 'common sense and experience' to a careful examination, although practicable methods of making such a study have now been available for some years. This attitude is perfectly reasonable provided that only a low standard of performance is demanded of the controls used; but as soon as we wish to attempt to find the best possible controller for a given job, exact analysis becomes essential. The apparent obscurity of control theory can be attributed largely to the early specialist control engineers who succeeded in surrounding much of their work with an aura of mystery, enhanced by an excessively illogical and confusing nomenclature…"

And then succinctly he lays out the aims in studying the theory of control:

> "…the first is to gain a clear understanding of the mode of operation of such controls; the second is to enable us to estimate the performance of a proposed control system; and the third is to gain some insight into the design of a control which will attain a specified performance."

And indeed the book does meet these aims.

The first five chapters, using simple mathematics, no more than that required to solve simple differential equations with constant coefficients, provide the reader with all that is needed to model and determine the stability, transient and harmonic response of simple control systems; and because he also explains, with a numerical example, Whiteley's standard forms, the reader has some guidance on the values of the coefficients needed in order to obtain an acceptable performance for systems of various class and degree. Throughout he uses, as an example, a hydraulic control system described by the equation

$$D(1+D/8)(1+D/12)\theta_o = K\theta \quad \text{where} \quad \theta = \theta_i - \theta_o$$

and I remember that when I was a student we did numerous examples on this system and it commonly appeared in examination questions.

In chapter 6 he covers the use of the harmonic response function and provides an introduction to the use of the Laplace transform. In the final chapter he gives a graphical method for finding the roots of the characteristic equation, the use of the harmonic response locus, the inverse harmonic response locus, amplitude and phase plots and, although he does not use the term, Nichols charts. In using the harmonic response locus he shows how to use M and N circles. The book ends with a short but informative historical survey, an extensive bibliography and appendices on complex numbers and the Laplace transform.

The book was reprinted in 1955 and it remained in widespread use until the middle of the 1960s.

In the academic year 1950-51 Macmillan was a Visiting Assistant Professor at MIT and in 1955 he gave a series of talks on Automation in the BBC programme "Talking about Science"; he also gave a series of popular lectures at the Royal Institution. These talks and lectures were subsequently published, in 1956, as a book *Automation: Friend or Foe*? [1]. Macmillan was appointed as Professor of Mechanical Engineering at the University of Swansea in 1956 and in the early 1960s joined the Motor Vehicle Research Association as its Director. He was an editor of the first volume of the series *Progress in Control* [2].

[1] R. H. Macmillan, *Automation: Friend or Foe?* Cambridge: Cambridge University Press, 1956.

[2] R. H. Macmillan, T. J. Higgins, and P. Naslin, "Progress in Control Engineering," in *Progress in Control Engineering*, vol. 1. London: Heywood and Company, 1962.

AN INTRODUCTION TO THE THEORY OF CONTROL IN MECHANICAL ENGINEERING

BY

R. H. MACMILLAN

Lecturer in the Engineering Department of the University of Cambridge

CAMBRIDGE
AT THE UNIVERSITY PRESS
1955

Cartesian co-ordinates; the resulting locus has μ as the variable frequency parameter. On such a chart one can draw contours of constant M and ϕ (just as on the Nyquist diagram); the form of these contours is shown in fig. 30.5, which is, of course, another transformation of the Nyquist or Y-plane.*

We can very easily determine the change of loop gain necessary to make the locus tangential to any required M contour, thus giving the desired maximum overall dynamic magnification. It is only necessary to shift the locus parallel to the λ axis by the required amount; in practice it is convenient to have the chart of fig. 30·5

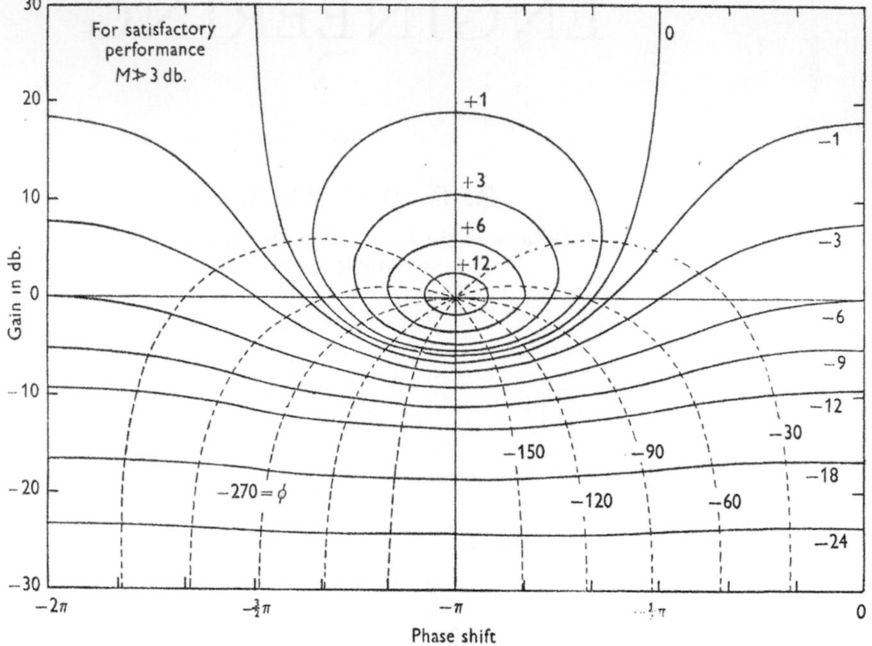

Fig. 30.5. M and ϕ contours on the $\lambda\psi$ diagram.

plotted on a transparent overlay, so that the appropriate gain can be found instantly by trial. The condition for stability on the Cartesian $\mu\psi$ chart is that the locus shall cross the $\psi = \pi$ line *below* the point $\lambda = 0$ (i.e. where λ is negative); if it does not do this $M_{max.}$ would become very large.

Yet one other way of plotting the harmonic response function has been used; here the $\lambda\mu$ and $\psi\mu$ graphs are replotted with λ and ψ as *polar* co-ordinates instead of Cartesian, as in the diagram which we have been considering. The result is a Nyquist locus with $|Y(i\omega)|$ to a logarithmic scale (i.e. attenuation rather than amplification is used for the radius); such a locus has certain advantages over the direct form of Nyquist diagram, but we shall not discuss them as the polar $\mu\psi$ locus is seldom used.

* It has recently been pointed out by MacLellan that if the scales used for the $\lambda\psi$ plot are such that equal lengths are used for 9 db. and 60°, then the graph is an *orthogonal* transformation of the $Y(i\omega)$ plane. When using these scales it is thus possible to construct from the $\alpha = 0$ contour an orthogonal grid of α and ω contours, to determine the transient response of the system.

Servomechanisms

West, J.C.
1953, London: The English Universities Press.

John Clifford West wrote this book as a text for undergraduate university students while a lecturer in the Electrical Engineering at the University of Manchester. He graduated from the University of Manchester in 1943 and served in the RNVR until 1946 when he returned to Manchester as a lecturer. In 1947 he gave a final year undergraduate course in control—the first undergraduate course in the subject in the UK. He was encouraged by the then Professor of Electrical Engineering, Frederic Calland Williams, to start a small servomechanisms laboratory and one of the first research topics Williams proposed was the position control of a rotating magnetic drum for use as a memory store for the digital computer.

Before the war, Williams, together with Arthur Porter, had worked on curve follower input devices for the Manchester University Differential Analyser. During war Williams, who was a practically minded inventor as well as an academic, worked for the Telecommunications Research Establishment on a number of aircraft tracking problems. One of his inventions was the Velodyne – a dc motor and tacho-generator combination which can be used for accurate position and speed control. It was used in position control systems for small radar sets and also as an integrator in electro-mechanical analogue computers. After the war, Williams became interested in the embryonic digital computing and he was a leading figure in the development of the Manchester computer. Commercialisation of the computer required the provision of back up storage mechanisms.

With this background it is not surprising that the West book introduces automatic control through electrical drives for position and speed control. The first five chapters proceed at a gentle pace with considerable emphasis on the student gaining an understanding of the physical behaviour. Very early in the book he introduces the student to the use of the phase-plane diagram—it is used in the first instance to explain the effects of coulomb friction. And throughout the book he reminds the student that the linear equations used in the analysis do not provide a full description of the behaviour of real systems, for example, he writes:

> "In practice there are limits and saturation effects. A servo-motor has a maximum torque, it has therefore a maximum acceleration. It has also a maximum safe speed which must not be exceeded in designing the system. The amplifiers also work within amplitude limits, beyond which there ceases to be a linear relation between input and output. The output power that the amplifier can deliver to the motor is also limited. These factors cannot enter into the simple linear equations, but they play an important part in the performance of the systems." [p. 72]

West briefly mentions the classical methods of solution of differential equations but quickly moves to a simple treatment of the servomechanism as a "network" and explanation as to how a network can be described by a transfer function $\Phi(p)$ where $p = \sigma + j\omega$. He then observes that in the rest of the book it will not "be necessary actually to solve the differential equations." There then follows a standard treatment of stability using Routh-Hurwitz, and the frequency response approach with an explanation of the Nyquist criteria. He also

SERVO-MECHANISMS

An Elementary Textbook
for Scientists and Technologists in
the Field of Physics, Mechanical,
Electrical and Production
Engineering

JOHN C. WEST
Ph.D., D.Sc., A.M.I.E.E.

Dean of the School of Applied Sciences and
Professor of Engineering, University of Sussex

explains the use of the gain-frequency diagram (he does not refer to this as the Bode plot although he references Bode's work). He explains clearly gain and phase margins and minimum phase networks.

The next chapters (11 and 12) lead the reader through the different methods of dealing with the problems of speed control, and in particular 'droop'. And with methods of reducing droop including integral action and the use of PID control. The effect of integral action 'wind-up' is described. Then follow chapters covering measuring devices, static amplifiers, rotary amplifiers and servo-motors. At first glance these seem to be standard treatments describing the devices and deriving appropriate transfer functions but a closer reading reveals a deep understanding of the practical issues and limitations of the devices when used in control systems. In chapter 18, dealing with steady-state accuracy of position control systems, he writes "[T]he best system can only reduce the error signal to zero, if the signals themselves are inaccurate, then the final positioning will be inaccurate. Thus the final limitation on the accuracy of position control systems rests on the accuracy of the measuring or monitoring devices."

The final two chapters of the book cover non-linear behaviour and give a brief introduction to methods of analysis and design of such systems, including treatment of 'bang-bang' control. And it is in the area of non-linear systems that West's major research contributions were made: In particular in 1956, with the dual input describing function [1]. He produced a book on non-linear control in 1960.[2]

In the Introduction he comments that the 1939-1945 war had led to the invention and development of many automatic control devices together with a rapid advance in the theory of control such that "...the whole subject is becoming of primary importance." It is possible, he argues "...to start from a few basic principles and build up a comprehensive theory of control" but "[T]he engineer...must be able to go from the abstract to the concrete, must be able to design a specific system for a given job." That the book succeeded in providing an excellent grounding for the student engineer needing to learn both the theory and how to move from the abstract to the concrete is one reason for its longevity—it remained in print until the late 1960s (the fifth reprint is dated 1965). Another reason, I think, is because it unfolds like a story of discovery. One can imagine John West in the lecture room encouraging the students to work out for themselves the next steps in the story.

John West moved to Queen's University Belfast in 1957 as Professor of Electrical Engineering and then in 1965 to University of Sussex as Dean of School of Applied Sciences. In 1979 he was appointed as Vice-Chancellor, University of Bradford and retired from this position in 1989. He served as the President of the Institution of Electrical Engineers in 1984 and was appointed as an Honorary Fellow of the Institution in 1992. John West has been active in retirement and arising from one of his many activities, in the year 2000, his name was added to the Roll of Distinguished Philatelists.

[1] J. C. West, "Forty years in control," *Institution of Electrical Engineers Proceedings-A*, vol. 132, pp. 1-8, 1985.

[2] J. C. West, *Analytical techniques for non-linear control systems.* London: English Universities Press, 1960.

curves, and is shown in Fig. 49. This is, of course, an approximate straight-line segmented characteristic, the true one having the corners rounded as shown dotted in the same figure.

The slope of the characteristic where it crosses the unity gain (zero db.) axis gives an indication of the phase angle at this frequency. Since the slope is 12 db. per octave for a small frequency range and 6 and

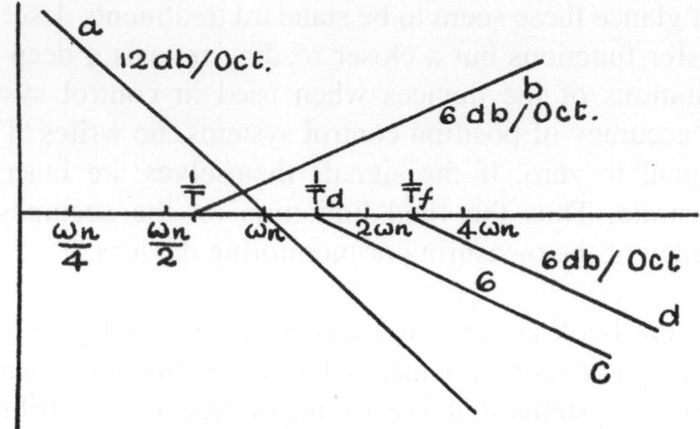

FIG. 48.—COMPONENTS OF EQUATION 139.

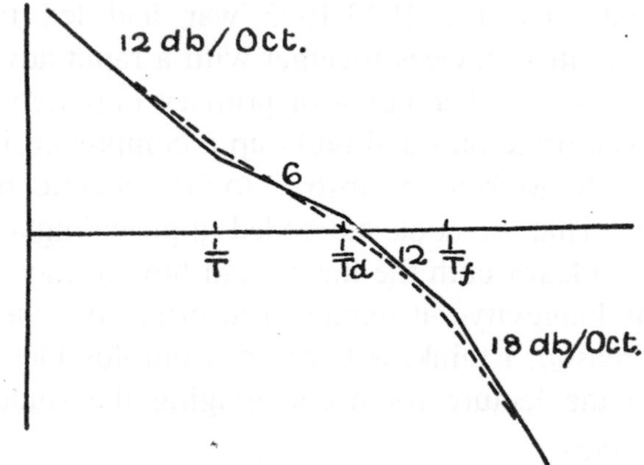

FIG. 49.—THE SUM OF CURVES *a*, *b*, *c* AND *d* IN FIG. 48.

18 db. on either side, the first estimate of the equation in approximate form (equation (137)) would indicate an angle of 180 deg. Thus the system, if not unstable, would be highly oscillatory. It was shown in Section 9.6 that the phase margin should be of the order of 30–40 deg., i.e., the phase angle should be 140–150 deg. at unity gain. This corresponds to a slope of 9–10 db. per octave. Hence the system can be improved over that of Fig. 49 if it is possible to reduce the slope. The advantages of this approximate form are, first, that it is easy to draw

EARLY CONTROL TEXTBOOKS
in the
UNITED STATES OF AMERICA

Contributed by

Peter Dorato

University of New Mexico
Albuquerque, NM, USA

Karl Johan Åström

Lund Institute of Technology
Lund, Sweden
(Engineering Cybernetics, H.S. Tsien)

Network Analysis and Feedback Amplifier Design

Hendrik W. Bode
Van Nostrand Company, Inc., New York, 1945.

Theory of Servomechanisms

Hubert M. James, Nathaniel B. Nichols, and Ralph S. Phillips
McGraw-Hill Book Company, Inc., New York, 1947.

Engineering Cybernetics

Hsue-Shen Tsien
McGraw-Hill Book Company, Inc., New York, 1954.

Automatic Feedback Control System Synthesis

John G. Truxal
McGraw-Hill Book Company, Inc., 1955.

Sampled-data Control Systems

John R. Ragazzini and Gene F. Franklin
McGraw-Hill Book Company, Inc., 1958.

Stability by Liapunov's Direct Method. With Applications

Joseph La Salle and Solomon Lefschetz
Academic Press, New York, 1961.

Adaptive Control Processes: A Guided Tour

Richard Bellman
Princeton University Press, Princeton, New Jersey, 1961.

Linear System Theory. The State Space Approach

Lotfi A. Zadeh and Charles A. Desoer
McGraw-Hill Book Company, New York, 1963.

Optimal Control

Michael Athans and Peter L. Falb
McGraw-Hill Book Company, New York, 1966.

Network Analysis and Feedback Amplifier Design

Hendrik W. Bode
Van Nostrand Company, Inc., New York, 1945

The *Bell Telephone Laboratories (Bell Labs)*, established in 1925, was a major center in the United States for the development of electronic technology. This classical text on feedback amplifier design is part of the *Bell Telephone Laboratory Series* of books that reported the technological innovations developed at the Laboratories. Researchers at the Laboratories made fundamental contributions to feedback control. In particular in 1927, H.S. Black introduced the concept of feedback to design high accuracy electronic amplifiers, and in 1932, H. Nyquist developed a frequency-domain criterion for the stability of feedback systems. Researchers at the Laboratories, such as R.M. Foster, also made fundamental contributions to electrical network synthesis. Network synthesis played a key role in the early implementation of feedback controllers, since the only electronics available at the time, e.g. 1940s, were passive electrical elements such as resistors and capacitors, and vacuum tube amplifiers.

Actually the major part of this book is devoted to passive network synthesis. As indicated in the title, Bode's text combines network theory and feedback design. The feedback design theory is based on Nyquist stability theorem, and the text introduces related frequency-domain plots, now known as *Bode plots* to expedite the feedback design. The frequency-domain design techniques presented in Bode's text dominated feedback design theory up to the 1960s, when state-space methods were first introduced. This text also introduces the *sensitivity function* to measure the sensitivity of closed-loop transfer-functions to plant parameter variations. The problem of feedback sensitivity design becomes a central control problem in the years following the publication of this text. In this text Bode also relates the gain and phase of a transfer function through his now *famous Bode gain-phase relations*.

The same year that Bode's text was published, a text by another Bell Labs researcher, *LeRoy A. MacColl*, appeared in the Bell Lab Series, with the title ***Fundamental Theory of Servomechanisms***. The MacColl text was not as detailed as the Bode text, but it covered much of the same frequency domain theory. Interestingly, it also had some discussion of sampled-data control and nonlinear control.

Network Analysis and Feedback Amplifier Design

By
HENDRIK W. BODE, Ph.D.,
Research Mathematician,
BELL TELEPHONE LABORATORIES, INC.

ELEVENTH PRINTING

D. VAN NOSTRAND COMPANY, INC.
PRINCETON, NEW JERSEY
TORONTO LONDON
NEW YORK

ment values in Fig. 17.7*A* and Curve II, to Fig. 17.7*B*. The gain and phase curves for the remaining structures are shown in Fig. 17.9. Curves

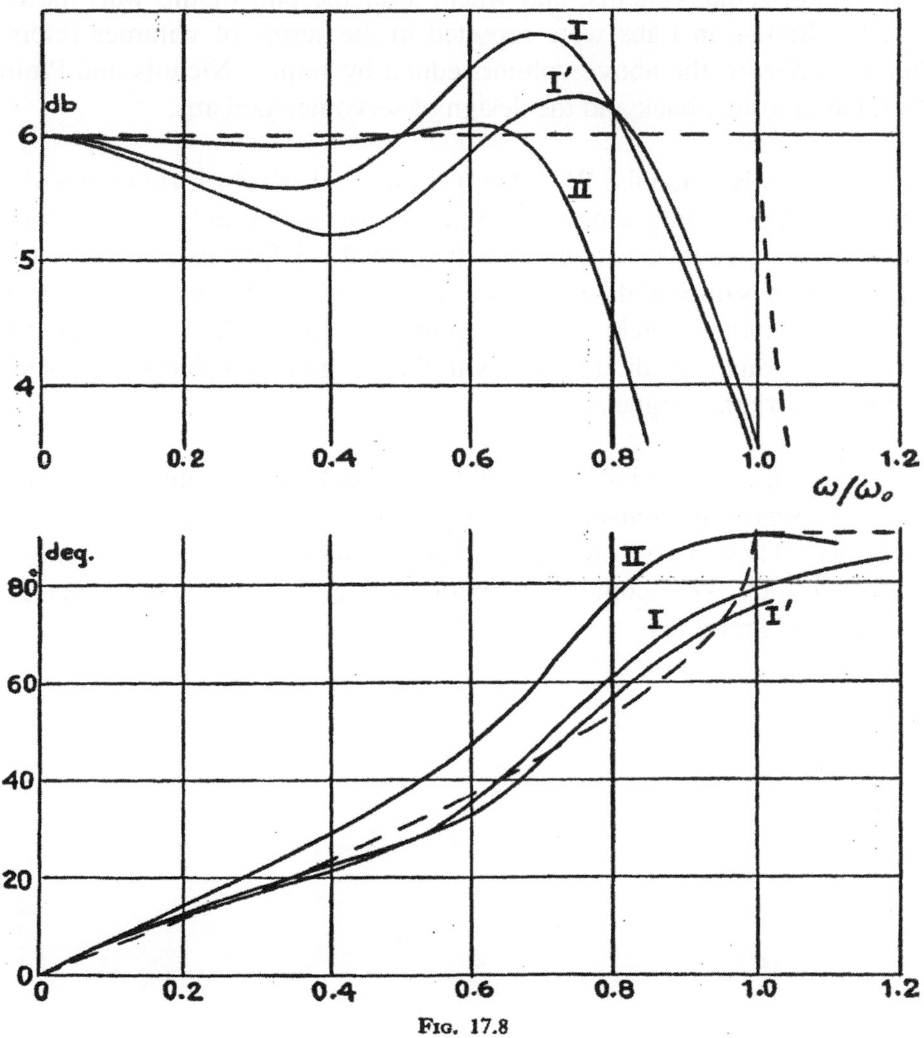

Fig. 17.8

III and IV refer respectively to the structures of Figs. 17.7*C* and 17.7*D*. The performance of the structure of Fig. 17.7*E*, which matches the ideal too closely to be shown by a separate curve, is indicated by the crosses.

This sample page illustrates an original "Bode plot" of magnitude and phase versus frequency.

Theory of Servomechanisms

Hubert M. James, Nathaniel B. Nichols, and Ralph S. Phillips
McGraw-Hill Book Company, Inc., New York, 1947.

During World War II, much new technology related to radar systems was developed at the *Massachusetts Institute of Technology's Radiation Laboratory*, established in 1940. Feedback was an essential part of radar systems, and when, after the end of the war, the results of the research done at the Radiation Labs were reported in the series of volumes referred to as the *Radiation Laboratory Series*, the above volume, edited by James, Nichols and Phillips, focused on those results relating to feedback and the design of servomechanisms.

In a chapter written by N.B. Nichols, W.P. Manger and E.H. Krohns, frequency-domain design principles are presented, including a new design plot, now commonly referred to as a *Nichols chart*. This volume also presents, for the first time in the USA, in a chapter written by W. Hurewicz, an analysis of sampled-data systems, including the concept of the *z-transform* for discrete-time signals. Another notable result reported in this volume was the application of Norbert Wiener's mean-square prediction theory to the design of feedback control systems in the presence of random disturbance signals.

Four years after the publication of the James, Nichols and Phillips volume, the two volumes of *Chestnut and Mayer,* ***Servomechanisms and Regulating System Design*** *(John Wiley & Sons, 1951)* were published. These two volumes, which included many of the results presented in James, Nichols and Phillips, were popular text books in the USA for control courses, because of their many examples and problems.

THEORY OF SERVOMECHANISMS

Edited by

HUBERT M. JAMES
PROFESSOR OF PHYSICS
PURDUE UNIVERSITY

NATHANIEL B. NICHOLS
DIRECTOR OF RESEARCH
TAYLOR INSTRUMENT COMPANIES

RALPH S. PHILLIPS
ASSOCIATE PROFESSOR OF MATHEMATICS
UNIVERSITY OF SOUTHERN CALIFORNIA

OFFICE OF SCIENTIFIC RESEARCH AND DEVELOPMENT
NATIONAL DEFENSE RESEARCH COMMITTEE

FIRST EDITION
THIRD IMPRESSION

NEW YORK · TORONTO · LONDON
McGRAW-HILL BOOK COMPANY, INC.
1947

decibel value.) For the ψ-contours the highest value of $|Y_{11}|$ is reached where the phase margin is equal to $90° + \psi \pm n\pi$ and in decibels equals $20 \log_{10} |1/\sin \psi|$.

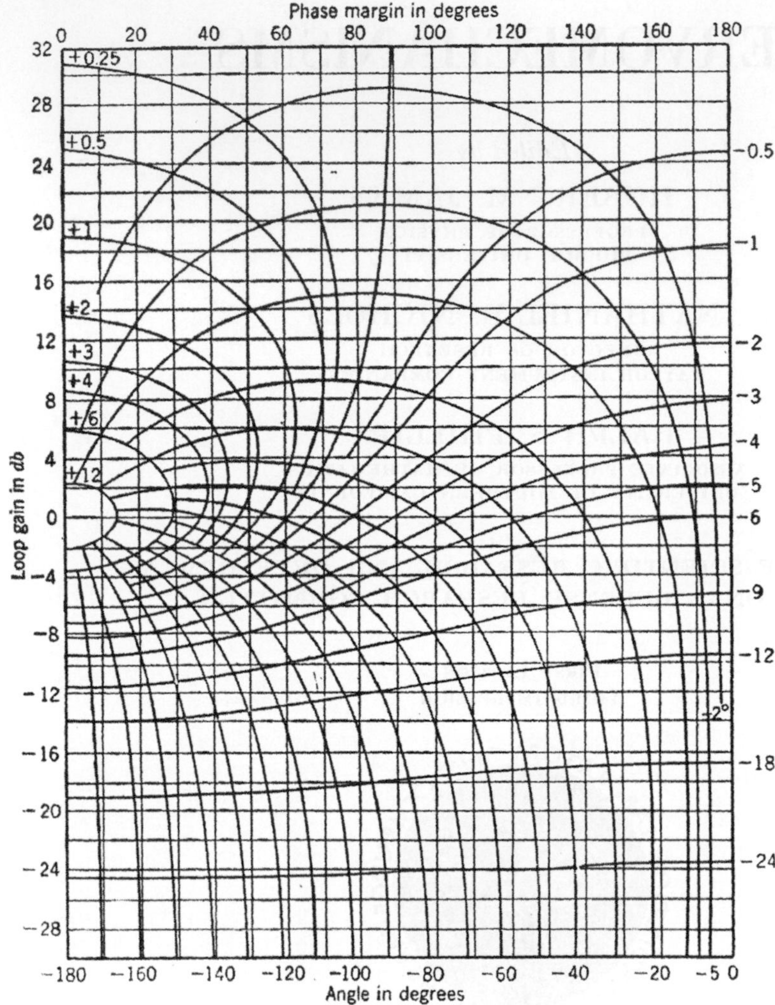

FIG. 4·27.—Constant-phase-angle and constant-amplification contours on the decibel-phase-angle loop diagram.

It is obviously possible to transfer the M-contours and the ψ-contours from this decibel–phase-angle diagram to the type of decibel–log-frequency graph discussed in Sec. 4·10. Then, after a study of the manner in which the attenuation curve crosses the M-contours, it is possible to alter the gain and, if necessary, the shape of the attenuation plot to

This page illustrates a "Nichols plot".

Engineering Cybernetics

Hsue-Shen Tsien
McGraw-Hill, New York, NY. 1954.

This is a remarkable book written by a remarkable person. Tsien graduated from Jiaotong University in Shanghai. He moved to the US where he got his PhD from Caltech in 1939. Tsien became a protégé of Theodore von Kármán, and an active researcher in the emerging aerospace field. He contributed to basic research as well as to early experiments with rocketry in what was to become the Jet Propulsion Laboratory (JPL). Tsien was also heavily involved with military projects; he joined von Kármán in Washington as a member of a small Scientific Advisory Group to the US Army. In 1946 he became professor at MIT, and in 1949 he returned to Caltech with consultancies at JPL and Aerojet Corporation. During the McCarthy era in the early 1950s he was accused of being a Communist and in 1955 he was deported to China where he became the leader of the Chinese missile program. Engineering Cybernetics was written in a transition period when Tsien had lost his security clearance, much of the aerospace research was closed to him and he turned his interest to research in other fields and to teaching at Caltech.

The early 1950s was also a transition period in the field of control. Development of classical control theory had reached a plateau, and the paradigm shift called modern control theory had not yet occurred. Recall that Bellman's book on Dynamic Programming, Kalman's paper on filtering and Pontryagin's book on Optimal Control were published in 1957, 1960 and 1962.

The book Engineering Cybernetics has 18 chapters. It gives an unusually broad coverage of the field of control. The first four chapters treat traditional servomechanism theory covering transfer functions, frequency response, the Nyquist stability theory and Evans root locus method in a nice concise way. The book then treats much material that was non-standard at the time. There is a short chapter on time varying systems that is motivated by linearization of rocket trajectories along a path. Non-interacting control with applications to turboprop engines, sampled data systems, and systems with time delays are covered in separate chapters. There is a good coverage of stochastic systems in two chapters which gives fundamentals of stochastic process and Wiener-Kolmogoroff filtering theory. A very interesting chapter on relay feedback describes classical results by Flügge Lotz, describing functions, the idea of minimum time control and switching surfaces developed by Bushaw. A chapter on nonlinear control covers jump resonance, frequency entrainment and parametric excitation. There are three chapters on optimal control. One of them suggests that control design should be formulated as an optimization problem, another solves it using adjoint variables, and the third deals with self-optimizing control developed by Draper and Li. The last two chapters deal with ultra-stability in the spirit of Ross Ashby and reliability.

A person lucky enough to have read Tsien's book in the late fifties would have been very well prepared for the exciting development of control that took place in the 1960s. The book *Thread of the Silkworm* by Iris Chang gives a fascinating tale of the career of Tsien Hsue-Shen.

ENGINEERING CYBERNETICS

H. S. TSIEN

Daniel and Florence Guggenheim Jet Propulsion Center
California Institute of Technology
Pasadena, California

McGRAW-HILL BOOK COMPANY, INC.
New York Toronto London
1954

rotating earth, as sketched in Fig. 13.1. The planar motion is possible due to the absence of the cross Coriolis force in the equatorial plane. The coordinate system is fixed with respect to the rotating earth, *i.e.*, it actually rotates with the angular velocity Ω, the speed of earth rotation. In the equatorial plane, the position of the vehicle at any time instant t is specified by the radius r and the angle θ from the starting point of the

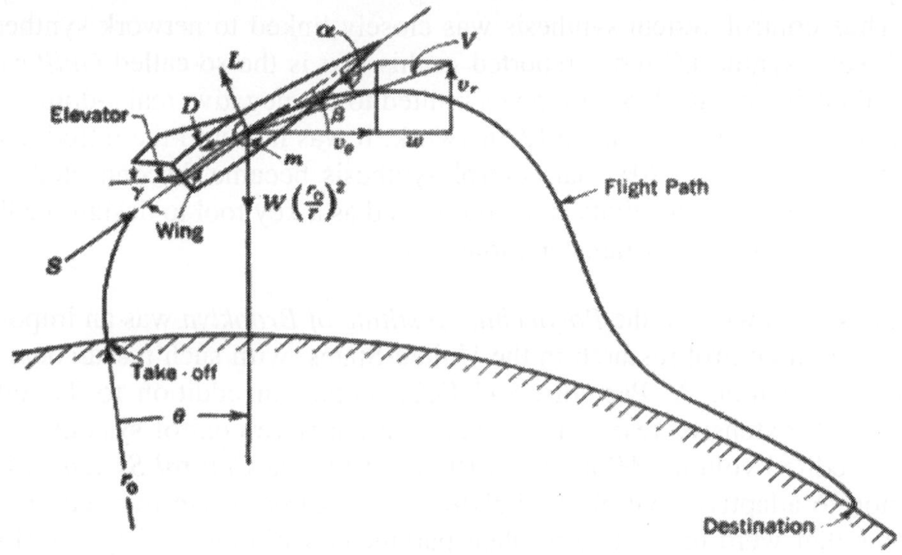

Fig. 13.1

vehicle. r_0 is the radius of the earth. g is the gravitational constant at the surface of the earth without the centrifugal force due to rotation. Let R and Θ be the force per unit mass due to thrust and aerodynamic forces acting on the vehicle in the radial and the circumferential directions, respectively. Then the equations of motion of the center of gravity of the vehicle are

$$\left.\begin{aligned} \frac{dr}{dt} &= \dot{r} \\ \frac{d\theta}{dt} &= \dot{\theta} \\ \frac{d\dot{r}}{dt} &= R + r(\dot{\theta} \pm \Omega)^2 - g\left(\frac{r_0}{r}\right)^2 \\ r\frac{d\dot{\theta}}{dt} &= \Theta - 2\dot{r}(\dot{\theta} \pm \Omega) \end{aligned}\right\} \tag{13.1}$$

where the plus sign in the second terms on the right will be valid for flights toward the east, and the minus sign for flights toward the west.

This page shows a good early example of an optimal control problem

Automatic Feedback Control System Synthesis

John G. Truxal
McGraw-Hill Book Company, Inc., 1955.

This text contains a very comprehensive presentation of most of the frequency-domain techniques for the design of control systems, available in 1955. It covers a wide range of topics, including Nyquist methods, root-locus methods. sampled-data systems, nonlinear systems, and statistical mean-square design methods. Due to the limitations in electronic hardware available at the time, a major focus in the text was on the design of controllers that can be implemented as RC networks. Thus control system synthesis was closely linked to network synthesis, as in the text of Bode. A key "synthesis" result reported in this text is the so-called *Guillemin* approach, which attempts to design controllers with poles limited to the negative real s-domain axis, so that the controller can be synthesized with an RC network. It was not until the introduction of operational amplifier chips in the 1970s, that control synthesis became disconnected from network synthesis. In this text Bode's sensitivity function is used as a key tool to design feedback systems that minimize the effect of plant parameter variations.

At the time this text was written, the *Polytechnic Institute of Brooklyn* was an important national center for systems and control research in the United States, with such researchers as L. Braun, I.M. Horowitz, E. Mishkin, A. Papoulis, and D.C. Youla, in addition to Truxal. Under the leadership of Truxal, extensive research was done on adaptive control systems, leading to the publication of the edited volume, *Mishkin and Braun, **Adaptive Control Systems** (McGraw-Hill, 1961).* In addition to adaptive control, the Polytechnic research group focused on the design of feedback systems that were insensitive to plant parameter variation. A key contribution in this area was made by Horowitz, with the introduction of sensitivity measures appropriate to large parameter variations. Details of Horowitz's frequency-domain design approach to feedback and uncertainty appeared in his text *Horowitz, **Synthesis of Feedback Systems** (Academic Press, 1963)***.** In the text of *Paul Frank, **Introduction to System Sensitivity Theory** (Academic Press, 1978)*, most of the major results associated with sensitivity design, that dominated feedback theory since the publication of Bode's text, were summarized. In the 1980s, the term *Robust Design* largely replaces *Sensitivity Design* in control text books.

A note on the term *Synthesis* used in the Truxal, and other control texts. It is a term that the control community inherited from electrical network theory, where "synthesis" meant, design theory with an existence theorem and a computable algorithm to find a design solution, when one existed. The term a*nalytic design* is generally understood to be synonymous with *synthesis*, as defined this way. Analytic design is in sharp contrast to design by trial-and-error. Design techniques based on shaping of the Nyquist plots are inherently trial-and-error techniques. Most of the results, presented in the texts of Truxal and Horowitz, cannot be said to be synthesis results in the strict network sense. It was not until the introduction of state-space techniques and H_∞ methods that the term *synthesis* could be used in its strict sense for control design. An example of a control text where the term is used properly in its strict sense, is the text of *Vidyasagar, **Control System Synthesis. A Factorization Approach** (MIT Press, 1985).*

AUTOMATIC FEEDBACK CONTROL SYSTEM SYNTHESIS

PROFESSOR JOHN G. TRUXAL

Department of Electrical Engineering
Polytechnic Institute of Brooklyn
Brooklyn, New York

McGRAW-HILL BOOK COMPANY, INC.

New York Toronto London

1955

2 and 3 is noteworthy, particularly in view of the apparently striking dissimilarity between the two system functions.

5.3. Determination of the Open-loop Transfer Function from the Closed-loop System Function. The second step in a synthesis along the lines suggested by Guillemin is a determination of the open-loop transfer function corresponding to the closed-loop system function chosen to meet the performance specifications. It is this step which was the stumbling block to successful logical synthesis of servomechanisms before Guillemin's work, for in its basic form this step involves the evaluation of the roots of a polynomial. The essential problem is illustrated in Fig. 5.16. $\frac{C}{R}(s)$ is known, with the numerator and denominator polynomials in factored form; the poles and zeros of the open-loop transfer function $G(s)$ are to be found.

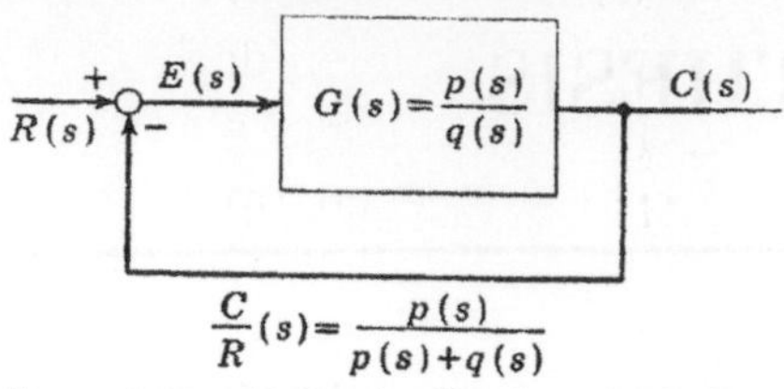

FIG. 5.16. Unity-feedback, single-loop feedback control system.

These unknown poles and zeros are related to the known $\frac{C}{R}(s)$ by the equation

$$\frac{C}{R}(s) = \frac{G(s)}{1 + G(s)} \tag{5.52}$$

If $\frac{C}{R}(s)$ is written as the ratio of polynomials, $p(s)/n(s)$, and $G(s)$ as $p(s)/q(s)$,

$$\frac{C}{R}(s) = \frac{p}{n} = \frac{p/q}{1 + p/q} = \frac{p}{p + q} \tag{5.53}$$

$$n(s) = p(s) + q(s) \tag{5.54}$$

$$q(s) = n(s) - p(s) \tag{5.55}$$

In other words, the zeros of the open-loop transfer function are identical with the zeros of the closed-loop system function, while the poles of the open-loop transfer function are the zeros of the polynomial $n(s) - p(s)$. In general, determination of these zeros involves solution of a polynomial of the degree of $n(s)$. It was this problem of zero determination which, until Guillemin's work, discouraged design through this method of working from the over-all system inward toward the compensation networks.

The foundation for the success of Guillemin's method lies in the procedure he presented to ensure a simple solution to the zero-determination problem. He proposed that the pole-zero configuration for $\frac{C}{R}(s)$ be chosen not only to meet the specifications, but also to ensure that all zeros of $q(s)$ lie on the negative real axis. If this condition is satisfied, the zeros of $q(s)$ can be determined graphically by plots of $n(s)$ and $p(s)$ (polynomials with known zeros) for real values of s and by graphical subtraction according to the equation

$$q(s) = n(s) - p(s) \tag{5.56}$$

This page illustrates the Guillemin synthesis concept of moving from closed-loop to open-loop characteristics.

Sampled-data Control Systems

John R. Ragazzini and Gene F. Franklin
McGraw-Hill Book Company, Inc., 1958.

This is one of the first texts written in the United States on sampled-data control systems. The authors were both former faculty members at *Columbia University*, where there was a strong research focus on sampled-data control, led by Professor John Ragazzini. Researchers at Columbia University in this area included, J.E. Bertram, B. Friedland, E.I. Jury, R.E. Kalman, G. Kranc, and L.A. Zadeh. This text presents much of the results developed by these researchers, e.g., the multirate sampling theory developed by Kranc. It is also one of the first texts to introduce the *digital computer* for the implementation of sampled-data controllers, although it was many years before efficient and inexpensive digital computers were to be realized. At the present time, digital-computer controllers have largely replaced operational-amplifier analog controllers.

In 1958, the text of another ex-Columbia University researcher Eliahu Jury was published. i.e. *Jury, **Sampled-data Control Systems** (John Wiley & Sons, 1958).* For a long time these two texts were the mainstay of the sampled-data literature in the USA. In 1958, the text of *Tsypkin, **Theory of Pulse Systems** (State Press, 1958)*, was published in the USSR and in 1972, the text of Jürgen Ackermann was first published in German, later to be expanded and translated into English in the text *Ackerman, **Sampled-data Control Systems. Analysis and Synthesis, Robust Design** (Springer-Verlag, 1985).* Since 1958 many texts have been written in the USA on the subject, with the title Sampled-data Control replace by Digital Control, an example being the recent text of *Franklin, Powell, and Workman, **Digital Control of Dynamic Systems** (Addison-Wesley, 1997).* This reflects the improvements that have occurred in digital technology, and the focus on digital implementation of controllers. Two other terms that have replaced the term Sampled-data Control are Discrete-time Control and Computer Control, e.g. *Ogata, **Discrete-time Control Systems** (Prentice-Hall, 1987)* and *Astrom and Wittenmark, **Computer Controlled Systems** (Prentice-Hall, 1984).*

SAMPLED-DATA CONTROL SYSTEMS

John R. Ragazzini
Dean, College of Engineering
New York University

Gene F. Franklin
Assistant Professor of Electrical Engineering
Stanford University

McGRAW-HILL BOOK COMPANY, INC.
New York Toronto London
1958

EXAMPLE

As an example of the design principle outlined in this section, a second-order plant with zero-order hold is considered. The pertinent transfer function of the continuous element is

$$G(s) = \frac{1 - e^{-sT/n}}{s} \frac{1}{s(s+1)} \tag{9.86}$$

From previous calculations (see Sec. 7.8) it is known that the pulse transfer function of this system has two zeros, so that $n = 2$. If, for

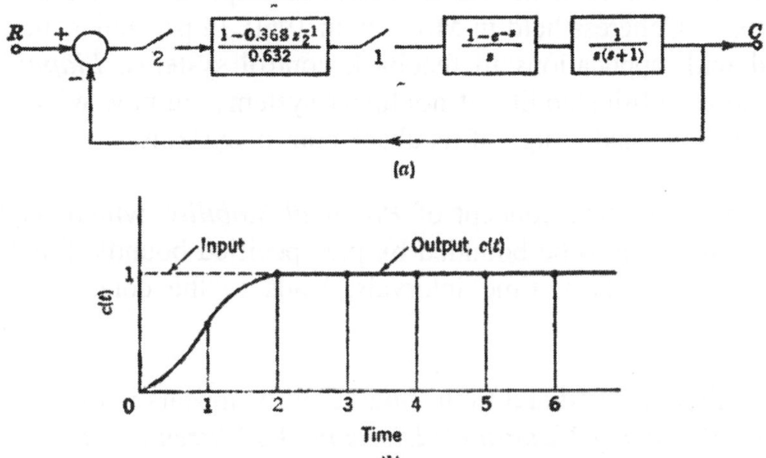

FIG. 9.14. (a) Block diagram and (b) step response of a multirate system designed for zero ripple.

purposes of this example, T' is taken to be 2 sec, then the previously calculated pulse transfer function can be used, namely,

$$G(z_2) = \frac{0.368z_2^{-1}(1 + 0.718z_2^{-1})}{(1 - z_2^{-1})(1 - 0.368z_2^{-1})} \tag{9.87}$$

Substituting from (9.87) into (9.85),

$$D(z_2) = \frac{(1 - z_2^{-1})(1 - 0.368z_2^{-1})}{0.368(1.718)(1 - z_2^{-1})}$$

$$= \frac{1 - 0.368z_2^{-1}}{0.632} \tag{9.88}$$

A block diagram of this system and the associated step response are shown in Fig. 9.14.

It is possible to generalize the interpretation of the basic design equation given as (9.82) considerably beyond the step-response case worked out in detail in (9.85). In some cases it is desirable, and for plants with

This page illustrates a multirate sampled-data system with a discrete-time compensator.

Stability by Liapunov's Direct Method. With Applications

Joseph La Salle and Solomon Lefschetz
Academic Press, New York, 1961.

La Salle and Lefschetz were both researchers at the *Research Institute for Advanced Studies (RIAS)*, a division of the Martin Company, located in Baltimore, Maryland, when this book was published in 1961. During the period 1955-1965, RIAS was one of the major systems and control research centers in the USA. The researches at RIAS included, in addition to La Salle and Lefschetz, J. Hale, R. Kalman, H. Kushner, G. Szego, L. Weiss and M. Wonham. Many fundamental results in linear, nonlinear, and stochastic systems originated at RIAS.

The above text presents some of the basic nonlinear results of Liapunov, that became known in the USA through the 1st IFAC Congress held in Moscow in 1960. In particular the text presents Liapunov's direct method with applications to feedback control systems. *Liapunov functions* used in the direct method to establish stability of nonlinear systems are now widely used in the control community for the analysis and design of feedback control systems.

This text also introduced the interesting concept of *Practical Stability*, which requires system variables to not just be bounded, but to be bounded by pre-specified bounds. Practical stability applied to systems operating over finite-time intervals, leads to the concept of *Finite-time Stability*.

A discussion of Liapunov's direct method and finite-time stability appears in the text of *Wolfgang Hahn, **Theory and Application of Liapunov's Direct method** (Prentice Hall, 1963)*, originally published in German in 1959. Liapunov's stability theory is now a standard component of most nonlinear control texts.

Stability by Liapunov's Direct Method

With Applications

JOSEPH LA SALLE

and

SOLOMON LEFSCHETZ

RIAS, Baltimore, Maryland
(A Division of The Martin Company)

1961

New York *ACADEMIC PRESS* *London*

§ 9. Liapunov's Stability Theorems

It is intuitively clear that if near an equilibrium state of a physical system the energy of the system is always decreasing, then the equilibrium is stable. Liapunov's theorems are a generalization of this idea. Liapunov functions are simply an extension of the energy concept. The central idea of the Liapunov method

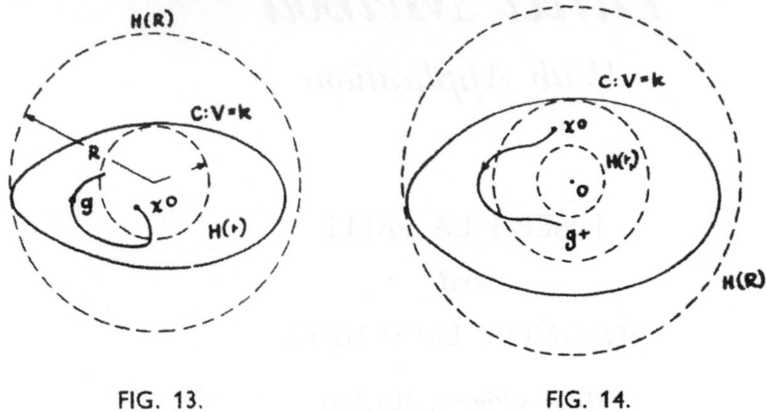

FIG. 13. FIG. 14.

is to detect stability for a system (FA) by means of properties of a Liapunov function $V(x)$ and to do this, not directly from a knowledge of the solutions, but indirectly from the system (FA).

Our proofs will be given as geometric a form as possible. However, for clearness in applying the theorems it is indispensable to formulate them in a precise analytic manner.

I. STABILITY THEOREM. *If there exists in some neighborhood Ω of the origin a Liapunov function $V(x)$, then the origin is stable.*

II. ASYMPTOTIC STABILITY THEOREM. *If $-\dot{V}$ is likewise positive definite in Ω, then the stability is asymptotic.*

The two theorems will be proved together. We rest them heavily upon the second geometric interpretation of a positive definite function $V(x)$. The sketch (Fig. 13) is conveniently made for $n = 2$ but the reasoning is valid for any n.

This page illustrates Liapunov's stability concepts and Liapunov functions.

Adaptive Control Processes: A Guided Tour

Richard Bellman
Princeton University Press, Princeton, New Jersey, 1961.

Richard Bellman was a researcher at the *RAND Corporation*, in Santa Monica, California during the period 1953-1956. During this period of time the RAND Corporation was one of the major centers in the USA doing research in systems and control. Some of the notable researchers at RAND working in the systems and control area included, J. Danskin, S. Dreyfus, I. Glicksberg, O. Gross, R. Isaacs, and R. Kalaba. Many RAND reports were published by the above researchers. The report most directly related to the above text is report R-313, *Some Aspects of the Mathematical Theory of Control Processes*, published in 1958 and authored by Bellman, Glicksberg, and Gross.

Bellman made many contributions to mathematics and control theory, but he is probably most famous for his invention of the *Principle of Optimality* and its application to optimal control (*Dynamic Programming*). Pontryagin's *Maximum Principle* and Bellman's *Dynamic Programming* provided solutions to many optimal control problems related to space programs and military applications (RAND was supported largely by the USA Air Force during the above period of time). In *Adaptive Control Processes*, Bellman covered a wide range of topics related to dynamic programming. In particular he discussed the calculus of variations, optimal stochastic control, differential games, and adaptive control. The style of presentation in this text is very informal, hence the subtitle, *A Guided Tour*. More details on dynamic programming may be found in an earlier text written by *Bellman,* ***Dynamic Programming*** *(Princeton University text, 1957)*. A very interesting sketch of Bellman's life and career may be found in his autobiographical book, *Eye of the Hurricane. An Autobiography, (World Scientific, 1984)*.

ADAPTIVE CONTROL PROCESSES: A GUIDED TOUR

BY

RICHARD BELLMAN

THE RAND CORPORATION

1961

PRINCETON UNIVERSITY PRESS

PRINCETON, NEW JERSEY

Markovian property described above. In this case, the basic property of optimal policies is expressed by the following:

PRINCIPLE OF OPTIMALITY. An optimal policy has the property that whatever the initial state and the initial decision are, the remaining decisions must constitute an optimal policy with regard to the state resulting from the first decision.

A proof by contradiction is immediate.

3.9 Derivation of Recurrence Relation

Using the principle cited above, we can derive the recurrence relation of (3.21) by means of purely verbal arguments. Suppose that we make an initial decision q_1. The result of this decision is to transform p_1 into $T(p_1, q_1)$ and to reduce an N-stage process to an $(N - 1)$-stage process. By virtue of the principle of optimality, the contribution to the maximum return from the last $(N - 1)$-stages will be $f_{N-1}(T(p_1, q_1))$. This is the Markovian property discussed in 3.5.

Hence, for some q_1 we have

$$f_N(p_1) = g(p_1, q_1) + f_{N-1}[T(p_1, q_1)]. \tag{3.22}$$

It is clear that this q_1 must be chosen to maximize the right-hand side of (3.22), with the result that the final equation is

$$f_N(p_1) = \operatorname*{Max}_{q_1} [g(p_1, q_1) + f_{N-1}[T(p_1, q_1)]]. \tag{3.23}$$

Although we shall discuss these matters in much detail in Chapter V which is devoted to the computational aspects, let us make this preliminary observation here. Observe that the usual approach to the solution of the maximization problem yields a solution in the form $[q_1, q_2, \ldots, q_N]$, the same one, of course, which is obtained as a result of our variational process. The approach we present here yields the solution in steps: first, the choice of q_1, then the choice of q_2, and so on.

The decomposition of the problem of choosing a point in N-dimensional phase space into N choices of points in one-dimensional phase space is of the utmost conceptual, analytic, and computational importance in all of our subsequent work. The whole book is essentially devoted to the study of various classes of multidimensional processes which can, by one device or another, be reduced to processes of much lower dimension, and preferably to one-dimensional processes.

3.10 "Terminal" Control

A case of some interest in many significant applications is that where we wish only to maximize a function of the final state p_N. In this case, the sequence of return functions is defined by the relation

$$f_N(p_1) = \operatorname*{Max}_{q_i} g(p_N). \tag{3.24}$$

This page illustrates Bellman's "optimality principle".

Linear System Theory. The State Space Approach

Lotfi A. Zadeh and Charles A. Desoer
McGraw-Hill Book Company, New York, 1963.

This is one of the early textbooks published in the USA that presented the state-space approach to linear systems, pioneered by R. Kalman in the late 1950s. Prior to the introduction of state-space methods, linear control theory in the USA was essentially limited to frequency-domain methods. This text covers topics such as stability, controllability, observability and realizability, for both continuous-time and discrete-time systems, and for both time-invariant and time-varying systems. For a number of years this text was the only in-depth text available on state-space methods, a subject that had become as important as frequency-domain methods in control theory. However after 1965, an increasing number of texts were published on the subject, e.g., *DeRusso, Roy, and Close, **State Variables for Engineers** (John Wiley & Sons, 1965), Schwartz and Friedland, **Linear Systems** (McGraw-Hill, 1965), Brockett, **Finite Dimensional Linear Systems** (John Wiley & Sons, 1970), Chen, **Introduction to Linear System Theory** (Holt, Rinehart and Winston, Inc., 1970), Kailath, **Linear Systems** (Prentice-Hall, Inc., 1980)*, and so on. Presently, almost every text on control theory contains some state-space material.

Linear System Theory

The State Space Approach

Lotfi A. Zadeh & Charles A. Desoer

Department of Electrical Engineering
University of California
Berkeley, California

McGraw-Hill Book Company, Inc.

New York / San Francisco / Toronto / London

We observe that the state equations *29* and *31* are not of the same form as the state equations of a system of the reciprocal differential operator type (*4.3.32*), nor are they of the canonical form

$$\dot{\mathbf{x}} = \mathbf{A}\mathbf{x} + \mathbf{B}\mathbf{u}$$
$$\mathbf{y} = \mathbf{C}\mathbf{x} + \mathbf{D}\mathbf{u}$$

Consequently, we cannot use Corollary *2.3.36*, as in Sec. 3, to show that $\mathbf{x}$ qualifies as a state vector of $\mathfrak{D}$.

Although it is an easy matter to verify that $\mathbf{x}$ as defined by *19* has the state separation property *2.3.19*, it will be instructive to demonstrate directly that $\mathbf{x}$ satisfies the four consistency conditions *1.6.5*, *1.6.11*, *1.6.15*, and *1.6.35*. This is what we shall do in the sequel.

Direct verification that x qualifies as a state vector

We have to verify that *21* is an input-output-state relation for the system characterized by the input-output relation

33 $$y(t) = (b_m p^m + \cdots + b_0)u$$

To avoid being influenced by notation, let us express *21* in the form

34 $$y(t) = \langle \boldsymbol{\phi}(t - t_0), \boldsymbol{\alpha} \rangle + M(p)[1(t - t_0)u(t)] \qquad t \geq t_0$$

where $\boldsymbol{\alpha}$ is an m-vector, $\boldsymbol{\alpha} = (\alpha_1, \ldots, \alpha_m)$, ranging over $\mathbb{C}^m$. [In effect, $\boldsymbol{\alpha}$ represents a point (state) in the state space Σ of $\mathfrak{D}$.] We employ the symbol $\boldsymbol{\alpha}$ rather than $\mathbf{x}(t_0-)$ in *34* because at this stage of our argument we have not yet demonstrated that the components of $\boldsymbol{\alpha}$ are the values of u and its derivatives at t_0-.

Since *34* is a general solution for *33*, it follows that *34* satisfies the mutual-consistency condition (see *3.8.24*). Furthermore, since the response is uniquely determined by $\boldsymbol{\alpha}$ and $u_{(t_0,t]}$, it follows that *34* satisfies the first self-consistency condition also. Thus, it will suffice to show that *34* satisfies the third self-consistency condition, since this condition implies the second self-consistency condition (see *1.6.39*).

We can simplify the argument without losing its essential features by carrying out the verification for the case where $m = 1$. Actually, this is all that we shall need at a later point (Secs. *8* and *9*) to develop an effective general technique for associating a state vector with a differential system of any finite order.

For $m = 1$, *33* and *34* become respectively

35 $$y = p\,u$$

36 $$y(t) = \alpha\delta(t - t_0) + p\,1(t - t_0)u(t)$$

where for simplicity we have set $b_0 = 0$, $b_1 = 1$. We assume furthermore that $u(t)$ is real-valued, which implies that α ranges over the real line $(-\infty, \infty)$.

This page illustrates state equations for linear systems.

Optimal Control

Michael Athans and Peter L. Falb
McGraw-Hill Book Company, New York, 1966.

The Calculus of Variations, Pontryagin's Maximum Principle, and Bellman's Dynamic Programming, theories expounded in the 1950s, as design techniques for optimal control, provided solutions to problems of special interest in the USA and the USSR. In particular, these methods provided the theoretical basis for the design of many control systems associated with space and military applications, applications that were of intense interest at the time, in both countries. The Maximum Principle and Dynamic Programming had special appeal, since these methods dealt directly with the problem of control signal constraints. In the USA texts on these methods began to appear in the 1960s. An early text, which presented the Calculus of Variations approach, was the text of *Kipinak, **Dynamic Optimization and Control. A Variational Approach** (M.I.T. Press, 1961)*. A more detailed and rigorous presentation of variational approach appeared a few years latter in the text of *Hestenes, **Calculus of Variations and Optimal Control Theory** (John Wiley & Sons, Inc., 1966)*. In 1962, an English translation of the Russian text of *Pontryagin, Boltyanskii, Gamkrelidze, and Mishenko, **The Mathematical Theory of Optimal Processes** (Interscience Publishers, Inc., 1962)* was published. This text included a detailed presentation of the Maximum Principle. However, one of the first texts written by USA authors on the Maximum Principle was the text of Athans and Falb, published in 1966. By introducing a sign change in the Hamiltonian function which appears in the Maximum Principle, Athans and Falb actually refer to the Maximum Principle as the *Minimum Principle*. The text by Athans and Falb is notable for its detailed discussion of some key design problems, such as time-optimal and fuel-optimal control. It also contained many Examples and Exercises, which made it attractive as a course text.

In the 1960s most graduate programs in the USA offered a course on *Optimal Control*. Some other notable early texts on the subject include the texts of *Leitmann, **An Introduction to Optimal Control** (McGraw-Hill Book Company, 1966); Lee and Markus, **Foundations of Optimal Control** (John Wiley & Sons, inc., 1967); and Bryson and Ho, **Applied Optimal Control** (Blaisdell Publishing Company, 1969)*. However optimal control theory had some important limitations. The optimization equations required the solution of difficult two-point boundary value problems. Also the theory, at least the deterministic theory, did not deal with uncertainties in system dynamics. Thus over time fewer graduate programs in the USA offered a full course on optimal control. However, one optimal control problem which endured was the Linear-Quadratic (LQ) control problem. Under the leadership of Athans, MIT became an important center for the study of LQ theory. LQ theory found many applications in industry because it applied to multivariable systems and software was available to solve the associated optimization equations (matrix Riccati equations). In the early 1970s, two texts appeared that focused on LQ theory, the text of *Anderson and Moore, **Linear Optimal Control** (Prentice-Hall, 1971)* and the text of *Kwakernaak and Sivan, **Linear Optimal Control Systems** (Wiley-Interscience, 1972)*.

OPTIMAL CONTROL

An Introduction to the Theory and Its Applications

MICHAEL ATHANS
Assistant Professor
Department of Electrical Engineering
Massachusetts Institute of Technology

PETER L. FALB
Associate Professor
Information and Control Engineering
The University of Michigan

McGRAW-HILL BOOK COMPANY
New York/St. Louis/San Francisco/Toronto/London/Sydney

We observe that the only difference between the two special problems is the difference between the forms of the target set S. In the next section, we shall state the minimum principle for these two problems.

Exercise 5-17 Formulate special problems 1 and 2 in *full* detail.

5-13 The Minimum Principle of Pontryagin

We shall state the minimum principle of Pontryagin for the two special problems in this section. We shall, of course, suppose that all the assumptions of Sec. 5-12 are in force, and we shall also use the notation of Sec. 5-12. Before stating the theorems, we shall introduce some additional terminology and notation. In particular, we have:

Definition 5-10 *Let $H(\mathbf{x}, \mathbf{p}, \mathbf{u})$ denote the real-valued function of the n vector $\mathbf{x}$, the n vector $\mathbf{p}$, and the m vector $\mathbf{u}$, given by*

$$H(\mathbf{x}, \mathbf{p}, \mathbf{u}) = L(\mathbf{x}, \mathbf{u}) + \langle \mathbf{p}, \mathbf{f}(\mathbf{x}, \mathbf{u}) \rangle \tag{5-391}$$

where $\mathbf{f}(\mathbf{x}, \mathbf{u})$ is the function which determines our system [that is, $\mathbf{f}(\mathbf{x}, \mathbf{u})$ is the right-hand side of the state equation] and $L(\mathbf{x}, \mathbf{u})$ is the integrand of our cost functional. We say that $H(\mathbf{x}, \mathbf{p}, \mathbf{u})$ is the Hamiltonian function (or, simply, the Hamiltonian) of our problem and that $\mathbf{p}$ is a costate vector.

We observe that, in view of our assumption CP1, the functions $H(\mathbf{x}, \mathbf{p}, \mathbf{u})$ and $\frac{\partial H}{\partial \mathbf{x}}(\mathbf{x}, \mathbf{p}, \mathbf{u})$ are continuous on $R_n \times R_n \times \bar{\Omega}$, where $\bar{\Omega}$ is the closure of Ω in R_m. Moreover, we note that $\frac{\partial H}{\partial \mathbf{p}}(\mathbf{x}, \mathbf{p}, \mathbf{u})$ is well defined and is given by

$$\frac{\partial H}{\partial \mathbf{p}}(\mathbf{x}, \mathbf{p}, \mathbf{u}) = \mathbf{f}(\mathbf{x}, \mathbf{u}) \tag{5-392}$$

Now let us suppose that $\mathbf{x}_0$ is our initial state and that t_0 is our initial time. If $\hat{\mathbf{u}}(t)$ is some admissible control and if we let $\hat{\mathbf{x}}(t)$ denote the trajectory of our system starting from $\mathbf{x}_0 = \hat{\mathbf{x}}(t_0)$ generated by $\hat{\mathbf{u}}(t)$, then, for *any* function $\mathbf{p}(t)$,

$$\dot{\hat{\mathbf{x}}}(t) = \frac{\partial H}{\partial \mathbf{p}}[\hat{\mathbf{x}}(t), \mathbf{p}(t), \hat{\mathbf{u}}(t)] = \mathbf{f}[\hat{\mathbf{x}}(t), \hat{\mathbf{u}}(t)] \tag{5-393}$$

In addition, if $\boldsymbol{\pi}$ is *any* n vector, then the *linear* differential equation

$$\begin{aligned}\dot{\mathbf{p}}(t) &= -\frac{\partial H}{\partial \mathbf{x}}[\hat{\mathbf{x}}(t), \mathbf{p}(t), \hat{\mathbf{u}}(t)] \\ &= -\frac{\partial L}{\partial \mathbf{x}}[\hat{\mathbf{x}}(t), \hat{\mathbf{u}}(t)] - \left(\frac{\partial \mathbf{f}}{\partial \mathbf{x}}[\hat{\mathbf{x}}(t), \hat{\mathbf{u}}(t)]\right)' \mathbf{p}(t)\end{aligned} \tag{5-394}$$

This page illustrates Pontryagin's "minimum principle".

$$H(x, p, u) = L(x, u) + \langle p, f(x, u) \rangle$$

$$\frac{\partial H}{\partial p}(x, p, u) = f(x, u)$$

[illegible]

[illegible]

Printed and bound by CPI Group (UK) Ltd, Croydon, CR0 4YY
03/10/2024
01040320-0018